Terrestrial Environmental Sciences

Series editors

Olaf Kolditz
Hua Shao
Wenqing Wang
Uwe-Jens Görke
Sebastian Bauer

More information about this series at http://www.springer.com/series/13468

Yonghui Song · Beidou Xi
Yuan Zhang · Kun Lei · Richard Williams
Mengheng Zhang · Weijing Kong
Olaf Kolditz

Editors

Chinese Water Systems

Volume 1: Liaohe and Songhuajiang River Basins

 Springer

Editors
Yonghui Song
Chinese Research Academy of
    Environmental Sciences (CRAES)
Beijing, China

Beidou Xi
Groundwater and Environmental System
    Engineering
Chinese Research Academy of
    Environmental Sciences (CRAES)
Beijing, China

Yuan Zhang
Riverine Ecology
Chinese Research Academy of
    Environmental Sciences (CRAES)
Beijing, China

Kun Lei
River and Coastal Environment Research
    Centre
Chinese Research Academy of
    Environmental Sciences (CRAES)
Beijing, China

Richard Williams
Pollution Science
NERC Centre for Ecology and Hydrology
Wallingford, UK

Mengheng Zhang
International Cooperation Centre
Chinese Research Academy of
    Environmental Sciences (CRAES)
Beijing, China

Weijing Kong
Laboratory of Environmental Criteria and
    Risk Assessment
Chinese Research Academy of
    Environmental Sciences (CRAES)
Beijing, China

Olaf Kolditz
Helmholtz Centre for Environmental
    Research—UFZ
Leipzig, Germany

and

TU Dresden
Dresden, Germany

ISSN 2363-6181   ISSN 2363-619X   (electronic)
Terrestrial Environmental Sciences
ISBN 978-3-030-09496-6   ISBN 978-3-319-76469-6   (eBook)
https://doi.org/10.1007/978-3-319-76469-6

Printed on acid-free paper

This Springer imprint is published by the registered company Springer International Publishing AG
part of Springer Nature
The registered company address is: Gewerbestrasse 11, 6330 Cham, Switzerland

EU-CHINA

EU-China Environmental Sustainability Programme:
Demonstration of Pollution Discharge Management for Water Quality Improvement in
the Songhuajiang-Liaohe River Basin (SUSTAIN H2O)
(DCI-ASIE/2013/323-261)

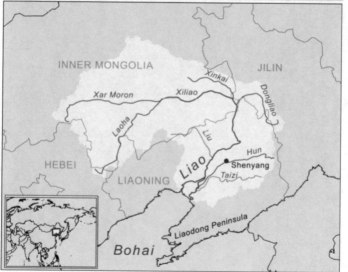

Preface

Rapid economic development and population growth in China go hand-in-hand with industrialization, increasing demand for energy and resources, intensified agriculture and increasing urbanization, involving growing mega-cities. These developments have caused and continue to cause severe pressures and risks to natural resources and the long-term provision of highly required goods and services based on natural resources. Pollution containing hazardous substances for environmental and human health, depletion and deterioration of water resources as a result of overexploitation and contamination, soil degradation and air pollution in mega-metropoles (such as Jing-Jin-Ji,[1] Taihu area) are increasing at an alarming rate. At the same time, the long-time neglected development of rural areas has to be tackled with corresponding environmental friendly master plans.

Consequently, to stop negative trends jeopardizing the economic and societal development in China, protection, remediation and productive management strategies as well as sustainable planning need to be developed and implemented for China's natural resources in a highly diverse, complex and dynamic environment. This offers most important opportunities for international collaboration in environmental science, technology and education.

Major Water Program

The Chinese government recognized the importance and complexity of the water situation and has initiated a programme entitled "Major Water Program of Science and Technology for Water Pollution and Governance" (2006–2020). While shortages resulting from regional resource depletion have led to projects of large-scale water transport from distant water-rich areas of China (Water Diversion Project), the water quantity and quality problems in other areas require efficient, flexible, and

---

[1] The national capital region of China (Beijing-Tianjin-Hebei).

site-specific solutions and overall management concepts. In April 2015, the Action Plan for Water Pollution Prevention (Clean Water Action Plan)[2] was published by the State Council of the People's Republic of China. It requires that by 2020, China's water environment quality will gradually improve; the percentage of severely polluted water bodies will be greatly reduced, and the quality of drinking water will be improved. The plan seeks to protect surface water in seven river basins: Yangtze, Yellow, Pearl, Songhua, Huai, Hai and Liao Rivers. It sets urgent, strict targets for water scarce regions such as Beijing-Tianjin-Hebei, Yangtze River Delta, and the Pearl River Delta.

In September 2013, China has formulated and implemented an in-depth Action Plan for the Prevention and Control of Air Pollution (Clean Air Action Plan)[3] in order to set up an evaluation system focusing on improving air quality and assessment results will be used for performance evaluation of the local leaders.

On 31 May 2016, China launched a new action plan to tackle soil pollution (Clean Soil Action Plan)[4] and China aims to curb worsening soil pollution by 2020 and stabilize and improve soil quality by 2030. These plans highlight the determination to control pollution, improve environmental quality, and protect the people's health.[5]

The Chinese government released its 13th Five-Year Plan (2016–2020)[6] on 17 March 2016. It promotes a cleaner and greener economy, with strong commitments to environmental management and protection, clean energy and emission control, ecological protection and security and the development of green industries. Specific objectives for environmental protection in the 13th Five-Year Plan period include: reduction of water consumption by 35% by 2020 as compared to 2013; estimated total consumption of primary energy in 2020 of less than 5 billion tons of standard coal; energy consumption per unit of GDP to be reduced by 15% in 2020 (compared to 2015); reduction of carbon dioxide emissions per unit of GDP by 40–45% by 2020 (compared to 2015 which is consistent with China's Plan for Addressing Climate Change (2014–2020)). On 3 September 2016, the presidents of China, Xi Jinping, and the USA, Barack Obama, announced the ratification of the Paris Agreement (of the 2015 United Nations Climate Change Conference) by their countries, respectively.

On 08 August 2016, Chinese government released its 13th Five-Year National Science and Technology Innovation Plan.[7] China will continue to support the national science and technology major projects, which include the major project in water

---

[2] http://www.mep.gov.cn/gkml/hbb/qt/201504/t20150416_299173.htm.

[3] http://www.gov.cn/zwgk/2013-09/12/content_2486773.htm.

[4] http://www.gov.cn/zhengce/content/2016-05/31/content_5078377.htm.

[5] http://www.mep.gov.cn/.

[6] http://www.gov.cn/xinwen/2016-03/17/content_5054992.htm.

[7] http://www.gov.cn/zhengce/content/2016-08/08/content_5098072.htm.

pollution control and treatment. The targets are: a number of key technologies shall be developed in terms of water circulation system restoration, water pollution control, drinking water safety, ecological service functions restorations as well as long-term management mechanisms. The comprehensive demonstration will be carried out in the region "Beijing-Tianjin-Hebei" and Taihu lake area. Comprehensive environmental information systems for water pollution control, environmental management and drinking water safety shall be established in order to set up the big-data-based platform for water environment monitoring and observation.

Sino-German Cooperation

German-Chinese governmental consultations are taking place on a regular basis, enabling discussion on recent topics for collaboration between the two countries at highest level. The first German-Chinese governmental consultations were held on 28 June 2011 in Berlin and provided the framework for the German-Chinese Forum for Economic and Technological Cooperation. During this first consultation, a joint declaration on the establishment of the bilateral "Research and Innovation Programme Clean Water" was signed. The second consultations between the two governments took place in August 2012 in Beijing, the third on 10 October 2014 in Berlin. The fourth consultations were held in June 2016. The major topic was how to link "made in China 2025" and "German Industry 4.0".

Recognizing the importance, opportunities and strength of Chinese-German bilateral research cooperation, the Federal Ministry of Education and Research published their "China-Strategy of the BMBF" in October 2015, a strategic framework for the cooperation with China in research, science and education.[8] The BMBF China-Strategy is dedicated to further improve the framework conditions for cooperation between Germany and China in science and research, networking and education. The main areas for cooperation are "Key Technologies", "Life Science", "Strengthening Social Sciences" and coping "Ecological Challenges".

The "Innovation Cluster Major Water" was established in 2016 in order to coordinate the German contributions to the "Major Water Program".[9] The cluster also provides actual information on Sino-German research projects, knowledge and technology transfer as well as training and education activities in the field of water science.

Sino-UK Cooperation

Britain and China have a long history of interactions commercially and governmentally. In modern times, since the creation of the People's Republic of China in 1949, Britain and China have sought to enhance historical ties particularly through trade and business. What was known as the "Group of 48" first started trade missions to China in the early 1950s, and was subsequently followed by missions from "The Sino-British Trade Committee" (formed in 1954). In the 1980s, these two groups merged to become what is now the "China-Britain Business Council" (although the 48 Group Club continues to operate as an independent business

---

[8] https://www.bmbf.de/de/china-strategie-des-bundesministeriums-fuer-bildung-und-forschung-20-15-2020-1882.html.

[9] http://sino-german-major-water.net/de/.

network promoting Sino-UK trade and academic interests[10]). Currently the China-Britain Business Council[11] acts as the delivery partner in China for the UK government Department of International Trade and covers sectors including agriculture, energy and education, amongst others.

The rapidly developing research and academic status of China, and the collaboration between Chinese and UK researchers was enriched when in 2007 a dedicated Research Councils UK (RCUK) office was established in Beijing, the first overseas RCUK office, capitalizing on existing collaborations and working to develop strong networks and future joint research and innovation activity between the two nations. The activities of RCUK China have supported the development of many initiatives such as the 2008 Water Availability and Quality programme, deriving funds from representative UK and Chinese funding bodies.[12] The RCUK China office continues to work collaboratively on behalf of the Research Councils with UK government and Chinese research bodies to develop opportunities and partnerships.

More recently, and formally, bringing together government and research ambitions for collaboration, in 2015 during a state visit from President Xi Jinping to Britain multi-million pound trade deals were agreed in several areas[13] including energy, medicine, infrastructure, transport, telecommunications and, significantly for environmental research, bilateral research funding under the Newton Fund (UK-China Research and Innovation Partnership Fund) with priority areas including Energy, Environmental Technologies, Food and Water Security, and Urbanisation amongst others.

---

[10] http://the48groupclub.com/the-club/about-the-club/.

[11] http://www.cbbc.org/.

[12] http://www.rcuk.ac.uk/documents/international/rcukchinaimpactbrochure-pdf/.

[13] https://www.gov.uk/government/news/chinese-state-visit-up-to-40-billion-deals-agreed.

Previous Works: Research and Education

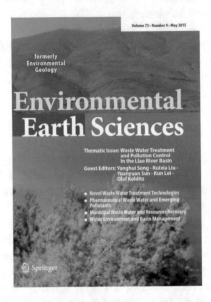

The Liaohe-Songhuajiang River Basin belongs to the priority areas of the Chinese Major Water Program. The Chinese central government launched a "Revitalizing Northeast China and Other Old Industrial Bases" campaign. The socio-economic development of Liao River Basin needs the supports of good water resource and environment. The "Major Water Program" took Liao River Basin as one of the most important demonstration basins, set up a full project under the river theme in the first stage of the programme (2008–2010), focusing on pollution source control technology development and continued such a project in the second stage (2011–2015), focusing on pollution load reduction and water environment restoration. The Topical Issue in Environmental Earth Sciences "Waste water treatment and pollution control in the Liao River Basin" compiles main results of the research of the "Major Water Program" within the 12th Five-Year-Plan dealing with wastewater treatment technologies, pollution control in the river basin, emerging pollutants and socio-economic studies.[14]

In addition to research work, educational material has been prepared for the Song-Liao River Basin. This tutorial presents the application of the open-source software OpenGeoSys (OGS) for hydrological simulations concerning conservative

---

[14]Yonghui Song, Ruixia Liu, Yuanyuan Sun, Kun Lei, Olaf Kolditz (2015): Waste water treatment and pollution control in the Liao River Basin. Environ Earth Sci, https://link.springer.com/article/10.1007/s12665-015-4333-7.

and reactive transport modelling. The tutorial was already applied on several international training courses on the subject held in China within the "SUSTAIN-H2O" project.[15]

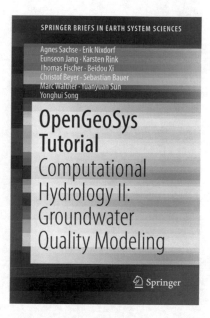

This tutorial is the result of a close cooperation within the OGS community (www. opengeosys.org). These voluntary contributions are highly acknowledged. The book contains general information regarding hydrological modelling of a real case study and step-by-step model set-up with OGS and related components such as the OGS Data Explorer. Benchmark examples are presented in detail.

Beijing, China	Yonghui Song
Beijing, China	Beidou Xi
Beijing, China	Yuan Zhang
Beijing, China	Kun Lei
Wallingford, UK	Richard Williams
Beijing, China	Mengheng Zhang
Beijing, China	Weijing Kong
Leipzig, Germany	Olaf Kolditz

[15] Sachse, A., Nixdorf, E., Jang, E., Rink, K., Fischer, Th., Xi, B., Beyer, C., Bauer, S., Walther, M., Sun, Y., Song, Y. (2017): OpenGeoSys Tutorial—Computational Hydrology II: Groundwater Quality Modeling. SpringerBriefs in Earth System Sciences. www.springer.com/us/book/9783319528083.

Contents

Editors and Contributors

Assistant Editors

Erik Nixdorf, Helmholtz Centre for Environmental Research—UFZ, Leipzig, Germany

Yuanyuan Sun, Chinese Research Academy of Environmental Sciences (CRAES), Beijing, China

Kexin Liu, Chinese Research Academy of Environmental Sciences (CRAES), Beijing, China

Shiguang Feng, Helmholtz Centre for Environmental Research—UFZ, Leipzig, Germany

About the Editors

Yonghui Song is Professor and the Vice-President of the Chinese Research Academy of Environmental Sciences (CRAES), which is affiliated to the Ministry of Environmental Protection (MEP) of China. He obtained his Ph.D. Degree in Environmental Science from the Research Center for Eco-Environmental Sciences of the Chinese Academy of Sciences in 1999 and his Dr.-Ing. in Environmental Engineering from the University of Karlsruhe (TH), Germany in 2003. His research focuses on water pollution control technologies, regional- and basin-level water environment management. He has undertaken over 20 national or ministerial/provincial scientific research projects as the principal investigator, published over 300 journal papers and obtained more than 20 patents. He received the "Outstanding Research Team Award" of the "11th Five-Year Plan" of the Ministry of Science and Technology (MOST) of China in 2012 and the "Award of MEP for Science and Technology" in 2016. He was selected as the Youth Innovation of Science and Technology Leading Talent by MOST of China in 2012 and was selected as the Innovation of Science and Technology Leading Talent by Organization Department of the CPC Central Committee General Office of China in 2013.

Beidou Xi is the Chief Engineer's Office Manager in Chinese Research Academy of Environmental Sciences, the Chief Specialist of the Innovation Base of Groundwater and Environmental System Engineering in CRAES, and Director of the State Environmental Protection Key Laboratory of Simulation and Control of Groundwater Pollution. His research fields include techniques and materials for the remediation of contaminated groundwater, solid waste resource utilization and secondary pollution control and water environmental protection. He got his Ph.D. from Department of Environmental Sciences and Technology, Tsinghua University. After he did the Post Doc research in University of Regina, he worked in Chinese Research Academy of Environmental Sciences until now. He is sponsored by National Science Fund for Distinguished Young Scholars. He is the member of Chinese Association of Environmental Sciences, Chinese Association of Energy Environment Technology and China Renewable Resources Recycling Association. He undertakes several national research projects such as "973", "863" and "Major Water Program". He is also the editor of several scientific journals such as "Environmental Science Research", "Environmental Pollution Prevention and Control". He has published over 150 SCI-indexed articles and obtained over 70 patents.

Yuan Zhang is the Deputy Director of the Water Environment Research Centre, Chinese Research Academy of Environmental Sciences (CRAES) and the Chief Specialist of the Laboratory of Riverine Ecological Conservation and Technology in CRAES. He is also an Adjunct Professor and Ph.D. Supervisor at Beijing Normal University (BNU). The area of expertise is related to river ecosystem restoration technology and catchment water environmental management sciences. He got his Ph.D. from Department of Environmental Sciences, Beijing Normal University. After he did the Post Doc research and continue to work in Chinese Research Academy of Environmental Sciences until now. He has chaired over 10 projects including "Program 973" projects, Major S&T Special Projects, "National Natural Science Foundation" general projects and international collaborative projects. He has published or contributed to four monographs, four patents and 194 scholarly articles, amongst which over 70 in SCI-cited articles published like Environmental Science and Technology, Water Research and Chemosphere.

Kun Lei female, born in 1973, served as Professor of Chinese Research Academy of Environmental Sciences (CRAES), graduated from Ocean University of China in 2001. Longterm she dedicated in watershed environment management, especially the establishment of the water target management technology system in watershed and control units areas, material flux estimation methods from watershed to the coastal zone, environment quality evaluation technique of the estuarine and coastal zone, environment evolution trend and the mechanisms of the coastal and estuarine ecosystem under the influence of human activities, estuarine and costal hydrodynamic and water quality simulation, pollutants environmental capacity and total pollutant load allocation, integrated coastal and watershed management. In recent years, she has served as the person in charge of the subject of watershed water quality target management technology system and the watershed water environment capacity evaluation and regulation, the 12th Five-Year Plan of costal environmental pollution prevention and treatment, studies on dynamic monitoring and evaluation technique of the estuarine and costal habitat in the Bohai Sea funded, Bohai land-sea flux estimate and total amount control technologies, which has lay the foundation for the watershed and coastal environment management, published more than 50 academic papers and 5 books (as a co-author or translator), got the second prize of the National Science and Technology in 2015, and the second prizes of the Ministry of Environmental Protection Science and Technology in 2009 and 2010.

Richard Williams is a Principal Scientist at the Natural Environment Research Council's Centre for Ecology and Hydrology (NERC-CEH), an organization he joined straight after receiving his degree in Chemical engineering from University College London, in 1980. His main area of research concerns the fate and behaviour of chemicals in the environment and in particular the application and development of mathematical models for helping to solve real-world problems related to river water quality. He has experience in modelling a wide range of potential contaminants of surface waters including nutrients, organic matter, micro-organic contaminants (including pesticides) and recently nano-particles. Currently, his research focuses on predicting the concentrations in rivers of "down-the-drain" chemicals e.g. personal care products, pharmaceuticals and steroid oestrogens. He uses GIS tools to make maps of contamination and combines maps with eco-toxicological effect levels to make risk maps. He makes these maps at a range of scales from small catchments up to the whole of the European continent. From 2013 to 2016, he led the NERC-CEH group on Water Resources Assessment. He was for 10 years a member of the Editorial Board of Pest Management Science (1999–2008). The Sustain H2O project was his first experience of working in China, although NERC-CEH has links with several universities and academies of science in China.

Mengheng Zhang has 25 years' working experiences in various capacities for international cooperation of climate change, ozone layer protection, sustainable development, marine protection, chemicals and air pollution. She has represented China as a chief negotiator at conferences of various multilateral environmental agreements, such as the Montreal Protocol on ozone layer protection, UNFCCC and Kyoto Protocol on climate change, Basel Convention on hazardous wastes and Rotterdam Convention on Chemicals. As a chief negotiator, she won USEPA's 2008 Ozone Layer Protection Award, which was presented to the Chinese Negotiating Team for the 19th Meeting of Parties. The award was recommended for her outstanding acumen shown in negotiating on the HCFC adjustments when she represented China at the 19th Meeting of the Parties to the Montreal Protocol. He has been the chairperson at the Intergovernmental Meeting of NOWPAP—a UNEP Regional Seas Programme. She has successfully coordinated the implementation of the project —"Reversing Environmental Degradation Trends in the South China Sea and Gulf of Thailand" which was the biggest UNEP/GEF project, covered seven countries and lasted five years. She has rich experiences in coordinating large and multilateral environment programmes, such as ESP SUSTAIN H2O: EU-China Environmental Sustainability Programme on Demonstration of Pollution Discharge Management for Water Quality Improvement in Songhuajiang & Liaohe River Basin (2014–2017), GLOCOM Project: Global Partners in Contaminated Land Management under FP7 (2011–2015).

Weijing Kong is the Professor from Chinese Research Academy of Environmental Sciences. His research interests are related to freshwater ecosystem health assessment, freshwater ecosystem restoration skills and freshwater ecosystem function management region delineation. He got his ecological Ph.D. degree in 2009 studying river scape system pattern and process. He has led several projects in freshwater ecosystem theory and management, e.g. ESP SUSTAIN H2O: EU-China Environmental Sustainability Programme on Demonstration of Pollution Discharge Management for Water Quality Improvement in Songhuajiang & Liaohe River Basin (2014–2017); the NSFC project "Mechanism of habitat patches spatial heterogeneity during riparian natural rehabilitation process of riverine nature reserve" (41201187, 2012–2015), the National Water Pollution Control and Treatment Science and Technology Major Project "Freshwater ecosystem function management region delineation in Liaohe River basin" (2012ZX0750100102, 2012–2015).

Olaf Kolditz is the Head of the Department of Environmental Informatics at the Helmholtz Center for Environmental Research (UFZ). He holds a Chair in Applied Environmental System Analysis at the Technische Universität in Dresden. His research interests are related to environmental fluid mechanics, numerical methods and software engineering with applications in geotechnics, hydrology and energy storage. Olaf Kolditz is the lead scientist of the OpenGeoSys project (www.opengeosys.org), an open source scientific software platform for the numerical simulation of thermo-hydro-mechanical-chemical processes in porous media, in use worldwide. He studied theoretical mechanics and applied mathematics at the University of Kharkov, got a Ph.D. in natural sciences from the Academy of Science of the GDR (in 1990) and earned his habilitation in engineering sciences from Hannover University (in 1996), where he became group leader at the Institute of Fluid Mechanics. Until 2001 he was Full Professor for Geohydrology and Hydroinformatics at Tübingen University and Director of the international Master course in Applied Environmental Geosciences. Olaf Kolditz is Editor-in-Chief of two international journals *Geothermal Energy* (open access) and *Environmental Earth Sciences* (ISI). He was initiating several Sino-German cooperation projects, e.g. the "Research Centre for Environmental Information Science-RCEIS" (www.ufz.de/rceis), the "Sino-German Geothermal Research Centre" (www.ufz.de/sg-grc), and collaborative project "Managing Water Resources in Urban Catchments—Chaohu" (www.ufz.de/urbancatchments). In 2015, He was awarded a visiting professorship under the CAS President's International Fellowship (PIFI).

Contributors

Yixiang Dend Centre for Ecology & Hydrology, Bailrigg, UK

Liang Duan Chinese Research Academy of Environmental Sciences, Chaoyang, China

Juntao Fan Chinese Research Academy of Environmental Sciences, Chaoyang, China

Francois Edwards Centre for Ecology & Hydrology, Wallingford, UK

Xin Gao Chinese Research Academy of Environmental Sciences, Chaoyang, China

Lu Han Chinese Research Academy of Environmental Sciences, Beijing, China

Olaf Kolditz Helmholtz Centre for Environmental Research, TU Dresden, DE, Leipzig, Germany; Helmholtz Centre for Environmental Research, TU Dresden, DE, Dresden, Germany

Weijing Kong Chinese Research Academy of Environmental Sciences, Chaoyang, China; Chinese Research Academy of Environmental Sciences, Beijing, China

Kun Lei Chinese Research Academy of Environmental Sciences, Beijing, China

Bin Li Chinese Research Academy of Environmental Sciences, Beijing, China

Kexin Liu Chinese Research Academy of Environmental Sciences, Beijing, China

Ruixia Liu Chinese Research Academy of Environmental Sciences, Beijing, China

Erik Nixdorf Helmholtz Centre for Environmental Research, DE, Leipzig, Germany

Jianfeng Peng Chinese Research Academy of Environmental Sciences, Beijing, China

Fei Qiao Chinese Research Academy of Environmental Sciences, Beijing, China

Yonghui Song Chinese Research Academy of Environmental Sciences, Beijing, China

Jing Su Chinese Research Academy of Environmental Sciences, Chaoyang, China

Yuanyuan Sun Chinese Research Academy of Environmental Sciences, Chaoyang, China

Ya Tao Chinese Research Academy of Environmental Sciences, Beijing, China

Qiang Wang Heilongjiang Provincial Research Institute of Environmental Science, Harbin, China

Siyu Wang Chinese Research Academy of Environmental Sciences, Beijing, China

Tong Wang Liaoning Academy of Environmental Sciences, Shenyang, China

Richard Williams Centre for Ecology & Hydrology, Bailrigg, UK

Jieyun Wu Chinese Research Academy of Environmental Sciences, Beijing, China

Beidou Xi Chinese Research Academy of Environmental Sciences, Chaoyang, China

Peng Yuan Chinese Research Academy of Environmental Sciences, Beijing, China

Ping Zeng Chinese Research Academy of Environmental Sciences, Beijing, China

Mengheng Zhang Chinese Research Academy of Environmental Sciences, Chaoyang, China

Moli Zhang Chinese Research Academy of Environmental Sciences, Beijing, China

Yuan Zhang Chinese Research Academy of Environmental Sciences, Beijing, China

Gang Zhou Chinese Research Academy of Environmental Sciences, Beijing, China

Chapter 1
Introduction to the Sustain H2O Project

Yonghui Song, Erik Nixdorf, Beidou Xi, Yuan Zhang, Lei Kun,
Weijing Kong, Mengheng Zhang, Richard Williams and Olaf Kolditz

1.1 Aims and Scope

In the framework of the EU-China Environmental Sustainability Program, the SUS-TAIN H2O project was launched in 2013 and ran over 36 months until 2016. Sustain H2O is an abbreviation which stands for "Demonstration of Pollution Discharge Management for Water Quality Improvement in the Songhuajiang-Liaohe River Basin (SLRB)". The SLRB represents China's old industrial base with a large number of water-polluting industries. It has been selected as it is a key river basin for water pollution control and water environment management with clearly defined challenges in terms of pollution emission reduction and water quality improvement. Some of these challenges/problems include excessive discharge of pollutants, no optimal load allocation mechanism of quantity control, poor water quality and severely damaged water ecology as well as high risk of environmental pollution accidental events that threaten the safety of drinking water sources. Moreover, the SLRB is a trans-national (transboundary) river catchment, flowing from China to Russia, and therefore is well suited as a joint Sino-EU action.

The overall objective was to develop and demonstrate management tools and practices for pollution reduction in SLRB and to support water quality improvement

Y. Song · B. Xi · Y. Zhang · L. Kun · W. Kong · M. Zhang
Chinese Research Academy of Environmental Sciences, Chaoyang, Germany
e-mail: songyh@craes.org.cn

E. Nixdorf (✉)
Helmholtz Centre for Environmental Research, DE, Leipzig, China
e-mail: erik.nixdorf@ufz.de

R. Williams
Centre for Ecology & Hydrology, Wallingford, UK

O. Kolditz
Helmholtz Centre for Environmental Research, TU Dresden, DE, Leipzig, Germany

© Springer International Publishing AG, part of Springer Nature 2018
Y. Song et al. (eds.), *Chinese Water Systems*, Terrestrial Environmental Sciences,
https://doi.org/10.1007/978-3-319-76469-6_1

in the demonstration areas to realize the goal of water pollution control in SLRB designated in the "12th Five-Year Plan" of China.

The specific objectives were:

1. To support sustained pollution reduction, optimize allocation of pollution load, and to improve wastewater discharge management systems based on ecological function zoning in the demonstration areas. It intends to carry out ecological function zoning, calculate the carrying capacity; optimize pollution load allocation, set pollution discharge limits on point sources from industries and intensive livestock and poultry, develop a unit control scheme and best practices including point and non-point sources; form a permit system and demonstrate it in SLRB.
2. To develop a methodology for river health assessment and water ecological restoration and management systems to promote water quality improvement in the demonstration areas. It tries to establish water ecological restoration technological systems for healthy rivers; develop a post-evaluation system of ecological restoration, thus promoting water quality improvement in demonstration areas like the Liaohe River Reserve.
3. To identify key risk sources and priority persistent organic pollutants in the demonstration areas, and develop a risk prevention management system for typical drinking water sources. It intends to establish a method of risk source identification of water environment in SLRB to prevent sudden water environment accidents; analyse the source and discharge of typical persistent organic pollutants, establish a list of key risk sources and put forward management scheme; identify sensitive water sources of groundwater at the demonstration areas of SLRB, build up risk reduction strategy.

1.2   The Project Team

The Sustain H2O project was led by the **Chinese Research Academy of Environmental Sciences (CRAES)**, which is the largest, multidisciplinary national environmental research institute in China. There are 835 staff at CRAES, including 5 academicians, 80 professors, 150 associate professors and 26 senior engineers. 275 of the scientific staff members hold Ph.D. degrees and 275 have master degrees. The expertise of CRAES covers all aspects of environmental science including water environment, atmospheric environment, ecology, climate change, environmental safety, cleaner production and circular economy, vehicle emission control, and development of environmental standards. Within the area of water environment, CRAES specialises in (1) simulation, mechanism, control, management and risk assessment of lake eutrophication and watershed protection; (2) biogeochemical process, mechanism, simulation, ecological restoration as well as comprehensive ecological management of river and coastal zone; (3) research on transition and transformation mechanism, quality evaluation, risk analysis, remediation technology as well as modelling and policy study of ground water; (4) comprehensive management, quality assessment, recycle and reuse, as well as landscape ecology of urban water environment. CRAES has many years of experience in environmental management at the provincial and national level including setting environmental standards, providing key advice to policy makers, and assisting the implementation of international conventions. CRAES has established cooperation partnership with national level institutions of EU, USA, Canada, Australia, Japan, Korea. Currently, CRAES is involved in 700 research projects with an annual funding of 790 million RMB and this amount is relatively stable over the past 5 years. CRAES has a long history of research and project implementation in SLRB, including emergency response during the 2005 Songhuajiang River nitrobenzene incident, 2010 National Environmental Risk Source Survey and Classification Exercise, and the 12th Five-year Plan of Liaohe River Basin Water Pollution Prevention. A number of national and international partners were included in the Sustain H2O project (Fig. 1.1)

- **Liaoning Academy of Environmental Sciences (LAES), China**: research and monitoring on pollution sources of the Liaohe River Basin, as well as verification of ecological restoration technology, principal stakeholder, test site owner and data provision;
- **Helmholtz-Centre for Environmental Research (Helmholtz - Zentrum für Umweltforschung GmbH - UFZ), Germany**: technologies and strategies on risk assessment and reduction of groundwater pollution, water ecological restoration technologies;
- **Heilongjiang Provincial Research Institute of Environmental Science (HRIES), China**: research and monitoring on pollution source of the Songhuajiang River, investigation, monitoring and risk assessment on drinking water sources of groundwater, principal stakeholder, test site owner and data provision;
- **Centre for Ecology and Hydrology (CEH) of Natural Environment Research Council (NERC), UK**: pollution load allocation and discharge permits, ecological status assessment and restoration;

The European partners are equipped with advanced river management experiences and technologies and understand the management rules and regulations of the EU such as the Water Framework Directive. They have participated in several large EU projects (such as AquaTerra, SWITCH, etc.) and have vast experiences in pollution load allocation, ecological restoration and non-point source pollution, which is vital in this action. The Chinese partners LAES and HRIES are major scientific research

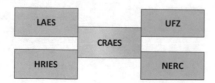

Fig. 1.1 Partners of the Sustain H2O project. Domestic and international partners are labelled in orange and green, respectively

and environmental management institutes of the pilot areas in China. They are familiar with problems in the SLRB and have accumulated substantial amount of data and experiences from the action areas, which will lay the foundation of the action. They provide direct expert input and policy oriented advice to local and regional government. Hence, LAES and HRIES will facilitate the implementation of strategy and plans in the SLRB. The Chinese partners will ensure wide dissemination and uptake of the action results.

1.3 Project Structure

The action focuses on the core theme of emission reduction and water quality improvement in particular demonstration areas (Fig. 1.2). Subsequently the workload was delineated by topics into 4 work packages, which also form the basis of the 4 major chapters of this book, plus a 5th work package for the project management.

- **WP1**. **Management methods and demonstration on pollution load of SLRB**: This WP supported sustained pollution reduction, optimize allocation of pollution load, and improve wastewater discharge management systems based on ecological function zoning in the demonstration areas. It intended to carry out ecological function zoning; calculate the carrying capacity; optimize pollution load allocation; set pollution discharge limits on point sources from industries and intensive

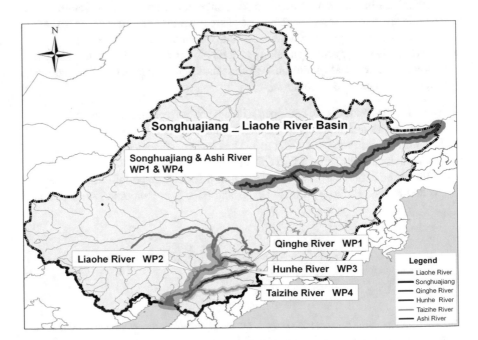

Fig. 1.2 Map of demonstration areas in the Songhuajiang-Liaohe River Basin

livestock and poultry; develop pollutant control unit schemes and best practices including point and non-point sources; form a permit system and demonstrate it in SLRB.

- **WP2. River health assessment, ecological restoration and management systems**: For the sustainability of water quality and ecosystem integrity in the demonstration areas of SLRB, the overall objectives of this work package are to develop methodologies for river health assessment, ecological restoration and management systems. WP2 will develop a methodology for river health assessment based on the ecological integrity. The key stress causing riverine ecosystem degradation will be identified in the pilot study area. The methodologies for selecting focal or indicator species, along with conservation and protection targets will be developed. Based on the protection target and stress causes, the research will establish water ecological restoration technological systems and carry out demonstrations along the main tributary of Liaohe River between Fudedian and Panjin. Finally, a post-evaluation system of ecological restoration will be constructed and carried out to evaluate the efficiency of water quality improvement in the demonstration areas.

- **WP3. Management technology and strategy for environmental risk sources and persistent organic pollutants (POPs) in Liaohe River Basin**: The overall objective of this work package is to identify key risk sources and priority persistent organic pollutants (POPs) in the demonstration areas of the SLRB as a basis for improvement of the drinking water risk management capacity of local government authorities. This includes establishing a method of risk source identification for the water environment in SLRB to prevent sudden water environment accidents, analyse the source and discharge characteristics of typical POPs, establish lists of key risk sources and POPs and develop a management strategy.

- **WP4. Sensitive groundwater sources identification and risk reduction management in SLRB**: The overall objective of WP4 is to identify sensitive groundwater sources and build up a drinking water management strategy for groundwater risk reduction. The specific objectives are to develop a risk assessment concept for groundwater sources in SLRB based on the hydrogeological characteristics and groundwater problems in the demonstration areas of Ashi river and Taizi River, to establish a sensitive groundwater source identification method in SLRB and to propose groundwater pollution risk reduction strategy for SLRB.

- **WP5. Project management and dissemination**: WP5 is devoted to build the framework to support the other WPs and managing the action during the entire project duration. One objective is to put in place and maintain the necessary framework, structures, and systems required for the management of the action to ensure effective functioning of the project. This entails setting up a common administrative and financial management framework, an effective dissemination system, as well as appropriate accountability arrangements. Additionally it ensures that all main activities are effectively carried out. WP5 will further promote the visibility of the activities, especially during field studies, applied research, demonstration, training, feedback and dissemination, to ensure wide dissemination and uptake of the action results for demonstration and promotion of a variety of outcomes from the action.

1.4   Summary of Key Results

During the SUSTAIN H2O implementation period, all work packages (WPs) with all the partners have been working on the action according to the work plans under the grant contract and achieved some good results which could be seen below. To be specific, WP1 developed and validated the basin hydrology model and water environment model suitable for water body characteristics of the control unit in the demonstration area and established the response relationship between the pollution load and water quality on the basis of hydrological condition. With the project Deliverable "Technical guidelines on pollution load allocation in control units", WP1 put forward the technical scheme of optimizing the pollution load allocation in the basin by control unit analysis and the calculation of the water environmental capacity of the model basin. Based on the water quality objective and best available abatement technology, it put forward the technology system of pollutant emission permit limits. For setting the industry pollutant emission permit limits in SLRB, it developed a demonstration platform of pollutant discharge permit license management in the Qing River Basin at the demonstration area of Tieling City.

WP2 focuses on river health assessment, ecological restoration and management system and achieved accomplishments in the following aspects:

- Review the river health assessment method from the EU WFD, assess the river health status in the Liao River Basin based on the related data collected, report it with the river health report card, outline the river health assessment guideline for big river based on the above review and assessment.
- Set the river protection targets by analysing the river system and ecological status in the demonstration area and comply the river protection target setting guideline.
- Analyse different types of wetlands and the restoration methods in the demonstration area, and build the management system.
- Based on the project activities and research result, submit two peer reviewed papers about the ecological status in the demonstration area to SCI journals.

WP3 mainly focused on the identification of key risk sources and POPs in the demonstration areas of the SLRB. WP3 established the water ERSI method, screened and identified the industrial pollution sources and POPs discharge in the demonstration area, developed the control strategy for POPs and water environment risk sources and submitted the result to the local Environmental Protection Bureaus for policy making.

WP4 mainly focused on the establishment of numerical groundwater flow model for Ashi River Basin and the development of groundwater sensitivity and risk maps in Songhuajiang-Liao River Basin. The groundwater flow model and contamination transport model of Ashi River Basin was developed by UFZ and several simulation results were obtained. A groundwater risk assessment system was developed for Song-Liao River Basin and each index was calculated and displaced in the manner of maps. The indexes were combined and groundwater risk maps of Songhuajiang-Liao River Basin were produced, which provide technical guidelines for sensitive groundwater source protection.

WP5 convened regular meetings, organised "Kick-off Meeting", "Midterm Review Meeting" and the "Final Meeting", three training sessions as well as two study tours to Europe. WP5 also finished complying regular progress reports of the action including standard reports required by EU according to the project reporting schedule, and did related work for dissemination, such as making fliers and setting up the project website. The main results are organizing all the milestone meetings of the project, the establishment of expert advisory group and management team, the project website and flyer as well as a smooth management of the project.

Chapter 2
Management Methods and Demonstration on Pollution Load Control of Song-Liao River Basin

Kun Lei, Ya Tao, Weijing Kong, Fei Qiao, Gang Zhou, Yuan Zhang, Yixiang Dend, Richard Williams, Kexin Liu and Jieyun Wu

2.1 Aquatic Eco-function Zoning Map

by Weijing Kong, Yuan Zhang

2.1.1 Introduction

This deliverables reports the "aquatic eco-function zoning map" delineation activities of the SUSTAIN H$_2$O project consortium during the project's reporting period.

As we know that Chinese water bodies are experiencing serious water environmental problems. To conserve aquatic environment Chinese government have implemented pollutant consistency control policy and object quantum control policy which both proved to be vain. And now capacity quantum control policy was thought to be an efficient way as it considers the capacity of water environment. Management based on regions is considered to be the most efficient way, but it was not considered fully for previous management policies. So in this project we hope to finish the eco-function zoning map in the demonstration area which have been proved to be efficient in developed countries.

In this report we first reviewed the literatures on ecoregion delineation, then we raised a 4 level framework for aquatic function management based on the review, and build the approaches for the delineation. We applied the framework and delineation approaches in Liao river basin, and finally the Liao river basin was delineated into 4 level 1 eco-function region, 14 level 2 eco-function region, and 50

K. Lei (✉) · Y. Tao · W. Kong · F. Qiao · G. Zhou · Y. Zhang · K. Liu · J. Wu
Chinese Research Academy of Environmental Sciences, Beijing, China
e-mail: leikun@craes.org.cn

Y. Dend · R. Williams
Centre for Ecology & Hydrology, Bailrigg, UK

© Springer International Publishing AG, part of Springer Nature 2018
Y. Song et al. (eds.), *Chinese Water Systems*, Terrestrial Environmental Sciences,
https://doi.org/10.1007/978-3-319-76469-6_2

eco-function region. In level 4 function of river reaches was assessed and the dominant function was chosen as the function of reaches.

Eco-function zone boundary offered a spatial unit for aquatic environment management, and different level meets different management demands. In this project, level 1ecoregion is suitable for the water quality standard establishment; for level 2 ecoregion it is suitable for reference site selection on river health assessment. For level 3, it is suitable for the pollution control planning, and level 4 ecoregion is suitable for river function rehabilitation effort, conservation target settlement, for each function type the natural status of physical, chemical and biological condition was different, and the management demands for each function type were also different.

2.1.2 Aquatic Ecosystem Function Management Region System

In this part we reviewed the aquatic ecosystem classification/delineation history, the approaches used in the world, and the water management region used in China. Based on this the aquatic eco-function management region framework was raised.

2.1.2.1 Review of Classification of Aquatic Management Units

Research Progress

In 1898, Merriam divided the United States using life zones (based on the distribution of mammals) and cropping data, which is the first instance of using biology as the basis of land regionalization, and it is thought to be the prototype of eco-region. In 1962, Canadian forest scientist Orrie Loucks proposed "ecological zones" using forest type and cover, landform, geology and soils. In 1967, Crowley described aquatic and terrestrial ecosystems which provided similar functions and called them ecological zones. This means that the traditional geographical zone research expended rapidly into the field of ecology. Subsequently, many scholars carried out in-depth studies, and proposed different types of ecological regionalization systems for different targets (Table 2.1). For example, U.S. Forest Service regionalization targeted forest ecosystems [66], the United States Environmental Protection Agency targeted the aquatic ecosystem [50], the Great Lake regionalization targeted lake riparian conservation, the marine ecological regionalization targeted marine ecosystems [11] and the World Wildlife Fund regionalization targeted biodiversity conservation.

It was found most useful to apply aquatic ecological regionalization in ecological management to all kinds of water bodies, such as rivers, lakes and wetlands, which provided a suitable spatial unit for aquatic management. According to aquatic ecological regionalization scheme [4], water bodies with the same properties should be managed uniformly, and have the same corresponding management standards, reference criteria for monitoring and recovery goals and feasible management measures. The North American aquatic ecological regionalization scheme of the

Table 2.1 Comparison of regionalization schemes in the world

Organization	Objectives	Regionalization characteristic indices	Partition methods	Partition system
United States Environment Protection Agency [85]	To propose ecological water quality reference standards of different water bodies, and provide basis for the establishment of water management objectives and regulations	Surface types, land use, soil and the potential natural vegetation types	Overlay the 4 thematic maps and delineate the boundary based on the weighting of the factors, depending on the expert knowledge	Partition Scheme: 15 level 1 functional regions, 52 level 2 functional regions and 84 level 3 eco-function regions
Austrian Standards Institute [10]	To provide scientific basis for the biodiversity conservation and resource management	Large scale: landform, climate and vegetation; Small scale: water quality and macro invertebrates	Bottom down method	Austria was divided into 17 freshwater ecoregions
Australia [24]	To carry out researches on the health of water bodies, and put forward the evaluation system	Physical geography, climate and vegetation	Used two main features of the US-EPA model	Victoria state was divided into 17 freshwater ecoregions
The European Union Water Framework Directive [37]	To make biological quality unit (the most important), water form quality unit and physical-chemical quality unit reach a "good state"	Landscape factors: geology, topography and climate; Factors controlling river communities: river channel morphology, water flow, riverbed morphology and riparian vegetation	Mainly "from top to bottom"	

United States Department of Agriculture (USDA) and The Nature Conservancy classification system of freshwater ecosystem are good examples.

The North American freshwater ecological regionalization scheme of USDA was established in 1995. It based on the 6 world animal geographic regions established by Darlington [31], Maxwell et al. [65] established multi-scale North American freshwater ecological regionalization hierarchies (with the support of USDA), based on the North American fish distribution. These hierarchies were animal geographical zones, animal geographical subzones, regions, subregions, basins, subbasins and even smaller partition units. At subzone, region and subregion scales, North American freshwater (except Mexico) were divided into 3, 12 and 58 aquatic ecoregions, respectively (Table 2.2).

Table 2.2 Freshwater ecological regionalization of North America and TNC

Level	Scale	North American freshwater ecological regionalization	TNC freshwater ecosystem classification
Animal geographical regions	$>1 \times 10^7 \, km^2$	Darlington Fauna, the Nearctic	–
Animal geographical subregions	$1 \times 10^6 – 1 \times 10^7 \, km^2$	Types of fish family	
Region	$1 \times 10^5 – 1 \times 10^6 \, km^2$	Patterns of fish community	
Subregion	$1 \times 10^4 – 1 \times 10^6 \, km^2$	Patterns of small fish community	Animal geographical unit, or large scale data on pattern, climate, geomorphology and geology of terrestrial basin
Basin	$1 \times 10^3 – 1 \times 10^5 \, km^2$	Fish species composition(Including native species)	Biological pattern, or natural geographical landscape, climate, freshwater ecosystems connectivity
Subbasin	$1 \times 10^3 – 1 \times 10^4 \, km^2$	Physical geography and species composition	
Watershed	$1 \times 10^2 – 1 \times 10^3 \, km^2$	Hydrological processes and fish genetic characteristics	Rivers hydrological geology, surface features, height of riverbed and other ecological factors
Subwatershed	$10 – 10^2 \, km^2$		–
River district	$0.1 – 100 \, km$	Geomorphological, climatic and hydrological process	–
Reach	$0.1 – 10 \, km$	River channel geomorphology	Habitat (river gradient, height, length, connectivity, geology, hydrological processes and riparian zone, etc.)
Channel unit	$<10 \, m$	The habitat features based on sample site	–

The American Nature Conservancy (ANC) classification system of freshwater ecosystem focused on the regionalization from region to reach. It has four levels: aquatic animal geographical units, ecological basin units, aquatic ecosystem and macro habitats (Table 2.2). These map across to the scales of subregions, basins, small basins and reaches, ranging widely. (Table 2.2).

Since the establishment of China P.R., researchers have developed a large number of schemes for regionalization from biological, physical geographical and ecological aspects, such as Chinese comprehensive physical regionalization [47], Chinese biological geographical divisions [62], Chinese ecological regionalization [18] and Chinese eco-function regionalization [89]. Different regionalization had different motivations and partitioning indicators (Table 2.3), and meets different management demands.

Research Approaches

Aquatic eco-function regionalization methods mainly include top-down and bottom-up approaches. Both include the screening technology of classification indicators, spatial classification, partition authenticating, controlling precision and mapping. These technologies basically cover the whole partition process.

The top-down approach reveals the internal differences existing in the large partition units from high to low level, and partitions lower units gradually. A top-down approach has strong flexibility, it can fully reflect the basins' nature, and the degree of data required is not high, it's suitable for partitioning regions where survey data are lacking [44]. The method has wide application in geography, especially in Chinese comprehensive physical regionalization and department regionalization. The method can give full play to the experience and knowledge of experts, especially the grasp of macro patterns at large spatial scales.

The bottom-up method directly uses ecological survey data for spatial clustering, and identifying freshwater eco-function areas according to aquatic organisms, habitat types, water chemistry and so on. In this method directly based on ecological characteristics, the uncertainty of partition results depends on the intensity of monitoring. Due to the need for survey data, it is relatively difficult to implement, thus this method is more suitable for eco-function regionalization at small scale [44].

Utility of Eco-function Region for Aquatic Environment Management

Aquatic eco-functional regionalization is river classification at the basin scale. River classification is the basis of river management, and it is widely used in water environment management of the United States [65, 85], the European Union [37], Australia [24], New Zealand [59] and other developed countries internationally. Aquatic ecosystem classifications (AECs) help governments manage and conserve aquatic resources at regional, provincial, and national scales. A variety of stream and lake classification methods are available, and such methods attempt to provide a

Table 2.3 Domestic regionalization schemes in China

Name	Indicators for regionalization	Regionalization's motivation	Management meaning	References
Chinese comprehensive geographical regionalization	Level 1: significant differences of physical geography; Level 2: climate, geographical factors; Level 3: Dry and wet conditions	Reflect the regional law of natural and geographical conditions	Basic unit for agriculture, forestry, animal husbandry, water and other businesses	[47]
Chinese ecological regionalization	Level 1: hydrothermal climate index and terrain; Level 2: climate index and the regional vegetation types; Level 3: geomorphology, ecological system types, human activity	Reveal the laws of similarity and difference among natural ecoregions and the rules of human activities disturbed the ecosystem	Provide basis for the construction of ecological environment and the formulation of management policy	[18]
Chinese biological geographical divisions	Level 1: various temperature indicators; Level 2: annual dry index and natural vegetation types Level 3: geomorphology	Reveal the regularity of regional differentiation, attach great importance to the interregional connection, improve the overall understanding of natural resources and natural environment	Provide basis for sustainable development and rational use of natural resources, improve the land production potential, analyse land management policy, introduce and promote advanced agricultural technology, renovate natural environment and select first level region in nature protection area	[92]

(continued)

Table 2.3 (continued)

Name	Indicators for regionalization	Regionalization's motivation	Management meaning	References
Chinese eco-function regionalization	Level 1: give priority to level 3 Chinese ecological environment comprehensive regionalization Level 2: based on the main ecosystem types and ecological service type; Level 3: give priority to the importance of ecological and service functions and ecological sensitivity index	Reveal the sensitivity of regional ecological environment and the importance of ecological service function as well as the similarities and differences of ecological environment characteristics	Provide a scientific basis for guiding the ecological environment protection and construction planning, maintaining the regional ecological security, promoting social and economic sustainable development; provide management information and management means for environmental management and decision-making departments	[69]
Chinese biological geographical regionalization	Cluster analysis was taken according to biological appearance information based on the biological species distribution similarity among administrative county boundary units	Reveal the differences of geographical distribution of animals and plants at different scales	Provide basis for the formulation of Chinese environmental protection policies, species conservation researches and strategies, and overall planning of national conservation areas	[62]
Chinese ecological hydrology regionalization	Level 1: geographical location and dry and wet conditions Level 2: basin location, vegetation types, hydrological features and influential intensity of human activities	Reveal the rules of similarity and difference among river basin ecological hydrology systems and how the laws of human activities disturbed the ecological hydrological system	Provide basis for guiding the classification of ecological environment protection, recovery and improvement	[90]

systematic approach to modelling and understanding complex aquatic systems at various spatial and temporal scales.

Aquatic ecological region is most widely applied in general water quality management and monitoring. Water ecoregion is mainly used to promote the theories and methods of quality management, or reflect a wide range of water quality management issues. Arkansas researchers applied water ecoregions to the water quality utility analysis (UAA), and found that United States Geological Survey (USES) water ecoregion was very effective in setting water quality standards, especially for dissolved oxygen and the related standards of indicated fishes, the study promoted the application of water ecological regionalization system in water quality management. Now, Arkansas starts to scientifically evaluate pH, water hardness and toxic contaminants based on aquatic ecological region.

In terms of river biological monitoring, water ecoregion is an effective means to select river reference areas with the minimum degree of disturbance. Reference areas are used to quantify the health of river ecosystems, and establish standards for comparison among rivers. In Florida, the EPA determined river biological monitoring reference areas based on water ecoregion, showing that it was an effective method. Hughes [49] deemed that the reference areas also provide a means for the comparative study of water ecosystems in different river basin.

Water ecoregion is an effective method to collate the data of groups of lakes and determine their nutrients standards. The productivity and biological communities of lakes are correlated with the water ecoregion, because lake productivity and trophic status are determined by climate, topography, soil, geology, land use and other factors. Due to the spatial variability of lake ecosystems, USEPA found that lake nutrient criteria cannot be used uniformly all over the country, and establishing a standard based on ecoregion is more suitable for lakes and reservoirs management. Therefore the United States enacted a national policy for formulating regional eutrophication benchmarks in 1998, and issued guidelines for lakes and reservoirs in 2000. Since then, it began to enact eutrophication quality benchmarks for lakes and reservoirs in 14 water ecological areas, which provided strong support for the effective management of lakes and reservoirs in the United States [6].

Water ecoregions play an important role in describing wetland characteristics and evaluating the influence of human activities on wetlands [15] combining the structure of water ecoregion and hydrological unit would be more conducive to carrying out research on wetlands. Bedford [14] pointed out that wetland management should be carried out at a large scale, as in his study of wetland standards, it was necessary to expand to landscape level from individual projects, and the water ecoregion was the most suitable unit to build the wetland standard at large scale.

Water ecoregion could be used to study the distribution of aquatic macro invertebrates and fish fauna. For instance, Whittier examined the relationship between water ecoregion and the aquatic fauna in the State of Oregon. By determining the degree of consistency between basic characteristics of the small watershed aquatic ecosystem and eight ecoregions, he found water ecoregion was an effective structure for classifications of basins based on aquatic organisms and management at large scale. While [40] divided Kansas into different fish ecoregions using multivariate statistical methods according to the characteristics of fish, his results showed that fish ecoregion have no correlation with USEPA water ecoregion. Although water ecoregion can't reflect species accretion and species occurrence, it can reflect the environmental characteristics of the aquatic ecosystem. For example, some studies found that most of the river physical environment parameters can be classified successfully by water ecoregion, and the physical and chemical properties of water had high congruence with water ecoregion. These studies further illustrated that water ecoregion is one way to reflect the distribution of aquatic organisms. However, the distribution of aquatic organisms was also influenced by some other factors, which couldn't be reflected easily by the existing ecological regionalization indicators, such as sediment load and hydraulic condition.

2.1.2.2   Water Environmental Management Region in China

For water environmental management methods, China has studied regionalization since the 1950s. Today, functional regions play an important role in environmental management. In China, water environmental management has centred on water quality goals for a long time, and established water environmental management strategies and methods [48]. Different management departments carried out different functional regionalization schemes, such as water functional regions, eco-function regions and principal functional regions etc. These regionalizing schemes were carried out according to the management need of different management departments. Considering the threats to the long term integrity of river basin ecosystems and the concerns over the sustainability of human development, there are more and more studies of basin eco-function regionalization. There are significant differences among functional regions in laws and policies, purposes, systems, indices, methods that related to regionalization and the roles they played in management (Table 2.4).

Aquatic functional region is the most important water environmental management scheme in China nowadays [61] to formulate the relevant standards of water quality protection for managing water bodies (Table 2.5). For 2069 rivers in 31 provinces, autonomous regions, municipalities and 248 lakes and reservoirs, 3397 level 1 eco-function regions were defined, and the area of level 1 region belonging to lakes and reservoirs was 28,949 km^2. According to the aquatic functional regionalization system, each province divided its basin into different water functional regions for management, and specified different management measures according to the condition of the river systems in the province. For example, since most part of Guizhou is plateau, 72% of the water function regions in the province were classified as reserved areas. In functional regionalization, the groundwater source of Liaoning Province was listed as the protect object due to its function as an important source of drinking water.

Table 2.4 Compare of aquatic ecosystem function management region and water function zone, ecological function region, and main function district

Topics	Aquatic eco-function regions	Water function regions	Ecological function regions	Principal functional regions
Concept	According to the spatial characteristics of structure and process of water ecosystem at different scales and the need to maintain the integrity of ecosystem, classified similar land and water as a geographical unit	According to the present condition of development and utilization of river basin or regional water resources and the different requirements that social economy had for water resource of different water sectors in different regions during a certain period. Also considering the sustainable utilization of water resources, defined the water with specific functions in rivers, lakes and reservoirs, and provided basis for the determination of water conservation objectives	Regions are defined according to the regional ecological environmental factors, the sensitivity of ecological environment and the laws of spatial differences among ecological service functions	On the basis of comprehensive analysis on present developmental density and development potential of environmental supporting capacity in different regions. The geographical unit with specific principal functions that is defined according to the spatial differences of natural environmental elements, development level of social economy and human activity types
Laws and policies	None	Water law of the People's Republic	The national ecological environment protection program	The People's Republic of China national economic and social development of the eleventh five-year plan outline
Management roles	To determine the water protection goals, health assessment, benchmark standards	The water quality protection standards determine each function level	Ecological function zones location	Industry entry requirements, regional economies development models

(continued)

Table 2.4 (continued)

Regionalization purpose	Reveal the spatial distribution pattern of different regions of aquatic organisms, clear water ecological function type and its importance, and determine the water ecosystem protection targets, To provide a scientific basis for river basin water ecosystem protection and restoration	According to the divisions over waters of the nature, combined with social demand, coordinate the relationship between the whole and partial, Determine the functions and features of the waters of the order. for developing and using water and provide a scientific basis for protection and management, in order to realize the sustainable utilization of water resources	Confirm all kinds of Ecological function zones' dominant ecological service function and ecological protections, Ecological function zones play a key role to the state and regional ecological security, To guide regional ecological protection, ecological construction resource utilization and economic and social development	To orient the development or conservation strategy in administrative level based on the regional environmental capacity, development potential and economic development status. As a basis to realize the coordinated development of the region, promote the formation and orderly pattern and overall coordination of space development
Regionalization system	Level 4 regional system. Lever I zone and level II zone are the aquatic ecological function regions; Lever III zone and level IV zone are the aquatic ecological function regions	Level 2 regional system. Level 1 functional zones are divided into 4 classes: reservoir reservations, exploitation and utilization of zone and buffer zone; Level 2 functional zones are divided into 7 classes: Drinking, industrial, agricultural, fishery, landscape, transition, pollution control	Level 3 regional system. Level 1 zone and the level 2 zone are the national ecological zoning, Level 3 zone is ecological function type, include: protecting the diversity of living things; water conservation; soil desertification, and control nutrients to maintain	Give priority to within country and province two stage build division system. Each level division system perform at 4 principal functional regions including the optimization of development, the key to develop, limit to development and prohibit development

Table 2.5 Standards of water quality protection

Classification indices for secondary regions	Environmental quality standards for surface water [4]
Drinking water source regions	Class II to III water quality standard
Industrial water regions	Class IV water quality standard
Agricultural water regions	Class V water quality standard
Fishery water regions	Class II to III water quality standard
Landscape recreational water regions	Class III to IV water quality standard
Transition regions	Choose corresponding quality controlling standards according to the requirements that water quality of outflow meet the water quality of adjacent functional region
Pollution controlling regions	Choose corresponding quality controlling standards according to the requirements that water quality of outflow meet the water quality of adjacent functional region

2.1.2.3 Aquatic Eco-function Regionalization (AEFR) Framework for Basins in China

Combining the domestic and world aquatic ecoregion regionalization schemes, and considering the scale of the basin, and the need for aquatic ecological management; a nested 4 level aquatic eco-function regionalization framework was put forward (Fig. 2.1). The system was an integrated regionalization system that combined the mosaic regionalization for terrestrial environmental factors in level 1 and level 2 AEFR, and the functional definition of river reaches in level 3 and level 4 AEFR.

Level 1 and level 2 is ecoregion showing the fauna distribution and community types respectively in basin and sub-basin scale. Level 3 reflects the pressure of land use on freshwater ecosystem. Level 4 defines the eco-function of the river types including ecological and functional demands for aquatic ecosystem.

Level 1 AEFR represents the fauna pattern which was influenced by climate, topology, geomorphology and geology, and it was reflected by the features of hydrology or water resources patterns at large scales. Level 2 AEFR represents the aquatic ecosystem types which were influenced by geomorphology and vegetation mostly, and it was reflected by the features topology at large or medium scale. Level 3 and level 4 regions were mainly classification of river subbasins, habitats and function regions, and reflected the characteristics of subbasins and the differences in ecological functions. Level 3 regions were divided mainly based on the types of human disturbance and river size, which reflected the disturbance background of land use, aquatic physical structure and the regional dominant aquatic ecological functions. Level 4 regions were definition of aquatic function in reach scale, reflecting aquatic ecological functions and service functions at reach scale.

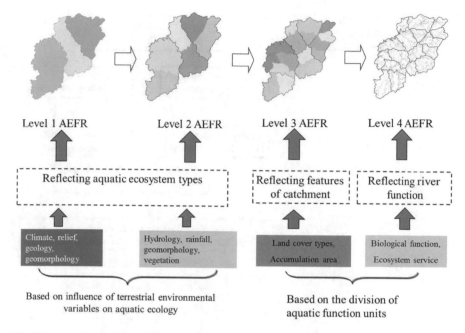

Fig. 2.1 Aquatic eco-function regionalization framework for Basin

2.1.3 Technical Routes of Aquatic Eco-function Regionalization

2.1.3.1 Technical Routes of Level 1 and Level 2 Delineation

Premise for level 1 and level 2 AEFR was that aquatic ecosystem was influenced by terrestrial landscapes. Level 1 and level 2 AEFR was delineation in large scale. Indicators were selected by analysis between aquatic community features and environmental variables (Fig. 2.2). Spatial data of environmental variables were used and qualitative spatial cluster analysis was conducted.

2.1.3.2 Technical Route of Level 3 Delineation

Level 3 AEFR was nested in level 2, and was carried out inside the boundary of level 2. It is the classification of rivers at catchment scale (Fig. 2.3). Indicators include land use type and accumulation area of catchment in catchment scale.

Fig. 2.2 Technical routes of
Level 1 and Level 2 aquatic
eco-function regionalization

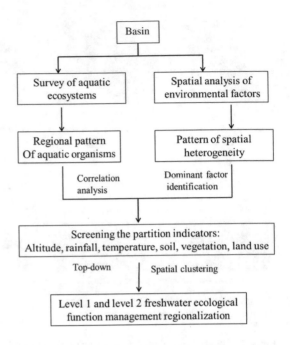

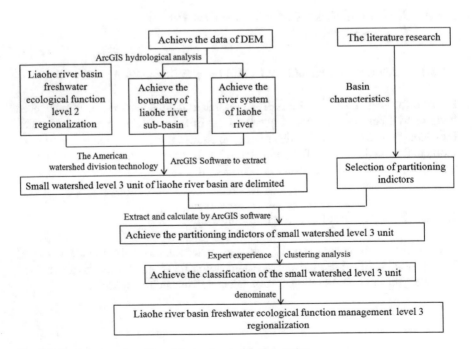

Fig. 2.3 Technical routes of Level 3 aquatic eco-function region

2.1.3.3   Technical Route of Level 4 Delineation

Level 4 AEFR was carried out at reach scale, and it was completed by classifying river habitats, and evaluating biological functions and river services functions (Fig. 2.4, Table 2.6). River function was determined by assessing the importance of functions.

Used SRTM DEM data of 90 m resolution, Liao river water system were extracted, and it was corrected combined with basin LADSAT TM remote sensing images. River systems were divided into reaches in the river nodes.

River types are related to ecological features of aquatic ecosystem, and were classified with indices in reach scale which represents the habitat types. Based on the literature review, the indices library in reach scale was established. And classification indices were selected on their applicability analysis. River reaches were classified by cluster analysis and frame diagram. Finally we got river types of reach scale, and we identified reaches with high biodiversity, and analyzed their ecological significance of river types.

Ecosystem functions include biological habitat functions and biodiversity supporting. Species distribution data were collected through field investigation and literature collection, especially the spatial distribution of rare, endangered and endemic species. Through field investigation, data on algae, macrobenthos and fish were collected, and calculated the biological diversity indices, identified the river reach with high diversity indices.

Ecosystem service functions include the function of drinking water source, water resources supply, contact and non-contact leisure activities and groundwa-

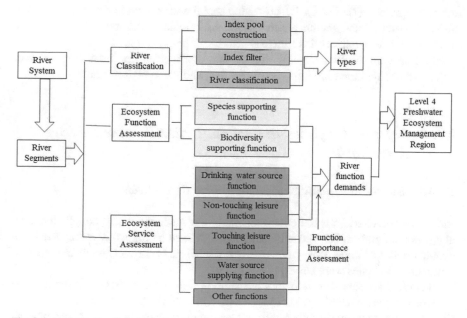

Fig. 2.4 Technical routes of Level 4 AEFR

Table 2.6 Functions of aquatic eco-function region

Issues	Function types	Function subtype
Ecological functions	Species support	Support rare and endangered species
		Support endemic species
		Support the important habitats for crayfish clam
	Biodiversity supporting	Support benthic animal diversity
		Support fish diversity
		Support the biodiversity in habitats similar to wetlands
		Support the biodiversity in habitats similar to reservoirs
		Support the biodiversity in habitats similar estuary
Ecological service functions	Related to human health	Surface water sources for drinking
		Contact entertainment
	Related to human use	Water supply
		Non-contact entertainment
		Groundwater recharge

ter recharge, etc. (Table 2.6). Based on the social and economic development data, channel structure and geomorphology data, we calculated the suitability of reaches to meet the ecological service function.

By assessing the function importance, and considering the importance of river types, we decided the function type of reaches, and named the reaches.

2.1.4 Aquatic Eco-function Regionalization Scheme for Liao River Basin

2.1.4.1 Survey of Aquatic Ecosystem in Liao River Basin

Indicators surveyed for water ecosystems include water chemistry, aquatic habitats, algae, macrobenthos, fish and aquatic plants. Different biological types represent the characteristics of water ecosystem at different levels in basin, catchment and reaches. Altogether 563 sites were surveyed (Fig. 2.5).

Technical schemes for field survey methods were adopted based on the method of river ecosystem health evaluation used by the USEPA. At selected sampling sites (Fig. 2.5), we measured temperature, water depth, transparency, dissolved oxygen

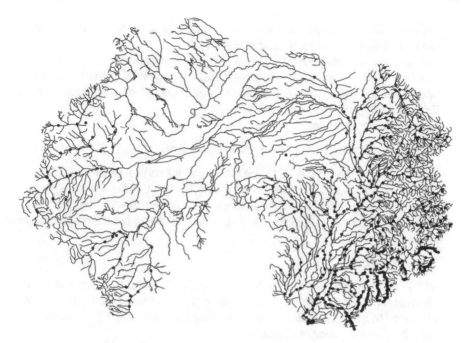

Fig. 2.5 Sampling sites in Liao River Basin for water ecological survey

and other physical parameters, collected water quality and sediment samples and acquired phytoplankton (plant), benthic animals, fish and aquatic vascular plants specimens, and evaluated habitat condition in the riparian zone. Thus we acquired physical, chemical and ecological indicators, then compared different rivers within basins, and the spatial differences from upstream to downstream, and identified the key constraints that formed differences in the ecosystems. The detailed research plan was as follows:

1. Sampling plan for water quality: Three sampling stations (right bank, middle and left bank) were set up in each river section to collect water samples. taking mixed samples if the water depth >2 m; taking mixed samples between 0.5 m under the surface and 0.5 m above the benthic surface if the depth ranged from 1 to 2 m; and taking samples in the central section of the water column when the depth <1 m.
2. Plankton sampling scheme: Taking 1000 ml water samples on the 0.5 m surface layer, put samples into the jar, and fix with 10–15 ml corus liquid. After taking water back into the lab, the bottles were shaken well and poured into 1000 ml circular or cylindrical separating funnel, and allowed to stand for 48 h. Supernatant fluid was drained with tiny siphon until the phytoplankton sediment volume reached about 20 ml, turning on the piston of separatory funnel, put the funnel into a 30 ml specimen bottle, then flushed funnel 2–3 times with a little precipitation supernatant fluid, and diluted with water to 30 ml. If the volume of quantitative sample water was larger than 30 ml, stand for more than 24 h again, then

drained redundant supernatant fluid. Each bottle labelled with sampling location, sampling date and number of sampling site.

3. Zooplankton sampling scheme: Taking 10 L water samples from the 0.5 m surface layer with WB-PM, samples were filtered through number 25 plankton net, gently opened the net top, poured the remaining liquid into 50 ml centrifuge tube and fixed by adding 4% formaldehyde solution.

4. Benthic animal sampling scheme:

   • Stream sampling (water depth is less than 1.2 m)

   Within 100 m, chose typical jet stream habitat, took three replicate samples with Surber net at each site, put samples into a sample bottle; in the coastal zone, took samples ranging 5 m (0.5 m × 10 times) with D shape brail. The sampling times in each habitat were allocated according to the number of habitat types and their relative area.

   • River sampling

   Within a 40–50 m sampling area at each sampling site, took samples with D form net for 10 times, every 0.5 m. Sampling times in each habitat were allocated according to the relative area of each habitat (water and static water area, water area and substrate composition, etc.).

5. Fish sampling: In wading rivers, fish was caught with backpack electric fisher, sampling for 30 min within 1 Km at sampling sites covering all kind of habitats of pool, riffle and run. In non-wadeable reaches, electro-fishing methods can be used in riparian shallow regions (water depth <1 m); while in central deep region, fish were sampled through net trawling from a boat, the travel distance at each sampling sites was less than 1000 m. After sampling, fish specimens were processed, and examined for scale and fin damage. After cleaning the fish, the body length and weight were measured, and the specimens were then labeled at the lower mandible or caudal peduncle. Only 10–20 specimens of rare fish and local special species were retained, all the rest were released back into the river. 5–10% formaldehyde solution was used to fix fish samples. For larger individuals, injected the right amount of fixed liquid into intraperitoneal. Fish samples were packed using gauze, which could be soaked with formaldehyde solution, to prevent surface drying, put into the collection box.

6. Physical habitat indicators: The physical habitat indicators include altitude, velocity, water depth, channel width and water temperature. Measure the substrate with scree, and classified substrates according to the following standards [32]: boulders (>256 mm), cobblestones (128–256 mm), the pebbles (64–128 mm), big round stones (32–64 mm), and small round stones (16–32 mm), thick gravels (8–16 mm), fine gravels (4–8 mm) and sand (<4 mm). Measuring cup was used to measure the volume of different types of substrate, bottom index was calculated on the basis of the following formula [52]: bottom index = 0.08% × boulders + 0.07% × cobblestones + 0.06% × pebbles + 0.05% × big round stones + 0.04% × small round stones + 0.03% × thick gravels + 0.02% × fine gravels + 0.01% × sand. The bottom index result is a mean size of the substrate.

2.1.4.2   Aquatic Eco-function Regionalization Scheme for Liao River Basin

Level 1 and Level 2 Delineation Schemes

Based on the review of literature around the world and knowledge on the relationship between aquatic ecological features and environmental variables, we built a delineation index system, and selected the indicators used for AEFR zoning, level 1 and level 2 regions were divided by spatial clustering, we verified the results of regionalization according to the results of biological clustering.

Establishment of Index System

Due to the different dominant environmental elements in each basin, the same index couldn't be used for all regionalization, so it was necessary to select the index which would most significantly reflect the characteristics of the ecosystem, and formed the final partition system. Based on the literature, we built a suitable environmental indicators library for regionalization at different scales (Tables 2.7 and 2.8), and analysed the ecological significance, types and availability of each indicator.

Table 2.7   Index system of Level 1 AEFR

Indicator types	Indicators	Manifestation of indicators
Climate	Average annual precipitation (mm), average annual temperature (°C), accumulated temperature, moisture index	Quantitative
Hydrology	Runoff depth	Quantitative
Geology	Lithology	Quantitative
Topography and geomorphology	Geomorphic types or elevation (m)	Qualitative, quantitative
Distribution pattern of aquatic organisms	Aquatic community types	Qualitative

Table 2.8   Index system of Level 2 AEFR

Indicator types	Indicators	Manifestation of indicators
Geomorphology	Geomorphic types, elevation (m), slope	Qualitative, quantitative
Vegetation	Vegetation types, vegetation index, coverage	Qualitative, quantitative
Soil	Soil types, soil constitution	Qualitative, quantitative
Land use	Land use types, the ratio of Human land use	Qualitative, quantitative
Distribution pattern of aquatic organisms	Aquatic community subtypes	Qualitative

Through analysing the sensitivity, independence, spatial heterogeneity of the alternative indicators, and the correlation analysis between fish species diversity index and environmental variables, combined with existing case studies and expert opinion, we explored the regional applicability of selected environmental factors, and further screened the index system for level 1 and level 2. Finally landform, annual average precipitation was chosen as indicators of level 1 aquatic eco-function regionalization; and elevation and rainfall are chosen as the indicators of level 2 aquatic eco-function regionalization.

Spatial Pattern of Indicators

River had different spatial pattern along the basin (Fig. 2.6). The aquatic eco-function region indicators showed significant spatial distribution pattern in the basin (Figs. 2.7, 2.8, 2.9, 2.10, and 2.11)

Results of Level 1 Regionalization in Liao River Basin

With k-means spatial cluster analysis combined with the boundary of sub-basins, the Liao river basin was divided into four level 1 aquatic eco-function regions (Fig. 2.12).

Spatial Cluster Analysis of Benthic Animal in Liao River Basin

Through study of the field monitoring results in August 2009 and June 2010 of the Liao river basin, we found that the spatial distribution characteristic of benthic animals showed clear heterogeneity pattern. The data for benthic animals were analyzed by cluster analysis in PC-ORD5, and tit showed distinct patterns (Fig. 2.13).

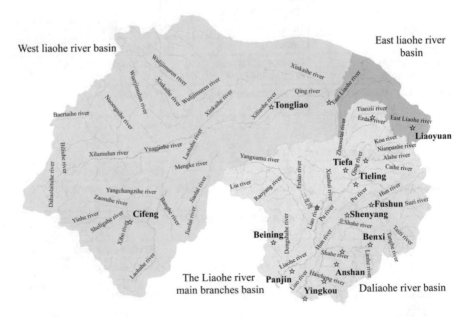

Fig. 2.6 Spatial distribution of rivers in Liao River Basin

Fig. 2.7   Spatial distribution of average annual precipitation in Liao River Basin

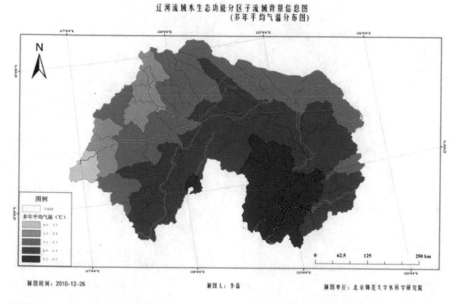

Fig. 2.8   Spatial distribution of average annual temperature in Liao River Basin

辽河流域水生态功能分区子流域背景信息图
（代表型DEM分布图）

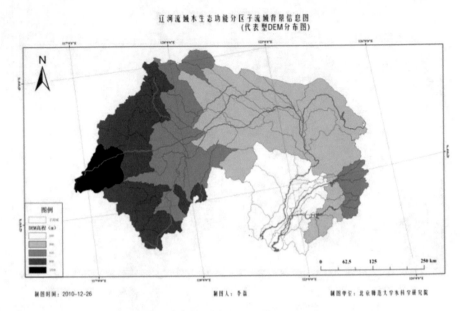

Fig. 2.9 Spatial distribution of altitudes in Liao River Basin

辽河流域水生态功能分区子流域背景信息图
（全流域坡度分布图）

Fig. 2.10 Spatial distribution of slope in Liao River Basin

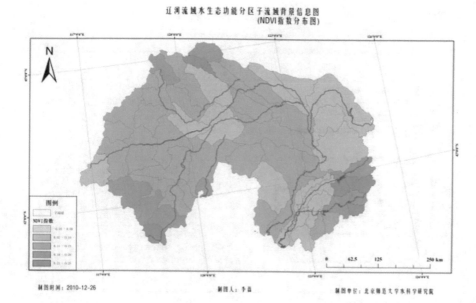

Fig. 2.11 Spatial distribution of vegetation index in Liao River Basin

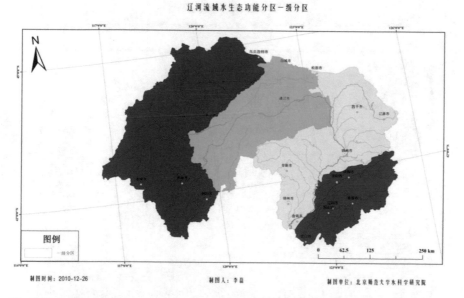

Fig. 2.12 Level 1 aquatic eco-function region in Liao River Basin

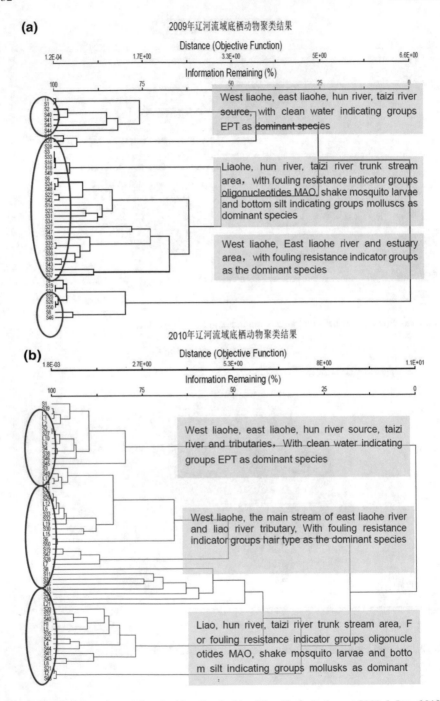

Fig. 2.13 The clustering results of zoobenthos in Liao River Basin (**a** August 2009; **b** June 2010)

Table 2.9 Characteristics of water ecoregions in Liao River Basin

Ecoregion	River system	Regional characteristics
Semi-arid water ecoregion in West Liao river upstream, plateau and hilly area	Including Xilamulun river upstream and midstream, and most part of Ur JiMuLun river, upstream from Goumenzi in Laohahe basin	Eastern Mongolia plateau, dominated by hills, about 500–1500 m height, 200–400 mm annual average precipitation
Drought water ecoregion in West Liao river downstream, desert area	Downstream of Xilamulun river and Laoha river, the Xinkai river and West Liao river	Relatively flat terrain, mainly sandy soil, annual precipitation is less than 200 mm, it is difficult to form surface runoff, river have dried up for 10 years
Semi-humid water ecoregion in hilly East Liao river, and Liao river mainstream, plain	East Liao river, Liao river mainstream and numerous tributaries in the west of Liao river downstream	East of the region is mainly hills, the elevation of downstream in Liao river mainstream is the lowest for the whole basin. Water rich, with an average annual precipitation about 600–800 mm
Semi-humid freshwater ecological area in Huntai river basin	Hun river, Taizi river and DaLiao river	Huntai river system is relatively independent compared to Liao river mainstream, so it was divided separately. Dominated by hills, annual precipitation are more than 1000 mm

Clustering results of benthic fauna showed that the spatial pattern of benthic animals were consistent with water level 1 eco-function regions (EFR), suggesting that level 1 eco-functional regions can reflect the characteristics of water ecosystem (Table 2.9).

Results of Level 2 Aquatic Eco-function Region in Liao River Basin

Level 2 regionalization in Liao river basin is the continual refining of level 1 regionalization and further reflects the influence of geomorphology, vegetation and other natural environment factors on aquatic ecological structure. Following the top-down method, we acquired spatial data based on the higher accuracy of climate factor, used statistics and spatial information technology for further subdivision of the aquatic ecoregion unit into different units according to the spatial heterogeneity of geomorphology, vegetation and soil environmental factors. The result was verified to be well matched with the community characteristics and bio-geographical pattern of macroinvertebrates (Fig. 2.14).

14 level 2 freshwater eco-function regions in Liao river basin, name the 14 regions according to the characters reflected by level 2 regionalization indicators, and the name for the 14 level 2 regions are in Fig. 2.14.

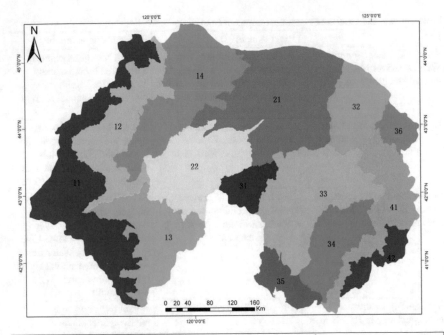

	The explantation of each aquatic eco-function region
11	the southern of Greater Hinggan Mountains, semi-arid, west Liao river, intermediate altitude and mesorelief mountain
12	the southern of Greater Hinggan Mountains, semi-arid, Silas wood Aaron river and Wulijimulun river, low altitude, rolling hills
13	the southern of Greater Hinggan Mountains, semi-arid, Laoha River and Jiaolai River, plateau and hilly
14	the southern of Greater Hinggan Mountains, semi-arid, Wulijimulun River, low altitude, hilly
21	the West Liao river, semi-arid, cultivated vegetation, lacustrine and alluvial plain; 22, the West Liao river, semi-arid, desert and alluvial plain
31	the Yangximu River, semi-humid, hilly and plateau
32	the East Liao river, semi-humid, low altitude, platform and plains
33	the Liao River, semi-humid, alluvial plains
34	the Huntai River, sub-humid, alluvial plains
35	the estuary of Liao river plains
36	East Liao River, sub-humid, low altitude, hilly
41	the south of Changbai Mountains, low altitude, small ups and downs region
42	the south of Changbai Mountains, taizi river, middle and lower altitude, middle ups and downs region

Fig. 2.14 Level 2 aquatic eco-function regions in Liao River Basin

Level 3 Aquatic Eco-function Regionalization Scheme for Liao River Basin

Level 3 eco-function regions reflect the disturbance pressure of land use characteristics, river catchment sixe and dominant ecological function at the scale of the river catchment. Level 3 regionalization was carried out by calculating indices for subbasins.

Establishment of Level 3 Regionalization Indicators System

The selection of level 3 regionalization indicators reflected the physical properties of catchment, structure and size of water bodies, and the influence of human activities on the river system. Combined with the literature research, land use, basin area and river structure (drainage density and level of rivers, etc.) were chosen as the regionalization indicators (Table 2.10).

Spatial Pattern Analysis of Indicators in Level 3 Regions

Land Use Pattern
Land use types in Liao river basin were interpreted from LANDSAT TM of 2009. Land use types include grassland, forest land, unused land, farmland, Urban and rural residents, industry and mining and water area (Fig. 2.15).

Table 2.10 The meaning and calculation method of indicators in Level 3 regions

Indicators	Meaning	Calculation method
Area ratio of land use types	The area percentage of land use types in each basic catchment units, including forest land, grassland, farmland, urban land and other construction land	Under the ArcGIS software, land use data layer was used to extract the area and calculate the percentage divided by the total area
Stream order	The stream order dominant and the largest stream order each basic catchment units	Calculated according to the Strahler's method which defined that tributaries at the top of river are the lowest level, increasing level according to the increasing number of tributaries
Number of nodes	The total number of nodes within each basic catchment units	One confluence of two rivers was one node, not counted as two, if 3 rivers meet at the same point, recorded as 2 nodes; 4 river meet at the same point, recorded as 3 (N − 1), and so on
Drainage area	The catchment area of each basic catchment units	In ArcGIS software, selected all the level 3 regions and all watershed units in upstream, calculated the area using the attribute table

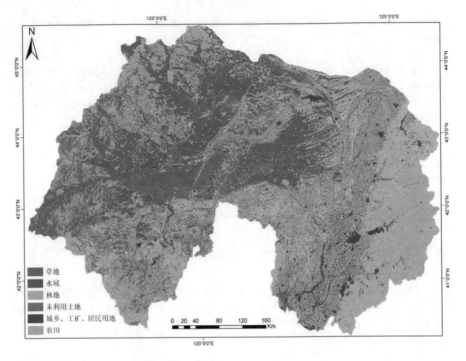

Fig. 2.15 Land use types in Liao River Basin

Different land use types showed different pattern in Liao river basin (Fig. 2.15) and they distributed among regions and rivers (Table 2.11). Farmland occupied the most percentage of the land area and distributed mainly in main area of Liao river main stem (Fig. 2.15 and Table 2.11). Natural vegetation of grassland and forest land distributed mainly in the upper part of the basins, and the two types occupied 46% of the total area. The 3 types occupied 83% of the total area, and they are the main land use type in Liao river basin.

The land use types in the whole Liao river basin were analyzed and results showed that East Liao river is covered mainly by farmland; West Liao river is covered mainly by grassland, farmland and some unused land; Liao river mainstream is mainly covered by farmland; the land use types of Hun river and Taizi river are similar, their upstream is covered mainly by woodland, and downstream mainly farmland, and a few urban and rural residents. The east of entire Liao river basin is covered mostly by farmland and woodland, the west mainly grassland, large reservoir occurred in the whole river basin.

Characteristics of Stream Order in Liao River Basin

The distribution features of stream order in the whole basin reflected that the area with more tributaries had the higher stream order. The stream order distribution of Liao river basin is shown in Fig. 2.16.

Table 2.11 The proportion of land use types and the main distribution regions

Land-use type	Types included	Proportion (%)	Main distribution regions
Farmland	Paddy field and dry field	37.15	The main stream of Liao river and most of its tributaries, most of East Liao river basin, most of DaLiao river basin, minority in West Liao river, minority in Hun river downstream and Laoha river, parts of Jiaolai river basin, few regions in Lubei river, Xinkai and Shali river
Urban and rural residents, industry and mining	Town, residential area and other construction lands	3.87	Small parts of Hun river midstream and Taizi river downstream
Unused land	Sandy land, the Gobi, saline-alkali land, wetland, bare land, bare rock fertile land	7.31	Parts of East Sha river, Laoha river and Xilamulun river basin. Small parts of Xinkai river upstream basin
Forestry land	Scrub, open woodland and other forestry land	18.09	Most of Hun river and Taizi river basin, parts of tributaries in Laoha river upstream
Water area	Canals, lakes and reservoirs pits, beach, permanent ice snowfield	2.12	Its distribution area is very small, mainly include several large reservoirs, such as the Qing river reservoir, Chessboard hill reservoir, Dahuofang reservoir and so on
Grassland	Grassland with high, medium and low coverage	28.2	Most of West Liao river, most tributaries of the Xilamulun river upstream and Xinkai river upstream

The spatial distribution characteristics of stream order was that East Liao river had 4 grades of stream order, Liao river mainstream and West Liao river had 6 grades, Hun river and Taizi river had 5 grades (Fig. 2.16). Cluster indices were based on 14 level 2 water eco-function regions using Ward's clustering method. The method ensured that the interclass variance was minimized, and the variance between groups was maximized, and thus separated the objects well. The method is a widely used clustering analysis, and it was completed in R software.

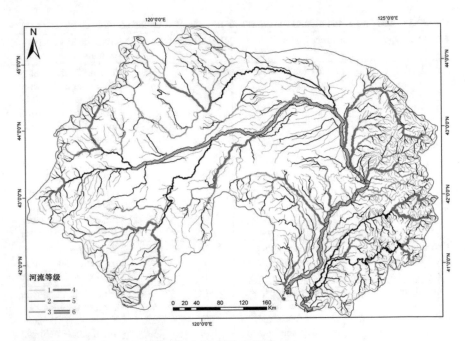

Fig. 2.16 Distribution of stream order in Liao River Basin

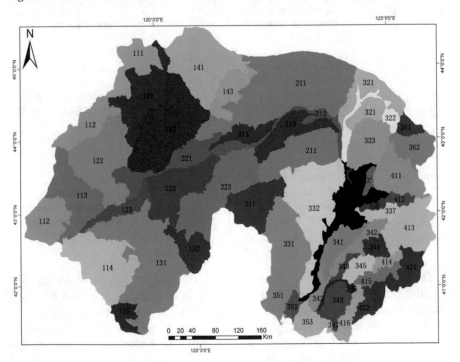

Fig. 2.17 Level 3 freshwater eco-function regions in Liao River Basin

Results of Level 3 Regionalization in Liao River Basin

According to the clustering results, combined with the analysis of basin characteristics, the whole Liao river basin was divided into 50 levels 3 AEFRs eventually (Fig. 2.17). The total 50 level 3 freshwater eco-function regions was named according to the characters of level 3 regionalization indicators. And the name for each level 3 AERR are as follows (Table 2.12).

Level 4 Aquatic Eco-function Regionalization Scheme for Liao River Basin

Level 4 regions are aquatic eco-function regions, for which the regionalization was completed by classifying water bodies and evaluating habitat function, biological function and service function.

Establishment of Classification Indicators

Through literature review, combined with expert opinion, the classification indicators at different scales were analyzed. The classification of Liao river basin was conducted at reach scale, hence the classification indicators must be able to reflect the size of reach, physical structure and natural form [57]. Stream order, confinement level, the number of channels, and sinuosity were selected as classification indicators of Liao river basin (Table 2.13).

Spatial Distribution Characteristics of River Classification Indicators

Four river classification indicators were quite different in spatial distribution and change trendamong tributaries and main stems. Liao river basin has 3158 river reaches. The stream order was classified as 1–6, which were divided into mainstreams and tributaries. There were 2935 tributaries, including level 1, 2 and 3 reaches, and a few level 4 and 5 reaches, and accounting for 93% of the total number of river reaches; Parts of level 4 and 5 reaches and whole level 6 reaches are main rivers, 223 in total, accounting for 7% of the total number of reaches.

According to the classification results of confinement level: confined river valleys included 306 reaches, accounting for 9.7% of the total number of river reaches, which are mainly distributed in Liao river source and Huntai river source; partly confined river valleys included 1308 reaches, accounting for 41.4% of the total number of river reaches, which were mainly distributed in the transition from the mountain areas to the plains; unconfined river valleys included 1544 reaches, accounting for 48.9% of the total number of river reaches, which were mainly distributed in the mid and downstream reaches of East Liao river, West Liao river and Huntai river and the Liao river plains.

According to the classification results of the channel number: single channels included 3076 reaches, accounting for 97.4% of the total number of river reaches, and were widely distributed; double channels included 82 reaches, accounting for

Table 2.12 Description of the numbering of eco-function regions in Liao River Basin

111	Wuerjimulun river, mid-altitude, middle ups and downs mountain, water conservation, source stream
112	Source of Silas wood Aaron River, mid-altitude, middle ups and downs mountain, water conservation, source stream
113	Upstream of Silas wood Aaron River, mid-altitude, middle ups and downs mountain, water conservation, medium river
114	Midstream and upstream of Yingjin river, mid-altitude, middle ups and downs mountain, covered by forests and grassland, water conservation, medium river
115	Laoha River, mid-altitude, middle ups and downs mountain, water conservation, source stream
121	Upstream of the tributary of Wuerjimulun River, mid-altitude, small ups and downs mountain, water conservation, seasonal river
122	Upstream of Silas wood Aaron River, mid-altitude, small ups and downs mountain, water conservation, medium river
123	Silas wood Aaron River, mid-altitude, small ups and downs mountain, water conservation, small tributaries
131	Midstream of Laoha river, platforms and hills, agriculture maintain, medium river
132	Jiaolai river, platforms and hills, water conservation, small tributaries
141	Teng's Guo Le river, low-altitude, hills, water conservation, seasonal river
142	Midstream and downstream of the tributaries of Wulijimulun river, low-altitude, hills, water conservation, seasonal river
143	Midstream of the main stream of Wulijimulun river, low-altitude, hills, water conservation, seasonal river
211	West Liao river, lacustrine deposit, alluvial plain, agriculture maintain, seasonal river
212	Downstream of the West Liao river, agriculture maintain, river
213	West Liao river, alluvial plain, agriculture maintain, seasonal river
221	Midstream of the main stream of the West Liao river, biodiversity maintain, trunk stream
222	Downstream of the Laoha river, biodiversity maintain, medium river
223	Midstream and downstream of the Jiaolai river, agriculture maintain, seasonal river
311	Source of Yangximu river, platforms and hills, agriculture maintain, medium river
321	Downstream of the East Liao river, platforms and plains, agriculture maintain, small river
322	Midstream and downstream of the main stream of the East Liao river, biodiversity maintain, trunk stream
323	Zhaosutai river, platforms and alluvial plain, city maintain, medium river
331	Midstream and downstream of the Raoyang river, alluvial plain, agriculture maintain, medium river
332	Tributary of the main stream of Liao river, alluvial plain, agriculture maintain, medium river
333	Main stream of Liao river, alluvial plain, biodiversity maintain, trunk stream
334	Downstream of Zhaosutai river, alluvial plain, biodiversity maintain, trunk stream

(continued)

Table 2.12 (continued)

335	Downstream of Qing river, alluvial plain, biodiversity maintain, trunk stream
336	Downstream of Chai river, alluvial plain, biodiversity maintain, trunk stream
337	Downstream of Fan river, plains, biodiversity maintain, medium river
341	Pu river, alluvial plain, city maintain, medium river
342	Midstream and downstream of the main stream of Hun river, alluvial plain, city maintain, medium river
343	Downstream of Taizi river, alluvial plain, city maintain, small river
344	Midstream of Hun river, water conservation, small river
345	Beisha river, alluvial plain, biodiversity maintain, small river
346	Downstream of Taizi river, plains, town and biodiversity maintain
347	Downstream of Haicheng river, alluvial plain, agriculture maintain, small river
351	Estuary of Raoyange river, plains, agriculture maintain, medium river
352	Estuary of the main stream of Liao river, plains, city maintain, trunk stream
353	Estuary of the main stream of DaLiao river, plains, city maintain, trunk stream
361	Upstream of East Liao river, low-altitude, hills, agriculture maintain, medium tributary
362	Upstream of East Liao river, low-altitude, hills, agriculture maintain, medium river
411	Midstream of Qing river, low-altitude, small ups and downs mountain, water conservation, medium river
412	Upstream of Chai river, low-altitude, small ups and downs mountain, water conservation, medium river
413	Upstream of Hun river, low-altitude, small ups and downs mountain, water conservation, medium river
414	Midstream of Taizi river, low-altitude, small ups and downs mountain, water conservation, small river
415	Midstream of the main stream of Taizi river, low-altitude, small ups and downs mountain, biodiversity maintain, medium river
416	Midstream of Taizi river, low-altitude, small ups and downs mountain, water conservation, medium river
421	Midstream and upstream of Taizi river, mid-altitude, middle ups and downs mountain, water conservation, small river
422	Midstream of Taizi river, low-altitude, middle ups and downs mountain, water conservation, tributary

2.6% of the total number of river reaches, which were mainly distributed in Laoha river midstream and upstream, one reach in Xilamulun river and 1 reach in Xinkai river midstream, 13 reaches in Huntai river, the rest distributed in Liu river, Xiushui river, Raoyang river, Zhaosutai river, Qing river basin.

According to the classification results of sinuosity: the sinuosity of total basin ranged from 1 to 4.2. 2168 reaches were low-sinuosity, accounting for 68.7% of the total number of river reaches, which were mainly distributed in Liao river, Huntai

Table 2.13 Statistics of river classification indicators

Indicators	Calculation method	Ecological significance
Stream order	Stream order was calculated in ArcGIS software using Strahler's method which defined that tributaries at the top of river are the lowest level, increasing level according to the increasing number of tributaries	Reflect the river extent and have connection with river morphology, river habitat and river flow
Confinement level	Observed by remote sensing image, compare the connection degree of channel and river valley, the connection degree of >90% were defined as confined river valley, the connection degree between 10 and 90% were partly confined river valley, <10% were unconfined river valley	Reflect the constrained degree of river valley to river; determine the space of river lateral movement and storage. The smaller the confined level, the more intense that river valley constrains the channel
number of channel	Observe channel number through high resolution remote sensing images., channels were divided into single and double channels	Related to the river bed stability and river habitat types
Sinuosity	S(sinuosity) $= Lr/Lv$, Lr is the length of measured river, L is the linear distance between two points in upstream and downstream. Both Lr and Lv were achieved in ArcGIS 9.0 software	Represent river lateral movement ability, decide the river longitudinal bending degree. Related to biodiversity, sediment transportation river morphology, and habitat diversity

source area and agricultural area; 658 river reaches with mid-sinuosity, accounting for 20.8% of the total number of river reaches, were mainly distributed in the transition to plains basin; 332 high-sinuosity reaches, accounting for 10.5% of the total number of river reaches, were distributed in the main channel of East Liao river, Huntai river and DaLiao river and parts of Liao river main channel.

Results of River Segments Classification

Liao river basin has 20 river types (Fig. 2.18). We named each river type, analysed the number of reaches, reach length and the percentage of the total length of the river reaches (Table 2.14). The results showed that the unconfined low sinuosity and partly confined low-sinuosity tributaries were the most widely distributed, with a length of 8186.3 km and 7717 m, respectively, accounting for about 23.1% and 21.8% of the total length of Liao river basin; The confined low-sinuosity and confined mid-sinuosity main channels, with a length of 4.6 km and 20 km, respectively, were the

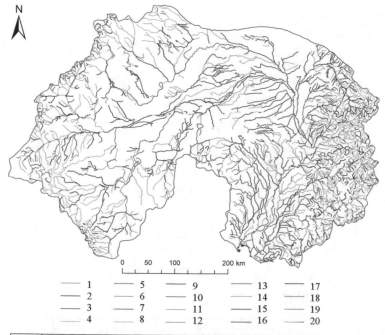

1	Partially restricted meandering main stream
2	Partially restricted meandering tributaries
3	Partially restricted low meandering main stream
4	Partially restricted low meandering tributaries
5	Partially restricted high meandering main stream
6	Partially restricted high meandering tributaries
7	Restricted meandering main stream
8	Restricted meandering tributaries
9	Restricted low meandering main stream
10	Restricted low meandering tributaries
11	Restricted high meandering main stream
12	Restricted high meandering tributaries
13	Nonrestricted meandering main stream
14	Nonrestricted meandering tributaries
15	Nonrestricted low meandering main stream
16	Nonrestricted low meandering tributaries
17	Nonrestricted duplex main stream
18	Nonrestricted duplex tributaries
19	Nonrestricted high meandering main stream
20	Nonrestricted high meandering tributaries

Fig. 2.18 Liao River Basin classification results

Table 2.14 River type statistics

The name of reach types	Number of reaches	Total length (km)	Proportion of total length (%)
Confined, low-sinuosity, main stream	2	4.6	0.01
Confined, low-sinuosity, tributary	218	1208	3.4
Confined, mid-sinuosity, main stream	3	20	0.06
Confined, mid-sinuosity, tributary	44	536	1.5
Confined, high-sinuosity, main stream	12	156	0.44
Confined, high-sinuosity, tributary	27	268.3	0.76
Partly confined, low-sinuosity, main stream	10	34.8	0.1
Partly confined, low-sinuosity, tributary	981	7717	21.8
Partly confined, mid-sinuosity, main stream	6	51.7	0.15
Partly confined, mid-sinuosity, tributary	211	2694.3	7.6
Partly confined, high-sinuosity, main stream	4	82	0.2
Partly confined, high-sinuosity, tributary	96	1445.5	4
Unconfined, low-sinuosity, main stream	99	856	2.4
Unconfined, low-sinuosity, tributary	811	8186.3	23.1
Unconfined, mid-sinuosity, main stream	42	992	2.8
Unconfined, mid-sinuosity, tributary	329	5974	16.9
Unconfined, high-sinuosity, main stream	34	932	2.6
Unconfined, high-sinuosity, tributary	147	3205.8	9
Unconfined, single channel, main stream	11	231	0.65
Unconfined, double channel, tributary	71	830.4	2.3

least widely distributed, accounted for only 0.01% and 0.06% of the total length of Liao river reaches, respectively.

Identify the Aquatic Biological Preferable Habitat Area

Aquatic biological habitat regions referred to regions with physical characteristics required by aquatic organisms for completion of their life history. They mainly included habitats required to propagate, grow, predate and migrate. Habitat protection is the most effective method of biological conservation. Understanding the habitats in which aquatic organisms are distributed can provide the basis for biological conservation, and level 4 aquatic eco-function regions can form the basis for aquatic ecosystem management and protection.

There were differences between the life histories of different species, and the niches were both overlapping and differentiated. For example, some fish spawn in stands of aquatic plant plexus, while some fish spawn in the benthic gravels. Therefore different organisms have different key habitats. Aquatic organisms occurring in habitat areas include fish, shellfish, benthic animals, algae, aquatic plants, birds and other organisms. According to the different life habits, and the different demands for living habitats by the same species in different life stages, habitats were divided into 11 types: endangered species breeding sites, endangered species feeding grounds, the migration route of endangered species, rare species breeding sites, rare species feeding grounds, the migration route of rare species, dominant species breeding sites, the migration route of dominant species, benthic animals sensitive habitats, benthic animals clean habitat. The identification of key habitats' for benthic animal and fish are introduced below.

Identification and Prediction of Macroinvertebrate Habitat Areas

Macroinvertebrate habitat areas were identified through field investigation. The distribution of benthic animals, community and environmental factors were measured in sampling sites. The distribution of macroinvertebrate habitats was determined using these indices by different analyses methods, and the potential distribution pattern was predicted (Table 2.15).

Identification and Prediction of Fish Habitat Areas

Fishes are the top of the food chain in aquatic ecosystems, which have a close relationship with humans, and their ecological status reflect the health of water ecosystems, so they can provide an important basis for aquatic ecological regions management.

Table 2.15 Indices of key macroinvertebrate habitats and the prediction methods

Indices surveyed and measured for macroinvertebrates	Indices of macroinvertebrate community pattern	The method for predicting potential distribution of macroinvertebrates
Hydrochemical indexed: (pH value, conductivity, dissolved oxygen, and total soluble substances, turbidity, suspended solids, total nitrogen, total phosphorus, BOD, COD, NH_4–N, etc.)	Diversity index	Using AUSRIVAS to predict potential species distribution based on the correlation of macroinvertebrate species characteristics and environmental variables
Physical habitat indices: (elevation, water depth, flow velocity, river width, river sediment particle size and composition)	Calculate index based on the species sensitivity	Using logistic regression to predict potential species distribution
Detection of habitat quality: (acquired by comprehensively scoring characteristics of benthic, habitat complexity and other habitat factors)	Macroinvertebrate biological integrity index (B-IBI)	The potential distribution of community indicators using logistic regression method

The identification of fish habitat included analysing the current situation and forecasting potential condition [21]. Analysing current situation mainly took history of fish fauna, composition of fish community structure, ecological characteristics and fish health into consideration. The measurement indices selected were similar to macroinvertebrates, including water chemistry, physical habitat, habitat quality monitoring and watershed land use interpretation [34]. The fish distribution was forecast based on the above evaluation indicators (Fig. 2.19). The method included the following steps:

1. Sclect reference sample. The selection of reference sample could refer to the sample selection criteria in calculating fish integrity. Take the conditions of water quality and habitat quality conditions into account or use some statistical analysis methods for selecting reference sample.
2. Choose fish community indicators. The selection of fish community indicators mainly considered the sensitivity and availability. Indicators that had a single response to environmental disturbance were selected generally, while considering the ease of calculation and the ecological significance.
3. The establishment of an index system for environmental variables. Environmental indicators including climate, geography, hydrology, soil and vegetation were classified according to the type and scale.

沙塘鳢现状分布

Fig. 2.19 The current distribution of Odontobutis

4. Model construction. To build the model reflecting the relationship between the fish community index and environmental factors, and verify the model. Generally, these methods include multiple linear regression models, generalized additive models, classification and regression tree models, etc.
5. Use the existing data to predict the unsurveyed regions of fish community characteristics.

Take Odontobutis obscurus in Liao river basin as example to predict the distribution of a single fish species (Fig. 2.20). After stepwise logistic regression analysis, the variables that were retained for the model were longitude, latitude, river level and average flow velocity, to establish the forecast model of Odontobutis:

$$y = \frac{1}{1 + \exp^a},$$

where $a = (-(1482.654 \times \text{longitude} - 1356.829 \times \text{latitude} + 9.313 \times \text{river level} + 5.616 \times \text{average flow velocity} - 910.489))$.

Odontobutis obscurus potential distribution was predicted using the prediction model for sampling sites with environmental data (Fig. 2.20). The result showed that the sites with potential distribution frequency of more than 85% were mainly

沙塘鳢潜在分布

Fig. 2.20 The potential distribution range of Odontobutis obscurus

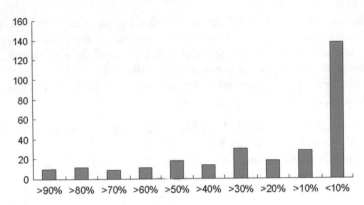

Fig. 2.21 The number of sites (individuals) where Odontobutis obscurus potentially **occured**

distributed in the north tributary of Taizihe, and this was consistent with the actual condition. 288 sampling sites in Liao river basin were predicted, 10 sites had Odontobutis frequency of more than 90%, 12 sites with frequency range from 80 to 90%, 9 sites with frequency range from 70 to 80%, 11 sites with frequency range from 60 to 70%, 18 sites with frequency range from 50 to 60%. The 60 points can be thought of as Odontobutis potential distribution areas and important habitats for protecting the species resources (Fig. 2.21).

Evaluation of River Service Function

River service function refers to the services that river systems provide for humans directly or indirectly [72]. In order to manage the basin water environment better, many countries implement water environment management methods that use a functional classification of water bodies [82]. Classic classification system are mainly divided into two categories, one was based on the classification and definition of ecosystem services of Costanza [27], subdivided from two aspects of ecosystem products and services; the other was based on the support, supply, adjust and culture function proposed by Millemium Ecosystem Assessment [68]. Today formulating management policy according to functions is the water environment management method generally used by developed countries. The United States has the most comprehensive water environment function classification system in the world. Referringed to the water functional classification system [86], the characteristics of Liao river basin and the type of level 3 regions, Liao river basin was classified into five major functional types: drinking water, protection of aquatic organisms, water resource supply, recreation and support services function. Each functional type was further divided into several sub functional types (Table 2.16).

Take fishing mentioned in recreational uses as an example to introduce the identification method for service function.

Table 2.16 Five functional types in Liao River Basin

Drinking water function		
Protection of aquatic organisms	Biological activity area	Endangered species breeding sites
		Endangered species feeding grounds
		The migration route of endangered species
		Rare species breeding sites
		Rare species feeding grounds
		The migration route of rare species
		Dominant species breeding sites
		Dominant species feeding grounds
		The migration route of dominant species
		Benthic animals sensitive habitats
		Benthic animals clean kind of habitat
	Habitat quality	High quality habitat
		Favorable quality habitat
		General quality habitat
	Habitat types	Cold fresh water fish habitat
		Warm fresh water fish habitat
		Estuary fish habitat
		Reservoir fish habitat
		Lake fish habitat
		Fresh water wetland habitat
		Estuarine wetland habitat
		Natural reserve
Public water supplies		Agriculture
		Industry
Leisure	Contact uses	Swimming
		Drifting
		Boating
		Angling
	Uncontact uses	Landscape
		Camping
Support services	Pubic water supplies	
	Sediment transport	
	Products of fish, shellfish	
	Water quality improve	
	Hydroelectric power	
	Sailing transportation	

Fishing function is the activity of capturing fish from the rivers, lakes and reservoirs using fishing rod, hook line and other tools. Human health and biological conservation were the goals of fishing function recognition, considering both social carrying capacity and social economic development needs, to follow the following principles:

1. This function belongs to contact recreation activities, requires water quality, and cannot be harmful to human health;
2. In fishing function recognition, all rivers have the representative fish regardless of the river size. But without consideration of water quality, human activities and other disturbance factors, the higher the river level the more the number of species and the higher the species richness;
3. Fishing function can be quantitatively measured based on the Shannon-Wiener diversity index (for example) when assessing reaches' fishing function;
4. Good habitat, not only conducive to the growth of fish, can also give a person aesthetic enjoyment when fishing, so the habitat index is one of the deciding indices for recognition of the fishing function;
5. As a leisure activity function, fishing function also needs to consider the distance between the reach and densely populated areas.

According to the above principles, combined with the specific situation of the Liao river basin, we established the following evaluation index system, so that only reaches that meet the following conditions can be defined as having fishing function:

1. Water quality must meet the Water quality standards for scenery and recreation area [2], and with reference to III class water quality standards in environmental quality standard for surface water [4];
2. River order greater than 2;
3. Rivers beyond 200 km distance from administrative cities higher than county;
4. Number of species which is greater than the arithmetic average number of species at the level 2 eco-region ;
5. The Shannon-diversity index value is higher than arithmetic average level;
6. Habitat quality reaches more than normal level whose total score is higher than 120.

According to the above index system, 635 reaches had a fishing function in Liao river basin, of a total length of 6245.48 km, representing 20.1% of total reaches (Figs. 2.22 and 2.23). Liao river main channel had a large water quantity, the number of fish species and the species diversity index were higher. There were more suitable regions for large fish survival near the Qing river reservoir and Chai river reservoir. Fish habitat quality in Huntai river was good, the rich benthic animals provided abundant bait for fish growth and also provided a good environment for wild fishing. Pu river, Haicheng river, Xi river and other reaches all have fishing function. The water in East Liao river main channel was plentiful, Erlong reservoir and Yangdachengzi reservoir could provide a good living environment for fish that grow for a long time, the related reaches could provide fishing function for humans. West Liao river had a

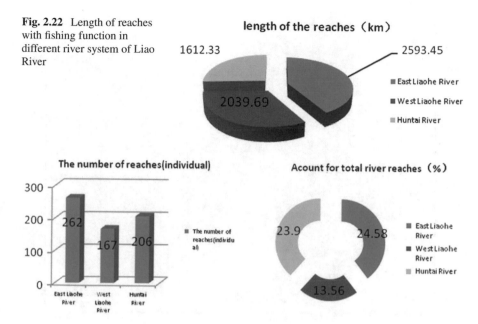

Fig. 2.22 Length of reaches with fishing function in different river system of Liao River

Fig. 2.23 Number of reaches with fishing function in different river system of Liao River

high number of fish species, but the fish species richness was not high, reaches with fishing function were distributed in Xilanulun river downstream and Zhaganmulun river downstream and Laoha river, Yangchangzi river, Xibo river of downstream from Hongshan reservoir (Fig. 2.24).

Aquatic Eco-function Regionalization Scheme in Liao River Basin

The division of the freshwater EFR level 4 was completed according to the level 4 technical route (Fig. 2.25). It was divided into 25 functional types, and the spatial distribution and change trend of different types were quite different. Analyses of the number, the river length, the proportion of the total length and the distribution of rivers for each function type were conducted (Table 2.17). the most widely distributed is the fish resources supply functional region, whose total length was 7265.5 km, accounting for 20.5% of the total length of the river; Undefined double channel was the least distributed, accounts for only 0.01% of the total length of the river.

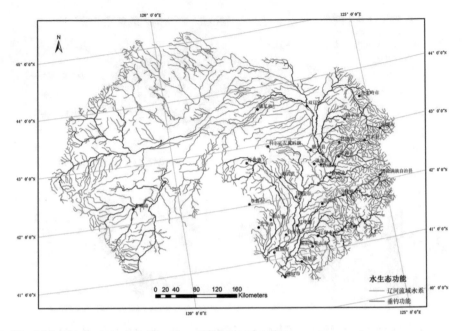

Fig. 2.24 Distribution range of reaches with fishing function

2.1.4.3 Conclusions

Management based on ecotypes have proved to be an efficient way, and eco-function based management zone is newly concerned management strategy for aquatic environment and ecosystem management in the world now.

After reviewing the literatures on ecoregion delineation, we raised a 4 level framework for aquatic function management, and build the approaches for the delineation. We applied the framework and delineation approaches in Liao river basin, and finally the Liao river basin was delineated into 4 level 1 eco-function region, 14 level 2 eco-function region, and 50 eco-function region. In level 4 function of river reaches was assessed and the dominant function was chosen as the function of reaches.

Ecoregion boundary offered a spatial unit for aquatic environment management, and different level is suitable for different management demands. In this project, level 1 ecoregion is suitable for the water quality standard establishment; for level 2 ecoregion it is suitable for reference site selection on river health assessment. For level 3, it is suitable for the pollution control planning, and level 4 ecoregion is suitable for river function rehabilitation effort, conservation target settlement, for each function type the natural status of physical, chemical and biological condition was different, and the management demands for each function type were also different.

1	Fish resources area
2	Fish diversity area
3	Fish migration area
4	Biodiversity area
5	Drinking water source area
6	Food chain top fish area
7	Unrestricted valley small river
8	Unrestricted duplex river
9	Non-touch entertainment function area
10	Restricted tributaries
11	Mussels, shrimp and crab protected areas
12	Partial restrictive tributaries
13	Algae diversity area
14	Biodiversity area
15	Rare and endangered species conservation area
16	Rare and endangered species functional area
17	Wetland habitat functional area
18	Water resource supply area
19	Water supply function area
20	Local endemic species area
21	Entertainment water function area
22	Zoobenthos diversity area
23	Mainstream river
24	Duplex river
25	Groundwater recharge area

Fig. 2.25 Level 4 freshwater eco-function regions in Liao River Basin

Table 2.17 Statistics of functional region types of river reaches

Functional types	Reach no.	Reach length (km)	Share (%)	Major rivers
Fish resources supply function region	561	7265.5	20.5	Most of the distribution in Xinkai river, Laoha river, Jiaolai river and East Liao river; few distribution in the source of Raoyang river, the estuary of DaLiao river and Beisha river
Fish community diversity support function region	6	56.8	0.16	Located in the junction of East Liao river and West Liao river
Fish migration channel function region	47	1051	3.0	Yangsha river, East sha river, West sha river, Badao river, Two longwan river, the downstream of the main stream of Liao river
High biodiversity support function region	1	18	0.05	a tributary of the upstream of the main stream of Liao river
Drinking water source function region	111	1011.7	2.9	The upstream of the main stream of Da Liao river and its tributaries, Sha river, Near the Dahuofang reservoir, the source of Laoha river, few in Pu river and Chai river
Large predatory fish support function reach	63	996.1	2.8	The mainstream of Hun river and Taizi river and their part tributary, the junction of East Liao river and West Liao river
Undefined small river valley	327	4436.3	12.5	The source of East Liao river and Zhaosutai river, The mainstream of Liao river and their part tributary, the downstream of West Liao and few in the midstream of Xinkai river
Undefined double channel	1	4.9	0.01	The downstream of Liangzhong river
Non-contact entertainment function region	69	1133.8	3.2	Bayantala river, Haihaerguole river, Tianshanxi river, part tributary of East Liao river and Xinkai river
Defined tributary	24	296.7	0.84	The source of Laoha river, Liuhao river,Beidi river
Important crayfish/clam habitat conservation region	14	272.1	0.77	The estuary of Shuangtaizi river, the tributary of the upstream of the main stream of Liao river
Partly-defined tributary	167	2254.5	6.4	The source of Xinkai river,and the tributary of the downstream, the midstream of East Liao river
Algal diversity support function region	262	2771.9	6.4	The two tributaries of the source of Xilamulun river, the source of Yingjin river, the source of Laoha river, the midstream of Beisha river, Haichenge river, Tang river and Kou river

(continued)

Table 2.17 (continued)

Functional types	Reach no.	Reach length (km)	Share (%)	Major rivers
Biodiversity support function region	209	1816.4	5.1	Most of the distribution in the source of Hun river, the midstream of Jiaolai river and Laoha river, few distribution in the source of the Xilamulun river and Raoyang river, Liao river plain
Rare and endangered species habitat reserves region	28	182.4	0.5	The mainstream of Hun river, near the Dahuofang reservoir and Suzi river
Rare and endangered species support function region of	494	3027	8.5	The source of Hun river, Taizi river, Qing river, Chai river
Wetland habitat function region	8	91.1	0.26	The source of Xisha river and Shuangtaizi river
Water supply function region	80	1317.4	3.7	The mainstream of Xilamulun river and its part tributaries, the downstrseam of the mainstream of the Liao river and its part tributaries
Water supply function region	70	811.4	2.3	The source of Shaolanghe river, Weitanghe river, Yangchangzihe river, Mengke river and Raoyang river
Native endemic species support function region	253	2208.2	6.2	The source of Xilamulun river
Contact recreational water function region	27	520.5	1.5	The source of Yangxumu river,the midstream of Jiaolai river,the near of Hongshan reservoir, and some tributaries of Xilamulun river
Benthic animal diversity support function region	298	3261	9.2	Xiushui river, the midstream and downstream of Yangxumu river, Liu river, the source of Kou river, the midstream and downstream of Qing river, Pu river and Chai river
The main stream	16	327.7	0.92	A reach of the mainstream of Xilamulun river, a reach of the mainstream of Jiaolai river,the rest distribution in the Qing river, Zhaosutai river and West Liao river
Double river channel	20	273.3	0.77	A tributary of the downstream of Xilamulun river, a tributary of the downstream of West Liao river, few in the tributary of Zhaosutai river and Kou river
Groundwater recharge region	2	21.6	0.06	Liangzhong river

2.2   Control Unit Schemes for SLRB Demonstration Areas

by Kun Lei, Ya Tao, Fei Qiao, Zicheng Li, Richard Williams, Gang Zhou

2.2.1   Introduction

This deliverable reports the "Schematization of control units" activities of the SUSTAIN H₂O project consortium during the project's reporting period.

According to water functional requirements, the condition of water pollution and the water risk level, a total of 114 priority control units out of a total of 318 control units in 7 key river basins of China have been determined. These units received particular attention for water pollution prevention during the 12th Five-Year Plan. The priority water quality control units were further divided into water quality maintenance units, water quality improvement units and water risk prevention units (form "the water pollution prevention and control plan in key river basin in 2011–2015", http://www.zhb.gov.cn/gkml/hbb/bwj/201206/t20120601_230802.htm).

As a water quality improvement priority control unit exemplar, the Ashi river basin of Harbin city in Songhua river basin has to undertake watershed total emission reduction to achieve water quality improvement, through comprehensive control measures. These measures strengthen the implementation of total pollutant total load control to cut pollutant emissions, to safeguard ecological river base flow, and to ensure water environmental quality improvements, notably in the surface water body. For example, the water quality goal is that the Ashi estuary section Ammonia Nitrogen concentration of 8.6 mg/L should be reduced below 7 mg/L and the rest of the indicators meet Class V class by 2015.

2.2.2   The Principles of Control Unit Divisions

According to "Specifications on Control Unit Division for Watershed Water Quality Objective Management" by CRAES, the main principle of Control Unit Division as following:

I Principle of Clean Boundary Segregation

According to the function characteristics of the river water body (water ecological function zoning, water environmental function zoning, water function regionalization), the outlet location of those river segments has been called the clean boundary when the river segment with higher river water body function and the higher water quality protection goal had been selected.

II Principle of Water Catchment Boundary Segregation

The control unit boundaries are defined hydrologically within the basin or sub-basin, so that all the pollutants received from that basin or sub-basin and not from waters from a different basin or sub-basin.

III Principle of Administrative Region Isolation

Given that the boundaries are defined on a hydrological catchment basis as the starting point, the control units should also keep the integrity of Local government administrative areas as far as possible. This ensures that there is a clearly defined administrative responsible main body for water quality objects and all management tasks.

IV Principle of Water Type Isolation

Consideration should be given to border selection based on rivers, lakes, reservoirs and estuaries as the boundary of different control units, thus linking to the different water quality targets set in management plans for different water body types.

2.2.3   Key Elements of the Division of Control Units

2.2.3.1   Spatial Distribution of Water System

Water systems have been defined as a combination of a main river and its tributaries at all levels in a watershed; it was influenced by certain geological structures and the natural environment. It can affect the hydrological process of runoff in a basin. Spatial distribution characteristics of water system have varied form, the common forms are: (1) the dendritic drainage; (2) the lattice water system; (3) Parallel drainage pattern; and (4) the radial and annular pattern.

2.2.3.2   The Aquatic Ecological Functional Division (Water/Water Function Areas)

The aquatic ecological function could provide the basis for control units and set the management goals. In practice, the aquatic ecological protection goals in different aquatic ecological functional areas may differ, so the water quality goal is set in the region without water ecological function division by reference to the Water Function Area or Water Environmental Function Zone.

2.2.3.3   Water Quality Monitoring Control Section

The Water quality monitoring control sections are set up to monitor and evaluate the influence of pollution from all sources. Those sections should give a real assessment of the condition of water quality and its spatial distribution. It should also help describe processes that control the concentrations of pollutants as they are transported along the river, including degradation rates and sorption to sediments.

2.2.3.4   The Administrative Area

The control units at different levels should be under the management of corresponding administrative departments, thus the boundaries of administrative area should be considered in the process of control unit division. The level of provincial, county or district administrative boundaries should be separately considered in the division of control units.

2.2.4   The Technique of Control Unit Division

2.2.4.1   Data Collection and Analysis

Water System Overview

The basic geographic information data of the research basin should be accessed firstly, including digital elevation model (DEM) data, remote sensing images, water system diagrams and the distribution map of water quality control section.

Based on the characteristic of the natural catchment, the water system needs to be generalized for different scales of sub-basin (watershed, catchment area). Then the corresponding relationships to aquatic function zones, sewage outlets and catchment areas should be established for different scales of sub-basin. The accurate boundaries of basin and sub-basins can be identified by GIS overlay approach.

It should be pointed out that the control units also could be divided based on the scope of administrative region in flat areas, due to the uncertain delineation of watersheds.

Section for Water Quality Monitoring

Monitoring cross sections for collecting detailed data of water quality and hydrology could be set up.

The specific methods as follows: set up two points at the inlet and outlet of a river in a catchment unit; set monitoring points monitoring above and below locations in rivers where pollution sources are concentrated on the basis of a pollution source survey; to filter and select the control section for control unit division, the existing state or provincial control sections, even the monitoring control section at county or district level should be preferred to be used, according to the actual situation.

Survey of the Pollution Sources

The pollution types (point source, non-point source and natural pollutant background), the number, amount and location of discharges and the nature of the discharges including existing rules and, pollution treatment technology should be confirmed through the existing data, report, and field investigation. Such a survey of the pollution sources and their impacts on water quality could be used as the basis for prioritizing control units. The above data would be the basis for pollution load calculations.

According to the distribution of river system, administrative center position, and hydrological experts specialized knowledge, a main sewage drainage river in the county (town) could be filtered and determined by the judgment method for sewage discharge direction in County (town). Usually, it was considered that a county (town) only has a corresponded main drainage river.

2.2.4.2 Control Unit Division Process

Data Preprocessing

(1) Data Acquisition and Preprocessing

Digital elevation model (DEM) including two kinds of data information: location and elevation. This information can be obtained by GPS and laser range finder or indirectly from the aerial or remote sensing images and obtained maps. If available, the highest resolution and therefore more accurate DEM data should be used. If these kinds of DEM data are not available, SRTM (Shuttle Radar Topography Mission) DEM data (90 m resolution or more precision data of any one geographical region in our country can be downloaded from the Chinese International Academy for science data services platform http://datamirror.csdb.cn/admin/datademMain.jsp). If there are depressions and peak in DEM raster data, to avoid reverse-flow phenomenon and get the grid terrain data without depression, this can be pre-processed by related commands of Arc Hydro hydrological.

(2) Drainage Network Extraction

The river network can be generated by related command of ArcHydro hydrological module of ArcGis. Firstly, according to the principle of "surface runoff flow from the high terrain to lower in the watershed space, and discharge from base outlet",the flow direction is determined. The watershed catchment area is determined based on the principle of "the higher terrain area may be the river basin watershed." According to the direction of river drainage, the river network can be extracted from confluence and converted into a vector network and water system layer (shape file format).

(3) River Generalization

In order to ensure that all river grade (major tributary is grade 1, tributary of a major tributary is grade 2 etc.) relationships are correct, the rivers are checked and the river grade is confirmed based on vector river data and topological relation.

The important water pollutant prevention and control river reach concerned by the department of environmental management, such as basin river main stream, key tributaries, key lakes, key water body and key pollutant tributaries in a city or small basin should be the focus of inspection and supplement after the generalized river system has been generated.

(4) Sub-basin Recognition

After finishing the river system generalized, the land areas of the sub-basins are partitioned in accordance with the principle of "one sub basin must have only one main river and all the rivers must flow into it". This combined with DEM data, aerial and satellite image information referring to the water resource partition and editing by hand, allows the actual watershed vector layer to be formed.

2.2.4.3   Control Unit Partition

(1) Basic Requirements

The control unit must reflect the natural catchment characteristics and administrative requirements. The natural river systems are the starting point; these are then reconsidered taking account of the integrity of administrative regions, and the same catchment area of administrative units to form the control unit.

1. Taking water to decide land. The control unit is the land area that drains to the water body according to the natural characteristics of the water system. The control unit is the water and land corresponding planar area. Natural water system is the land partition benchmark. According to the natural catchment characteristics, determining the land catchment area and forming the water and land corresponding unit.
2. Taking district and county as the smallest administrative unit. As the basic administrative unit of statistical and investigation of environmental data, district and county are the smallest space units for water environment management.
3. Covering the whole basin. The watershed integrity must be ensured, the control units are expressed as the non-overlapping units covering the whole basin.
4. Linking with water ecological function area (water function area and water environment function area). The control units should be linked with these function areas.
5. Dynamic adjustment. The partition of control units is the process of national and local concurrent linking and coordination. Absorbing ideas of the different levels of departments of water environment management and adjusting the control unit partition results.

(2) Control Unit Classification

The control unit levels could be defined based on the watershed classification, generally.

1. Firstly, defining the first level control unit based on the main stream of watershed.
2. Defining the second level control unit based on the first grade tributaries in first level control unit.

3. Defining the third level control unit based on the second grade tributaries.
4. Defining the fourth level control unit based on the third grade tributaries.
5. And so on. This is a kind of nested partitions. For a given watershed scale, this method gives the divided control units.
6. According to actual situation, any one independent water system (sub-basin) within the basin can be defined at the same level of control unit. In addition, according to water environment planning and management needs, any level of control unit can be divided into several sub-control units based on the relationship between the upstream and downstream, sensitive protection goals or on the boundary of administrative regions. The prime motivation is to maintain the boundary water quality targets to be consistent with each other.

(3) Control Unit Space Determination

The space of control units include water and land, the specific method are as follows:

1. Waters
   Analyze the water function and corresponding water quality target, focus on cleaning boundary separation principle, link with water ecological function area and administrative area; divide the water area into two or more sections. The premise of the division is to maintain consistent boundary water quality targets.
2. Lands
   Based on the water division, catchment area isolation principle, administrative region isolation principle were considered comprehensively, according to water catchment area of DEM extraction, base map information of aerial and satellite images, and water resources division. Editing by hand, get the reasonable land range division results of control units.
   For the district and county of single dominant control of emissions, it can be transferred into a control unit. For the district and county of non-single dominant emission control, it can be split into the different control units.

(4) Rationality Analysis

Retrospective analysis of partition results, combining the principle of simplification of pollution source management and a clear responsible person for water quality, analyze the control unit division results. If necessary, adjust the control unit boundary.

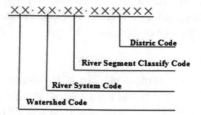

Naming the Control Unit

(1) Naming Convention

The results of the control unit are delineated by the form of tables and maps. The chart includes the name of control unit, the name of controlled section it contains, the river which the control section in, including administrative (to county, town), the area of the control unit etc. Name, should adopt a comprehensive and consistent naming method, which is "Watershed name + River + Reach classification + Administrative". Structure of the code:

As an example:

Naming the control unit: Liao River/Taizi River/The middle/Benxi city Benxi County

Structure of the code: BA/06/02/210521

Watershed Code/River Code/Reach classification Code/Administrative Code

(2) Expression Outcomes of the Control Unit Delineation

Results of the watershed water quality objectives management control unit, including text, tables and maps, those maps should be prepared using computer graphics.

(1) Form

Fill the delineation results of watershed water quality objectives management control unit form.

(2) Maps

The maps of the watershed water quality objectives achievement management control unit division mainly refers to the watershed—a control unit map and maps of various scales in the same area should be consistent. Coverage information includes: watershed boundaries, the control unit boundaries, major water bodies and names, control section identified with names, the provincial administrative center, regional administrative center and the county administrative center.

2.2.5 Qing River Watershed Control Unit Delineation

2.2.5.1 Basic Overview of Qing River Watershed

Qing river originates in Qingyuan county, Beiyingelinmen township, Sandaogoumiao ridge. The river brings together the Erdaogou, Nianpan and Kou Rivers and is a large tributary of the Liao River, which has a basin area of $5674.3\,km^2$ and a main river length of 217.1 km. It flows through Kaiyuan city, Lijiatai township and Shangqinghe village before entering Tieling city (Fig. 2.26).

Climate Meteorology

Qing river watershed has a north-temperate with southwest-monsoon continental climate, where the summer is hot and rainy and the winter cold and dry. The frost-free period is approximately 140 days: the first frost usually occurs in late September and the last frost in late April the following year. Annual precipitation is around 700–800 mm d. Watershed yearly average temperature is $6.0\,°C$: minimum temperature is usually in January, the lowest temperature recorded was $-41\,°C$. The highest recorded temperature was $35.7\,°C$. The highest temperatures usually occur in in July or August.

Fig. 2.26 Schematic for Qinghe watershed

Hydrology

Due to its geographical location, terrain, topography and weather factors, Qing river is a large water resources tributary of the Liao river basin and Qing river is the main source of flood water in the Liaohe river: annual runoff of is 1.64 billion m^3. It is the spatial and temporal distribution of rainfall that determines the spatial and temporal distribution of runoff. Because it is affected by a wide rainfall distribution during the year and considerable inter-annual variability, runoff itself is highly variable within and between years. Storm centers in the Qing river basin are generally located upstream of the reservoir (see **Qinghe Reservoir** below). The very uneven distribution of runoff during the year means that runoff in the flood season accounts for 70% of the annual flow as measured at the Gengwangzhuang hydrological station for example.

Qing river flooding is formed from rainstorm confluence, and flood features are constrained by watershed characteristics and storm characteristics. The main source of storm rainfall is water vapor picked up from the Yellow Sea. The vast majority of the basin is part of the Changbai mountains area of low hills and the spatial distribution of heavy rain is broadly consistent with changing topography: the large storm center generally appear in the hilly area of Qing river reservoir upstream, where the elevations are above 200 m. Qing river is a major storm center of the Liao river, affected by Pacific subtropical high pressure with most rainstorms occurring in late July to early August. The river basin is in the northern part of the subtropical high-pressure airstream and a second heavy rainfall generally occurs within two days and lasts no more than three days. The resulting flood peaks are high as is the flood volume reflecting the intensity of the storms steeply sloping ground and rapidly flowing stream. Flood duration is generally 2–4 days, according to statistics from the Gengwangzhuang and Bakeshu hydrological stations. On one occasion one day of flood water accounted for more than 70% of the annual flow.

The vegetation of Qing river basin is good; sediment yields are relatively low, and consistent with the regional distribution of sediment with floods, mainly around the Nianpan River and Lijiatai Bakeshu, which is the Qing river upstream station. These sediment sources account for about 80% of the sediment above the Qing river reservoir. The distribution of sediment loads during the year is even more concentrated than runoff: flood-sand accounts for more than 95% of the annual sand load, and is concentrated in the several major flood events.

Qinghe Reservoir

There is a river reservoir in the central section of the Qing River: Qinghe reservoir. As noted above, the Qing river basin is located in the Changbai mountain low mountain hill, spring drought, summer floods are the basic characteristics of watershed. Qinghe reservoir was completed in 1960 and lies 89 km from the headwaters of the Qing River with a catchment area of 2376 km² (42% of the total drainage area). It is a large reservoir with multiple functions, including; flood control, power generation, irrigation, tourism, aquaculture and other comprehensive benefits. Because of the distance from Qinghe into Liaohe port is short, Qinghe reservoir plays a crucial role in the Liao river runoff regulation and Liao river flood control.

Above the Qinghe Reservoir, the water quality is Grade III standard. From the reservoir to its confluence with the Liao River, the Qing river flows through Yangmu, Babao and other eight towns, as well as two industrial cities in Qing river district and Kaiyuan, it also accepts sewage effluents from Xiaoqinghe and Mazhonghe. There are two urban centralized drinking water sources of groundwater in the northern of Qinghe district and Kaiyuan district, Qinghe reservoir is an alternate water source for Tieling city.

2.2.5.2 Data Collection and Analysis

Relevant information was collected for the control unit delineation as described below.

River Network Data

The 1:4 million digital river network (Fig. 2.27 (left)), including both primary and secondary tributaries within the range of the Qinghe watershed control unit was used in the correction work of extracting the control unit river.

Terrain Data

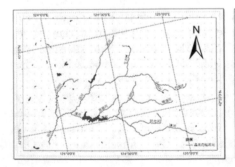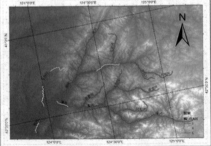

Fig. 2.27 River maps (left), DEM maps (right)

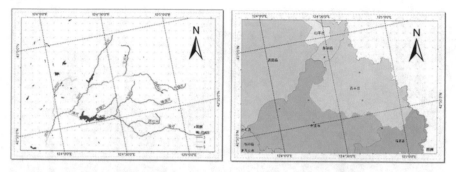

Fig. 2.28 Water environment function zones map (left), administrative map (right)

DEM data at a resolution of 90×90 m (Fig. 2.27 (right)) was used for digital river network extraction and sub-basin delineator.

Water Environment Functional Area Information:

Major tributaries of the Liao river and their tributaries were the units used to set river water quality objective (Fig. 2.28 (left)).

Administrative Data:

There were 6 districts in the Qing river basin (Fig. 2.28 (right)), which were used for the division of management unit.

2.2.5.3   Division of the Control Unit

Watershed Delineator

Using the information on the distribution of water environment function zones, from the mainstream to the tributaries, the different rivers and the river water environment function area water quality objectives, each water feature was divided into catchment or sub-basin. The Qing river basin (in the scope of dotted line) is divided as follow (Fig. 2.29).

Division of the Control Unit

Based on the division into sub-basin and the superimposed administrative regions, eventually 18 control units were formed in which to consider water quality objectives, drainage distribution and with administrative control (Fig. 2.30).

The Main Control Unit List

Based on the watershed, districts and river network, some information was added to the control units, including watershed, river, name, water quality objectives, control section, and so on (Table 2.18).

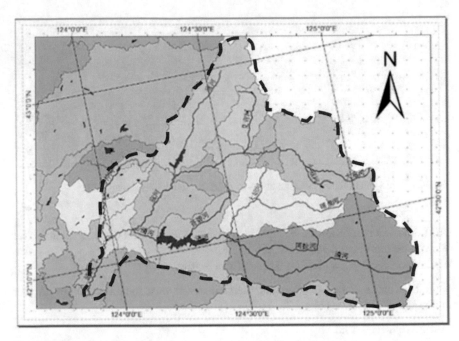

Fig. 2.29 Watershed delineator map

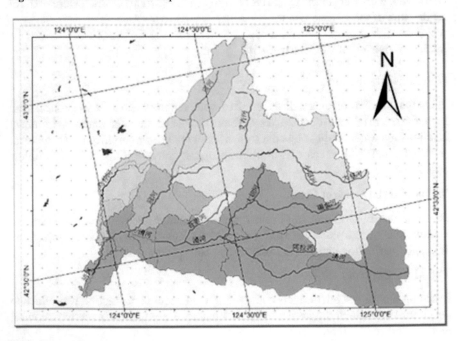

Fig. 2.30 Control unit division map

Table 2.18 The main control unit list

Water-shed	River	The control unit		Water quality objectives	Control section	Sewage area		
		Name	Level			Provinces (autonomous regions)	City	District and county
Qinghe	Qinghe (Kou River)	Kou River_Kaiyuan_1	3	II	Outlet of Kou river in Kaiyuan	Liaoning	Tieling	Kaiyuan
Qinghe	Qinghe (large Kou River)	large Kou River_Xifeng	4	III	Outlet of Large Kouhe	Liaoning	Tieling	Xifeng
Qinghe	Qinghe (Kou River)	Kou River_Xifeng	3	V	Outlet of Kou river in Xifeng	Liaoning	Tieling	Xifeng
Qinghe	Qinghe (Kou River)	Kou River_Kaiyuan_2	3	III	Outlet of Kou river in Kaiyuan_1	Liaoning	Tieling	Kaiyuan
Qinghe	Qinghe (Mazhong River)	Mazhong River_Changtu	3	V	Outlet of Mazhong river in Changtu	Liaoning	Tieling	Changtu
Qinghe	Qinghe (Mazhong River)	Mazhong River_Kaiyuan	3	IV	Outlet of Mazhong river in Kaiyuan	Liaoning	Tieling	Kaiyuan
Qinghe	Qinghe	Qinghe_Kaiyuan_1	2	III	Outlet of Qinghe in Kaiyuan_1	Liaoning	Tieling	Kaiyuan
Qinghe	Qinghe (Taibi River)	Taibi River_Xifeng	4	III	Outlet of Taibi River in Xifeng	Liaoning	Tieling	Xifeng
Qinghe	Qinghe	Qinghe_Kaiyuan_2	2	IV	Outlet of Qinghe in Kaiyuan_2	Liaoning	Tieling	Kaiyuan

(continued)

Table 2.18 (continued)

Water-shed	River	The control unit		Water quality objectives	Control section	Sewage area		
		Name	Level			Provinces (autonomous regions)	City	District and county
Qinghe	Qinghe	Qinghe_Fushun	2	III	Outlet of Qinghe in Fushun	Liaoning	Fushun	Fushun
Qinghe	Qinghe	Qinghe_Kaiyuan_3	2	III	Outlet of Qinghe in Kaiyuan_3	Liaoning	Tieling	Kaiyuan
Qinghe	Qinghe	Qinghe_Kaiyuan_4	2	IV	Outlet of Qinghe in Kaiyuan_4	Liaoning	Tieling	Kaiyuan
Qinghe	Qinghe	Qinghe_Kaiyuan_5	2	III	Outlet of Qinghe in Kaiyuan_5	Liaoning	Tieling	Kaiyuan
Qinghe	Qinghe	Qinghe_Xifeng	2	III	Outlet of Qinghe in Xifeng	Liaoning	Tieling	Xifeng
Qinghe	Qinghe	Qinghe_Kaiyuan_6	2	IV	Outlet of Qinghe in Kaiyuan_6	Liaoning	Tieling	Kaiyuan
Qinghe	Qinghe (Nianpan River)	Nianpan River_Xifeng	3	III	Outlet of Nianpan River in Xifeng	Liaoning	Tieling	Xifeng
Qinghe	Qinghe (Kou River)	Kou River_Lishu	3	III	Outlet of Kou River in Lishu	Jilin	Siping	Lishu
Qinghe	Qinghe (Mazhong River)	Mazhong River_Changtu	3	III	Outlet of Mazhong River in Changtu	Liaoning	Tieling	Changtu

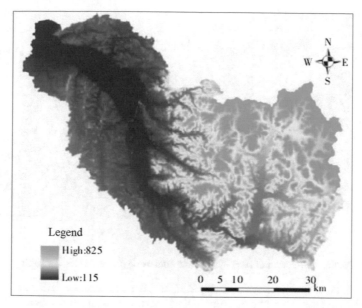

Fig. 2.31 The DEM spatial distribution of Ashi River Basin

2.2.6   The Control Unit Division of Ashi River Basin

Ashi river basin is located in the south of Heilongjiang province at longitude 126° 40′–127° 42′ east, latitude 45° 05′–45° 49′ north. It is bounded on the north by Songhua River and on the south by Mangniu River and has a basin area of 3,545 km², which accounts for only 0.65% area of Songhua River basin, which is the fifth longest river in China and covers a drainage area of 545,000 km² (Fig. 2.31).

2.2.6.1   Hydrology Conditions

Ashi River is the primary tributary on the right bank of Songhua River, which flows through Harbin city. The Ashi river originates from Maoer mountain town in Shangzhi city, flowing through Maoer mountain town, Pingshan town, Xiaoling town, Yuquan town, Jiaojie town, Shuangfeng town, Aching district, Lindian township and Xiangfang district in Harbin, before flowing into Songhua river in the area around Harbin Yatai cement company.

The Ashe river system had 79 tributaries, the total river segments length in Ashe river system is 1278 km with a drainage density of 0.36 km/km². The main stem of the Ashi River runs for about 257 km long and has a river bend coefficient of 1.6.

There are 12 primary tributaries of the Ashi river: Aching river, Dashi river, Yuquan river, Haigou river, small yellow river and Dongfeng ditch run into on the right, and Huangni river, Willow river, Fanjia river, Miaotai ditch and Xinyi ditch run into on the left, (See Fig. 2.32, Table 2.19).

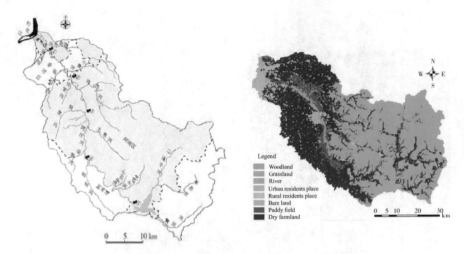

Fig. 2.32 Spatial distribution of river system and land use types in Ashi River Basin

Considering the influence of human activities on the water functions in Ashi river basin, the land use types was defined as an indicator for control unit division. Land use types in Ashi river basin are shown as Fig. 2.32.

Ashi River is representative of mountainous rivers in northern China; the river runoff supply primarily reflects the rainfall distribution and precipitation. The average annual runoff averaged over multi-years was 4.6×10^8 m^3, the maximum and minimum annual runoff was 9.0×10^8 m^3 and 1.1×10^8 m^3 respectively. In the frozen period, from November to the following March, the natural flow of the river low to zero, the observed "runoff" in the downstream of river is comprised almost 2.0×10^5 t/a of sewage (Fig. 2.33).

2.2.6.2 Water Pollution Problems

Ashi river basin was the agricultural base of Harbin city, and with large area of land designated to agricultural production and high farmland area per capita. It is not surprising to find out that one of the major sources of pollution is from agricultural farmland (herbicides, insecticides, and nutrients).

At the same time, point source pollution (human and industrial effluent) is also an important source of water pollution (COD/BOD, Ammonia, Heavy Metals and Toxic Organic Pollutants). One of the main challenges and limitation on water quality in the Ashi River is the large ratio of wastewater to natural runoff.

Most of water quality monitoring sections in the downstream parts of the river is in Class V or worse [1]. The major pollution indicators were COD (37248.9 t/a) and NH$_3$–N (4160.7 t/a), and the main pollution sources were domestic sewage and Industrial wastewater from Harbin and Acheng District (Fig. 2.34).

The Ashi River currently meets Class III standards upstream of the city of Acheng, specifically upstream of the Acheng hydrological monitoring station just downstream

Table 2.19 Relation table between main stream and branch of Ashi River

River Name	Headwater location	Estuary location	Length (km)
Ashe River	Jian shanzi in Maoer mountain	Cement plant in Harbin	213
Haigou River	Nan pojiangan ditch	Hai heyan in Lindian town	68
Yuquan River	Yi cuomao mountain in Yuquan street	Yue ji village in Yagou town	30
Dashi River	Qi liban ditch in Xiaoling town	Wang jintun in Xiaoling town	26
Huaijia Ditch	Xi fa village in Alamo town	Gao jiatun in Limin village	24
Xinyi Ditch	Yi fayuan village in Dawn town	Xingfu village in Xiangfang district	18
Small Yellow River	Feiketu in Aching	Xi nantun in Donggang villang	30
Miaotai Ditch	Xinli street in Aching	Infall of water conveyance main canal in Chenggaozi	25

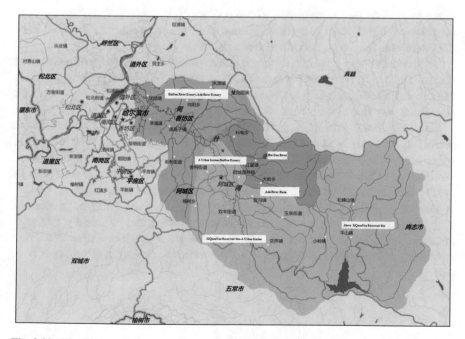

Fig. 2.33 Distribution of water resource of Ashi River Basin

Fig. 2.34 Distribution of main sewage outlet and primary tributary outlet in Acheng District

of the town. The long term aim is to reach Class II. It seems that the main reason for not reaching Class II is agricultural non-point source pollution. The area is an important area for the production of rice and maize.

The river has very poor quality downstream of Acheng hydrological monitoring station. The poor water quality is due to high levels of COD, which point to the likely cause being the discharges from various point sources just downstream of the weir and all the way to the confluence with the Songhua River.

The weir on the Ashi River, just downstream of Acheng keeps river levels high above this point partly to provide irrigation water for the local area. The amount of water passing through the weir looks very low. The impact of the current irrigation abstractions and back water on the final river flows allowed to pass the water and its quality need to be considered in an assessment of the pollution problems.

Serious pollution has been observed along the Ashi River. There are numerous outfalls of industrial wastewater and domestic wastewater and sufficient wastewater gathering and treatment services are lacking. The water pollution influences living conditions and crop production greatly as well as the public health of people in river side.

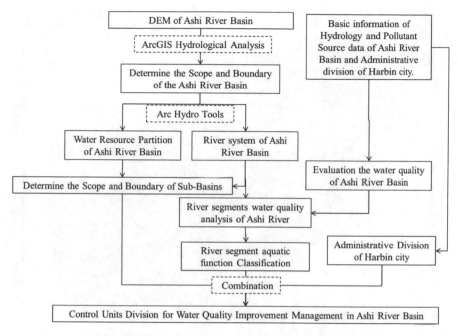

Fig. 2.35 Scheme of the control unit division for water quality improvement of Ashi River Basin

2.2.6.3 The Scheme of Control Unit Division

1. Drainage basin boundaries were delineated on the basis of DEM and locations of implied vallies using ArcGIS software. Used the SRTM DEM images with 30 m resolution to extract Ashi river water system and sub-basins; and based on this the river nodes were chosen to divide river system into reaches.
2. Combined literature research, basic information and sub-basin features to select partition factors; analyzed the pollution loads and distribution of sewage outlet and evaluated the water quality condition of the Ashi River Basin.
3. Used expert experience to complete the division of the watershed. In addition, the Water Resource Partition of the Ashi River Basin and Administrative Division of Harbin City also should be considered in the Control Unit Division of the Ashi River Basin.
4. Named the classification results according to river basin features to complete Control Unit Division.

Figure 2.35 shows the Scheme used to establish the Control Unit Division for water quality improvement of the Ashi River Basin.

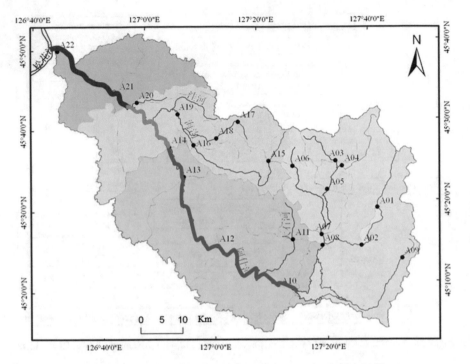

Fig. 2.36 Watershed water quality objectives management control unit of Ashi River

2.2.6.4   The Result of Control Unit Division of the Ashi River Basin

The Ashi river basin mainly includes the administrative regions of Acheng district and Harbin municipal District and the distributions of pollution sources mainly are concentrated in the middle and lower areas of the Ashi River. Considering the water resource of the Ashi River, the main stem of the Ashi river could also be divided into two parts with the boundary located at Acheng hydrology station.

According to the distribution of industrial enterprises and rural settlements in the Ashi River Basin, and considering the distribution of sewage discharge outlet, water resource partitioning, and administrative divisions, the Ashi river main stream has been divided into 4 control units: 2 water quality maintenance control units, 1 water quality improvement control unit and 1 water risk prevention control unit in accordance with the principle of the control unit division, as shown in Fig. 2.36 (Table 2.20).

Comment:—means there have not exact information in this term.

1. **Water quality maintaining control units**: The sub-watershed of Xiquanyan reservoir (A1–A10).

   In this control unit, to adopt the comprehensive strategy of water quality maintaining overall: strictly control the management of forest use and conservation, implement reasonable and efficient use of limited human activity and develop-

Table 2.20 Ashi River Watershed water quality objectives management control unit basic fact sheet

Watershed	River	The control unit		Water quality objectives	Control section	Sewage area		District and county
		Name	Level			Provinces (autonomous regions)	City	
Ashi river basin	Xiquanyan reservoir (upstream segment)	Acheng district	3	II	Xiquanyan reservoir outlet	Heilongjiang	Harbin	Acheng district
Ashi river basin	Ashi river (upstream segment)	Acheng district	3	II	Acheng hydrological station	Heilongjiang	Harbin	Acheng district
Ashi river basin	Ashi river (middle reaches)	Acheng district	3	IV	Acheng district outlet of Ashi river	Heilongjiang	Harbin	Acheng district
Ashi river basin	Ashi river (downstream segment)	Harbin district	3	V	River mouth of Ashi river into the Songhua river	Heilongjiang	Harbin	Harbin district

ment, strengthened the pollution prevention to improve or protect water sources for drinking water and aquatic ecosystems conservation.

2. **Water quality maintaining control units**: Ashi River upstream, from Xiquanyan reservoir outlet (A10, E127.2634°, N45.222°) to Acheng hydrological station (A14, E126.9924°, N45.5692°).

   In this control unit, to adopt the comprehensive strategy of water quality maintaining overall: strictly control the management of agriculture irrigation and back water impoundment (at Ascheng weir), implement reasonable and efficient use of fertilizer, pesticides and irrigation water, strengthened the pollution prevention to improve or protect water sources for drinking water and aquatic ecosystems conservation.

3. **Water quality improvement control unit**: Ashi river middle reaches and the middle and lower reaches, from Acheng hydrological station to the Acheng district outlet of Ashi River (A20, E126.9239°, N45.6767°).

   In this control unit, to generally adopt mainly pollution treatment strategy, to strengthen regulation on sewage outlet and strictly control the pollutant emission amounts, to effectively reduce the load and concentration of main pollutants, to safeguard river ecological base flow, to improve the water quality of the main stem and tributaries of the Ashi River System.

4. **Water risk prevention control unit**: Ashi River downstream, which was belong to Harbin district, from the Acheng district outlet of Ashi river to the river mouth of Ashi river into the Songhua river (A22, E126.7227°, N45.8199°).

   In this control unit, the control strategy of combination of Water quality improvement and water risk prevention should be adopted, to strengthen regulation on sewage outlet and strictly control the pollutant emission amounts and concentration, to strengthen the construction of emergency mechanisms for water sudden pollution accidents, and to actively prevent the water pollution in Administrative Region transboundary and chemical industry areas.

2.3   Technical Guidelines on Pollution Load Allocation in Control Units

by Kun Lei, Ya Tao, Gang Zhou, Yixiang Dend, Richard Williams

2.3.1   Compilation Basis/Application Range

1. The Technical Guidelines are formulated to regulate the method of pollution load distribution in a basin based on control units.

2. The Guidelines are applicable to the basins based on freshwater ecological function region, and can be referred to for the basins which are not yet under freshwater ecological function region.
3. The total water environment capacity is calculated by regarding the basin as a whole, the control units as the basis, and the overall water quality target as the target, by comprehensively considering about the hydrodynamic relation between the hydrologic process and the receiving water body and also that of upstream and downstream and left and right banks in the basin, to ensure the local and overall harmonious relations.
4. The scientific and reasonable calculation method of the total water environment capacity is determined according to the water quality target management demands, water types and basic data; the hydraulic conditions are scientifically designed according to the combination principle of water quality and quantity and based on the hydraulic and water engineering regulated features, and the total load calculation and distribution is conducted on the basis of the stipulated design risks.
5. Regarding the pollution load distribution in a basin, a step-by-step distribution relationship of "receiving water body – pollution discharge outlet of a river – land pollution source" is established on the principle of "determining land by water".

2.3.2 Basic Technical Contents of Pollution Load Allocation Based on Control Units

The technical contents of pollution load allocation based on control units can deepen and implement the concrete control requirements of water quality target management in a basin; the decomposition of control indexes reflects the layered control philosophy of water quality target management, the layers are connected with each other and implemented one layer after another, and the water quality target management is finally implemented in each and every executable control unit. The basic contents and steps of pollution load allocation based on control units are as follows:

(1) Division of Control Units

The division of control units can decompose complex system problems of a basin into independent problems of units, and make it much easier to calculate the water environment capacity and divide the discharge quantity distribution. The basic units of water pollution prevention are divided with water body and water quality standard attainment as the target and in accordance with freshwater ecological function region, catchment features and administrative division.

(2) Water Environment Problem Diagnosis

Water environment problem diagnosis is to determine the water quality target of each control unit in accordance with the freshwater ecological function region, water quality requirements in water function region, then investigate, analyze and evaluate the physical and chemical features of the control units, determine the types and degrees of damage, identify the main over-standard elements or elements with huge over-standard risks, and define the key points of water environment management.

(3) Pollution Load Evaluation of Control Units

Pollution load evaluation of control units is to investigate the land utilization patterns, pollutant generation volume and discharge forms in a basin, analyze the types and features of pollution (proportions of point and non-point source pollution) and spatial distribution features of the control units in the statistical or basin load model methods, study and analyze the influences on the water body from the pollution sources within the control units, and define the pollution sources which needs to be emphatically controlled.

(4) Analysis of Water Environment Capacity

Analysis of water environment capacity is to select appropriate receiving water body and water quality models in accordance with the pollution types, pollution source discharge rules, pollution indexes, and the hydraulic conditions of the receiving water body and other elements, study the pressure-response relation between the pollution load and water body and water quality, in accordance with the water quality target requirements, scientifically design the corresponding hydrological conditions, establish the water environment planning model, and calculate the water environment capacity of the receiving water body in the control units.

(5) Total Pollution Load Distribution of Control Units

The total pollution load allocation of control unit is to define the pollution load allocation principle in accordance with the social, economic and environmental demands within a basin, and calculate the maximum permissible discharge amount of various pollution sources in the control units based on the water environment capacity of the receiving body of each control unit. It also includes the two-layer distribution:

1. The basin control unit—estuary load distribution, in accordance with the water quality target requirements of the control cross-section, the positions of the discharge estuary cross-sections and other conditions, distribute the water body environment capacity of the control units to each and every estuary discharge outlet, and obtain the maximum permissible discharge quantity of each estuary discharge outlet.
2. Estuary—pollution source load distribution. In accordance with the maximum permissible discharge quantity of each discharge quantity and by combining the pollution source structures in each discharge area and estuary coefficients of different pollution sources, distribute the total pollution load to different pollution sources within the control units.

(5) Evaluation of Pollution Load Distribution Effect of Control Units

Based on the pollution load allocation of each control unit and by combining the pollution discharge reduction methods which adapt to the social and economic development level, the step of evaluation s to formulate definite pollution load allocation

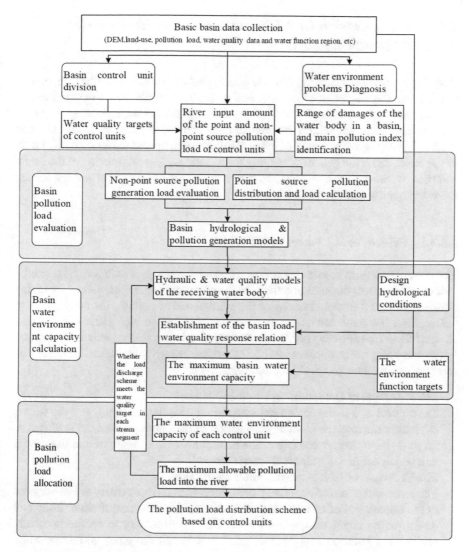

Fig. 2.37 The pollution load allocation technical flowchart based on control units

schemes for various kinds of pollution sources within the control units and ensure the attainment of the target of the total amount reduction and water quality improvement of the control units.

The pollution load allocation technology roadmap of control units is shown in Fig. 2.37.

2.3.3 Basin Control Unit Division and Water Environment Problem Diagnosis

Dividing control units can divide a complex basin into several mutually independent but also interdependent units, and makes it easier to calculate the water environment capacity and make a refined distribution of total pollution load. Solving the water environment problems within the control units and coping well with the units can realize the water quality targets of the control units and the entire water quality target of the basin, reach the goal of protecting the aquatic ecological function of the basin, and then makes it more convenient to management the basin system and implement the water quality management schemes.

2.3.3.1 Principles of Control Unit Division

The control units simultaneously reflect the natural catchment features and administrative management demands, on the premise of natural water systems and by giving consideration to the completeness of the administrative regions, the administrative units within the same catchment area are combined as control units. The division should follow the following principles: determining land by water, independent water body type, complete catchment area, definite administrative responsibilities, and harmony with aquatic ecological function region, etc.

1. **The Principle of Determining Land by Water**
   The control unit is the equivalent area region of water and land, and the waters catchment range is the boundary limitation of land division, and the land spatial range is determined according to the natural attachment features and the spatial unit which combines water and land is formed.
2. **The Principle of Independent Water Body Type**
   The cross-sections of river-lake, river-reservoir, and river-estuary are considered as the boundaries of control units to connect different types of water body and water quality target management schemes. It is necessary to comprehensively analyze the influences on hydrological catchment from gates and dams while dividing the control units in the gate and dam regulating region.
3. **The Principle of Complete Catchment Area (Basin/Sub-basin)**
   According to the basin hydrological cycling features, a control unit may include one or more sub-basins, and the sub-basins should keep complete within the control units. The principle considers the basin or sub-basin boundaries as the isolation boundaries among control units to ensure that the pollution load within a control unit don't exchange with that in other control units and the pollutants of the receiving water body are completely from the very control unit.
4. **The Principle of Complete Administrative Division**
   The control unit division should not break the boundaries of administrative units as much as possible, that is, the completeness of the administrative units should

be ensured in the division process of control units, this can guarantee that various tasks and measures of water quality target management of control units are finally implemented in the administrative regions to assure the realization of task breakdown and implementation, supervision management, pollution source calculation and discharge reduction. The complete administrative units within the control units are beneficial for clarifying the responsible body for pollution source management and environment quality improvement.

5. **The Principle of Harmony with Freshwater Ecological Function Region**
   The control units should be connected with the freshwater ecological function region, the water (environment) function region, and water resource region as much as possible. Control units are divided according to the freshwater ecological function region in the regions which have been defined as the freshwater ecological function region. And control units can be divided by referring to the water (environment) function region in the regions which have not yet been defined as the freshwater ecological function region.

2.3.3.2 Control Unit Division Technology Scheme

The basin control unit division should be based on the systematic exploration of the hydrological and water environment situations in the basin and conducted by adopting the GIS hydrological analysis technology. In accordance with the basic principles of control unit division, under the designated water quality target conditions, the natural catchment range within the basin are under overlapping combination with the administrative units to determine the water and land spatial ranges of each control unit. The control unit division technology roadmap is shown in Fig. 2.38, and the technology roadmap process can be summarized into the following steps:

The collected control unit division data shall include but be not limited to:

1. The basin digital elevation model (DEM), remote-sensing image, land utilization, soil types, river and water system distribution;
2. The water resource region, freshwater ecological function region, water (environment) function region, etc.;
3. The administrative region, hydraulic engineering distribution, the positions and discharge loads of the industrial enterprises, livestock farms and other fixed pollution sources;
4. The positions and flow information of hydrological/water level stations, etc.;
5. The positions of the water quality control cross-sections, the pollution index monitoring values, etc.;
6. And other social and economic development states and future planning data of the basin, etc.

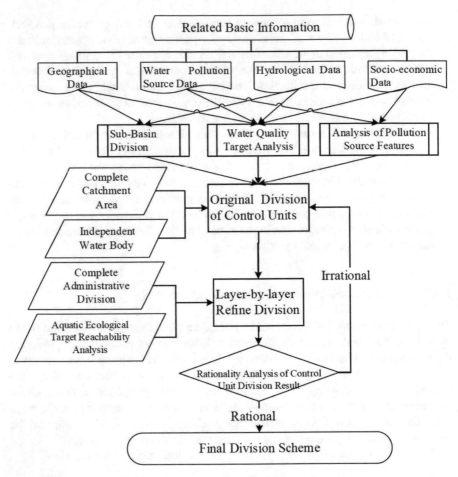

Fig. 2.38 Technology roadmap of basin control unit division

Basin Boundary Identification and Sub-basin Division

The basin surface hydrological conditions are analyzed by digital elevation model (DEM) to obtain the river net and sub-basin division information including the spatial topological information. After basin division by using appropriate software, such as ArcGIS, it is necessary to refer to the measured water system chart to make the extracted river net consistent with the measured water system and identify the accurate basin boundaries and sub-basin boundaries at various levels.

Water Quality Target Analysis

For the sub-basins, in accordance with the function division (freshwater ecological function region/water function region/water environment function region) of the corresponding main stream segments, the water body functions, the corresponding water quality targets and sensitive ecological protection targets of all the stream segments are analyzed and then the basin water quality target is determined.

Basin Pollution Source Feature Analysis

It briefly analyzes the types, distribution and discharge directions of the pollution sources (industrial, domestic, intensive livestock and poultry breeding pollution sources, etc.), and the rural population and land utilization conditions in the basin; by combining the basin water system features, it establishes the point source and non-point source pollution distribution and the relation between the estuary discharge outlets and the pollution receiving rivers to understand the general situations of pollutant generation, collection and discharge into rivers.

Main Stream Segment Division and Control Region Determination

Based on the analysis of the water body functions and corresponding water quality targets, the step firstly starts from the main stream of the basin, and then conducts segment division along the main stream. In general, the step, in accordance with the Principle of complete catchment area, considers the boundaries of the upstream, medium-stream and downstream boundaries in the basin as the boundary of the main stream division for larger seas and rivers according to the basin terrain, topography, and hydrological situations, and the corresponding catchment areas are considered as the tier-one control area; and the step is not required for smaller-dimension basins. At the same time, if a basin involves several administrative regions, the boundaries of these administrative regions can be considered as the main stream cross-sections in accordance with the principle of complete administrative regions, and the corresponding catchment area is regarded as the tier-one control area.

Later, various tier-one control areas are further divided into tier-two control areas. At the very moment, if there exist high-functioning water bodies in the main streams, the principle of priority protection of high-functioning water bodies should be followed, and the upper and lower interfaces of high-functioning water bodies are considered as the interfaces to divide the main stream segments. At the same time, in accordance with the principle of independent water bodies, the main streams in the tier-two control area can be divided into several segments, and corresponding catchment areas are divided as the tier-two control area range.

Various tier-two control areas can be further refined into the main stream segments in accordance with the principles of administrative boundaries, types of water bodies, and clear management responsibilities, and the catchment area in each and every segment is considered as the tier-three control area. What calls special attention is that, during the main stream segment division process, the water quality targets of the interfaces in all the segments should coordinate with each other to ensure the coordination of the water quality targets of the interfaces between two control units, so that the two control units can be under independent pollution control planning. If the water quality targets of the upper and lower cross-sections are uncoordinated, it can be considered that the two segments are combined, and the coordination between the water quality targets of the cross-border water quality cross sections is solved by virtue of the coordination between the administrative subjects in the pollution control planning. After the main stream segments are divided, it is necessary to determine the corresponding catchment area of each main stream segment as the tier-three control area.

Determination of Control Units

A basin can be decomposed into several independent control areas, there exists no cross-border water quality target conflicts among these control areas, and it can also ensure that the pollutants within the control areas all flow into the main stream segments, these control areas will be considered as the spatial units for formulating the sub-basin water quality target management schemes, total pollutant control schemes, or water pollution prevention and control planning, the measures and tasks determined within such range will be finally decomposed and implemented in each and every administrative region, such as, districts and counties, villages and towns are all within the control units. Therefore, the sub-basin boundaries should be overlapped with the administrative region boundaries to determine the distribution and quantity of the administrative regions within the control units, and the administrative regions within the sub control areas are considered as the control units. In this way, the administrative regions accept the pollutant reduction indexes and tasks (e.g.: total discharge reduction) from different control areas, and the sum of the tasks of different control areas is the total pollution reduction task of the basin.

Rationality Analysis of Control Unit Division Effect

By making a retrospective analysis of the division result and combining the principles which are conducive to simplifying pollution management and defining the responsibility subject for environment quality, the rationality of control unit division effect is analyzed. The control unit boundaries need to be adjusted if necessary.

1. Test of water area and water quality target unity of control units. Each control unit should only have one single water area and water quality target.
2. Test of sub-basin completeness of control units. One control unit can contain or more sub-basins, but should not be across the sub-basin boundaries.

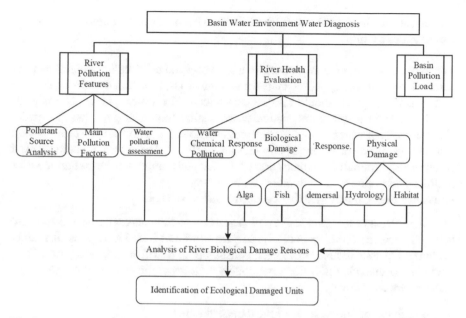

Fig. 2.39 Diagnosis technology roadmap of water environment problems of control units

3. Test of administrative unit completeness of control units. Each control area boundary should be overlapped but not be crossed with the administrative region boundary at the corresponding level.

2.3.3.3 Diagnosis Technologies and Methods of Water Environment Problems of Control Units

In order to stably keep its planned design aquatic ecological environment function, it is necessary to prove the ecological completeness of the basin control unit from the perspectives of chemical, biological and physical completeness. The chapter, starting from completely mastering the damage information of control units, incorporates the chemical index, physical index and biological index into the diagnosis of water environment problems of control units.

Through making field survey and monitoring of the basin water environment, the environment problems for the main water bodies in the basin are diagnosed, the basin pollution features analyzed, the ecological health state of rivers evaluated, and the water ecological environment damage situations of all the control units within the basin identified. The diagnosis technology roadmap of water environment problems of control units is shown in Fig. 2.39.

Water Quality Evaluation and Identification Methods of Damaged Environment Units

The types of over-standard pollutants and over-standard time are determined via actual measured water quality analysis; and the types, time and degrees of damage of the damaged units should also be determined. The water quality state is evaluated according to the cross section water quality monitoring results; the spatial change trends of water quality for key cross sections are analyzed; and the pollution type change trends of featured pollutants are analyzed. The current commonly used water quality evaluation methods are: Classified and comprehensive pollution index method, and pollution load duration curve method.

Classified and Comprehensive Pollution Index Method

The Classified and comprehensive pollution index method makes it easier to compare different regions, different control units, and different function regions. It mainly evaluates the four indexes of oxygen consumption, nutritive salt, heavy metal and bacteria and then identifies the water quality features of the control units. The basic contents are as follows:

1. Identify the pollution types of the damaged units;
2. Determine the degrees of damage of classified pollution of the damaged units;
3. Diagnose the trends of classified pollution of the damaged units;
4. Diagnose the transformation trends of pollution types of the damaged units.

The classified pollution index of the damaged water body > 1, and that of the undamaged water body < 1.

Pollution Load Duration Curve Method

The duration curve is a frequency curve whose parameter value is larger or equal to a given parameter value in a specific parameter series, and is a curve where the percentage of the quantity of other parameter which are no smaller than the specific parameter value among the total is the X-coordinate, and the corresponding parameter values the Y-coordinate (logarithmic coordinates generally adopted) after sorting the parameter values from big to small/ when the parameters are the flow-related parameters, it is a flow duration curve, and the load flux is obtained when the flow-related parameters multiply the specific pollutant standards, and the duration curve which considers the load flux as a new parameter is the load duration curve for such index. The analysis steps are as follows:

1. Establish the daily flow data of the water quality cross sections, and establish the flow duration curve;
2. Determine the water quality targets of the actually measured cross sections (the permissible concentrations of different pollutants);
3. Obtain the permissible load duration curve by multiplying the flow duration curve date of the water quality cross sections by the water quality targets, and draw the permissible load duration curve chart;

4. Transform to the actual load by multiplying the actually measured water quality by the flow rate of the current day;
5. Obtain a daily load of the monitoring cross section by multiplying the daily flow and the concentration of the water quality sample, and draw the load points on the load duration curve chart. The points above the curve indicate the deviation from the permissible load and those below the curve indicate they meet the standards and the actually measured water quality satisfies the give using functions.
6. Determine the Over-Standard Recurrence Period According to Over-Standard Points.

   The load duration curve can identify the following water quality features of control units:

   Identify the water quality over-standard phenomenon, distinguish the point and no-point source pollution problems, determine the pollution control period, determine the over-standard recurrence rate and over-standard frequency, and determine the eliminating rate required for pollutant standard attainment.

   The analysis of damaged control units: the over-standard recurrence period of the damaged water body < 3 years, and that of the undamaged water body > years.
7. Evaluation Method of Water Pollution Load Contribution Types

   The method of water pollution load contributor types is adopted to identify the non-point source pollution discharge control area, and the specific evaluation indexes are shown in Table 2.21 (Fig. 2.40).

Biological Evaluation and Identification Methods of Damaged Environment Units

The biological completeness is evaluated by analyzing the diversity index, uniformity index and richness index of phytoplankton, zooplankton and zoobenthos.

The Plankton Evaluation Method

Table 2.21 Analysis of water pollution load types

Water pollution load contribution types	Point source pollution contribution proportion (r)	Economic reduction measure
Point-source oriented	$r \geq 80\%$	Point-source reduction
Point-source dominated	$60\% \leq r \leq 80\%$	Point-source reduction oriented
Hybrid	$40\% \leq r \leq 60\%$	Point-source reduction oriented
Non-point dominated	$20\% \leq r \leq 40\%$	Non-point-source reduction oriented
Non-point oriented	$r \leq 20\%$	

The biological state of the region is evaluated in accordance with the richness index, diversity index and uniformity index of phytoplankton and zooplankton.

1. Margalef Richness Index:

$$D = (S - 1) / \ln N \qquad (2.1)$$

2. Shannon-Wiener Diversity Index:

$$H' = -\sum_{i=1}^{s} P_i \ln P_i \qquad (2.2)$$

3. Formula of Pielou Uniformity Index:

$$J = \frac{H'}{\ln S} \qquad (2.3)$$

where: S is the total number of species; in $P_i = N_i / N$, P_i is the proportion of the individual number of the ith specie among the total number of individuals; And N is the total number of individuals of all the species.

The water quality pollution situation is evaluated by using Shannon-Wiener Diversity Index for zooplankton, and then a pollution classification standard is established as shown in Table 2.22.

The Zoobenthos Evaluation Method

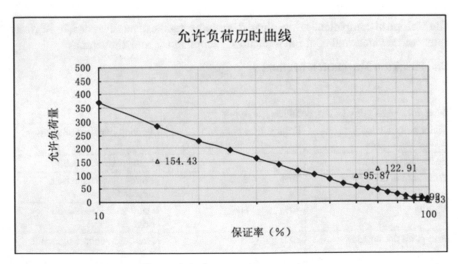

The x-axis is the guarantee percentage.
The y-axis is the allowable load.

Fig. 2.40 Permissible load duration curve

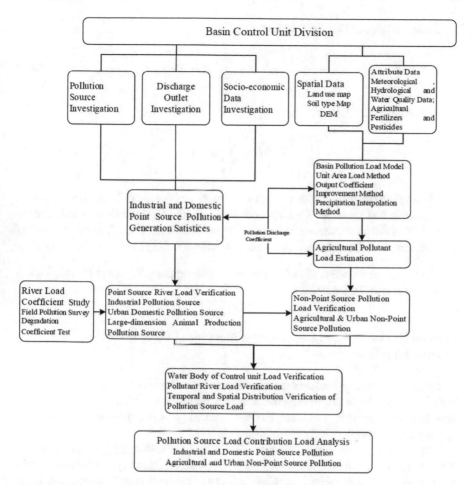

Fig. 2.41 Technology roadmap of basin control unit pollution load evaluation

The biological state of the region is evaluated according to the richness index, diversity index and biological index of the zoobenthos:

1. Margalef Richness Index:

$$D = (S - 1) / \ln N \qquad (2.4)$$

Table 2.22 The diversity index classification standard for zooplankton

Diversity index	3.1–4.5	2.1–3.0	1.1–2.0	0–1
Pollution grade	Slightly polluted (Unpolluted)	Lightly polluted	Moderately polluted	Heavily polluted

2. Shannon-Wiener Diversity Index:

$$H' = -\sum_{i=1}^{s} P_i \ln P_i \tag{2.5}$$

3. Biological Index (BI):

$$BI = \sum_{i=1}^{s} n_i a_i / N \tag{2.6}$$

where: S is the total number of species; in $P_i = N_i / N$, P_i is the proportion of the individual number of the ith specie among the total number of individuals; and N is the total number of individuals of all the species; N_i is the number of individuals of the ith specie; a_i is the pollution receiving value of the ith specie.

The water quality classification grades of zoo-benthos by using BI (Biological Index) are shown in Table 2.23:

Physical Evaluation and Identification Methods of Damaged Environment Units

The physical damage evaluation method mainly evaluates the habitats for aquatic organism individuals, populations or communities by analyzing the steady-state river flow and sand content and other indexes according to the flow and sand data of hydrological stations.

The physical damage evaluation method mainly conducts analysis according to the flow and sand data of hydrological stations, and it can also expand the index range to increase the comprehensiveness and validity of evaluation and identification.

Its main thought is to segment the historical data, consider the early stage with fewer human disturbance activities as the background state, then make comparison with the data in the recent period, and judge the degrees of change or damage.

Ecological and Environmental Flow Indexes and Evaluation Methods

Calculate the river steady-state design flow indexes of the recent ten years, each index is given a certain value and the full mark is 100, and obtain the score of the recent 10 years.

Table 2.23 Water quality classification grades of zoobenthos by BI

BI	0–5.48	5.49–6.53	6.54–7.59	7.60–8.64	8.64–10.00
Pollution grade	The cleanest	Clean	Slightly polluted	Moderately polluted	Heavily polluted

Table 2.24 Ecological and environment flow indexes and evaluation methods

River steady-state design flow			Individual score P_i	Total score
Name of design flow	Code	Significance		
The harmonic average flow BxQhy	30B3	Biological safety—traditional pollutants design dilution flow	20	100
	4B3	Biological safety—toxic substances design dilution flow	15	
The extreme arithmetic average flow HxQay	30Q5	Human Health—non-carcinogens design dilution flow	15	
The geometric average flow BxQgy	30G5	Bacteria dilution flow	10	
The perennial daily harmonic average flow Qh	Long-term	Human health—carcinogens Design dilution flow	15	
The perennial daily arithmetic average flow Qa(ADF)	Long-term	Water resource quantity index	10	
The minimum ecological flow QeTennant method (10% Qa)	30	The minimum flow for ecological safety	15	

Table 2.24 shows the 7 indexes used in evaluation, the full point for each index is Pi and the full mark is 100.

$$P = \sum_{1}^{7} \frac{Q_t}{Q_0} P_i \qquad (2.7)$$

where, Q_0 is the background flow value, Q_t is the flow value during the evaluation period, and P_i is the score during the evaluation period, and should be more than 100.

The Sand Content Index and Evaluation Methods

Analyze the changes of sand contents in wet, normal and dry seasons, and give a certain value to the sand content and the full mark is 100.

Comprehensive Evaluation and Identification Methods of Damaged Environment Units

Standardize the evaluation results of water quality, aquatic organisms and physical habitats, weigh each index according to the features of the research regions, and do a weighted sum to obtain the comprehensive evaluation result of the water environment of the control units.

The comprehensive evaluation adopts the 100% progressive five-tier evaluation system, and multiple indexes return to the five-tier 100% evaluation system in which a higher score is superior and lower one inferior via weighted index after entering into the next layer.

1. Physical Habitat Evaluation

   The weights of 0.6 and 0.4 are respectively given to the river steady-state design flow and sand content, and get the final weighted score of physical habitat evaluation.

2. Chemical State Evaluation

   Each of the five types of comprehensive pollution indexes is weighted 20 scores, and the full mark is 100, and the chemical state value is obtained according to the following formula:

   When $0 < x \leq 0.5$, $y = 20$; when $0.5 < x < 2$, $y = -13.333x + 26.667$; when $x \geq 2$, $y = 0$.

   (Where, x: The Classified and comprehensive pollution index: y: the index score)

3. Biological State Evaluation

   In accordance with the biological diversity index evaluation, the classification standard is shown in Table 2.25. Phytoplankton, zooplankton and zoobenthos are respectively given the weights of 0.2, 0.4 and 0.6 to calculate the biological state scores.

4. Comprehensive Evaluation

   The physical, chemical and biological states are respectively given the weights of 0.2, 0.3 and 0.5, and the comprehensive evaluation score is obtained according to their own weighted scores; the classification standard of the river degrees of damage is shown in Table 2.26:

Table 2.25 Biological state evaluation and classification standards

Biological diversity index	$D > 3$	$2 < D \leq 3$	$1 < D \leq 2$	$0 < D \leq 1$	$D = 0$
Pollution grade	Unpolluted	Slighting polluted	Moderately polluted	Moderately-heavily polluted	Heavily polluted
Score	100		$Y = 30D$ $(0 < D \leq 3)$		

Table 2.26 Classification standards of degrees of damage

Damage situation	Unpolluted	Slighting polluted	Moderately polluted	Moderately-heavily polluted	Heavily polluted
Score	80–100	60–80	40–60	20–40	0–20

2.3.4 Pollution Load Evaluation of Basin Control Units

The pollution load estimation of basin control units is an important link to control unit water quality and improve the management work, and can provide various kinds of pollutant source discharge loads for formulating the total pollutant control and reduction discharge schemes. According to the pollution discharge features, the pollution sources in a basin are divided into point source and non-point source. The technology roadmap of pollution load verification of control units is shown in Fig. 2.41.

2.3.4.1 Point Source Pollution Load Calculation

There are mainly 4 methods of sewage discharge from point pollution sources: the sewage is directly discharged into the surface water body through the discharge outlets in plants; the sewage is discharged into the surface water body after mixing with urban domestic sewage; the sewage is discharged into the nearby surface water body through sewage channels; and the sewage is discharged into the underground water body through seepage pits and wells. The fixed point source pollution load evaluation within the basin is mainly calculated by analyzing the data provided by pollution source census and monitoring system in statistical method. The point source pollution discharge load can be quantitatively estimated in several mature methods, for example, it can be quantified in the methods of monitoring the sewage amount and concentration from pollutant discharge enterprises, measuring the materials used in their production processes, etc.

Data Collection and Investigation Methods

The water environment quality is investigated comprehensively in the methods of data collection, field investigation and actual monitoring. In the practical operation, the method of sampling investigation can be employed to make a field investigation of the urban domestic sewage production, discharge and treatment situations.

The field inspection and monitoring of pollution sources include mastering the positions and dimensions of pollution sources and pollutant discharge and treatment, mastering the types, physicochemical and biological features, discharge patterns and rules of discharged pollutants, and calculating the discharge intensity.

In verifying the industrial point source pollution load, the main investigation contents include the sewage discharge amount from enterprises, pollutant weight and discharge direction and also the water function regions to which they belong, and also supplementary investigation or monitoring of the industrial and mining enterprises which lack certificates. COD, ammonia nitrogen. ammonia nitrogen (TN) and total phosphorus (TP) are the investigated mandatory pollution indexes, and investigation items can be added for industrial and mining enterprise with different natures according to the demands, such as BOD_5, SS, volatile phenol, petroleum and derivatives, total mercury (T-Hg), total cadmium (Cd), etc.

In verifying the urban domestic pollution load, the main investigation contents include: the urban pollution, the urban floating population and population growth situations of towns and cities, the residents water consumption, and sewage discharge and treatment amount. The other aspects include the local industrial structure, and the per capita income of urban residents; the urban sewage pipeline network layout situation, the domestic sewage treatment ratio and treatment reusing ratio, and the operation situations of urban sewage treatment plants; the water quality states of different monitoring cross sections, etc.

Indirect Estimation Methods

The indirect estimation methods of industrial pollutant discharge amount mainly include material balance method, pollution discharge coefficient method, etc.

Material Balance Method

The amount of pollutants transformed is calculated according to the fuels and materials used in the enterprise manufacturing process and also the consumption within a unit time as well as the related compo contents contained in products and by-products.

Pollutant Discharge Coefficient Method

The pollutant discharge amount is obtained according to the empirical discharge coefficient of unit product in the production process, and the calculation formula is:

$$L_i = K_i * W \tag{2.8}$$

where, L_i is the discharge amount of a certain pollutant, kg; K_i is the pollution discharge coefficient of unit product, kg pollutant/t product, and can be changed with the technological advancement and management level improvement; W is the total product weight, ton.

Furthermore, the industrial sewage discharge amount can be obtained by simply multiplying industrial water consumption, sewage production coefficient and discharge coefficient, and then the pollutant discharge amount from enterprises can be estimated by determining the concentrations of main pollutants contained in sewage in production processes.

The estimation method of pollutant discharge amount from enterprises is:

$$L_i = Q * \gamma_1 * \gamma_2 C_i \tag{2.9}$$

where: L_i is the discharge amount of a certain pollutant, kg; Q is the water consumption of enterprises, 10,000 t; γ_1 is the industrial sewage production coefficient; γ_2 is the industrial sewage discharge coefficient; and C_i is the sewage pollutant concentration, g/m^3.

Population Coefficient Method

The urban domestic pollution load is obtained by the per capita sewage discharge amounts of different regions multiplying their estimated future population and their average pollutant concentrations minus the pollutants reduced by sewage treatment. Shown in the following formula:

$$W_{ijt} = P_{it}q_{it}c_{ijt}(1 - \alpha) + P_{it}q_{it}c'_{ijt}\alpha \tag{2.10}$$

where, W_{ijt} is the pollution load of the jth pollutant in the tth year in the i region, P_{it} is the population in the tth year in the i region, q_{it} is the per capita sewage discharge amount in the tth year in the i region, c_{ijt} is the sewage concentrate of the jth pollutant in the tth year in the i region, α is the sewage treatment coefficient, and c'_{ijt} is the discharge concentration of the jth pollutant after sewage treatment in the tth year in the i region.

2.3.4.2 Non-point Source Pollution Load Estimation Methods

The basin non-point source pollution load estimation is relatively complex, considering the differences in non-point source pollution production mechanisms and features of different types of underlying surfaces, there exist great differences in non-point source pollution estimation methods for different types of underlying surfaces. Due to the differences in basic data sources, types of underlying surfaces and verification methods, the verification methods from simple to complex are adopted, and different verification methods can check and verify each other.

The non-point source pollution production process is obviously influenced by the rainfall runoff and other hydrological processes, and different types of underlying surfaces (land utilization) have obvious temporal and spatial difference distribution features. Therefore, for the non-point pollution load verification, it is required to not only determine the total pollution load at the basin outlets and the change relations with the passage of time, and also determine the non-point source pollution loss spatial distribution regulations and also the key non-point pollution areas. Meanwhile, it is also required to verify different types of agricultural non-point source pollution (farm production pollution, rural domestic pollution, livestock and poultry breeding pollution, aquiculture pollution), and the pollution load contribution ratio to the receiving water body of control units of the urban runoff pollution, and determine the main pollution sources and pollutants to provide grounds for formulating the total pollutant control and discharge reduction schemes.

The non-point source pollution load quantified methods applicable to all research regions are called the general methods, such as, Unit Areal Load Method, Export Coefficient Model, Mean Concentration, rainfall deduction method, the load-runoff relation model, etc.

Unit Areal Load Method

Unit Areal Load (UAL) Method is the simplest and most widely applied method to estimate the non-point source pollution load. It calculates the total pollution load within a basin. Such kind of model can identify the influences on the non-point source pollution load from different types of land utilization, and has some scientificalness and representativeness, and play a certain guiding role in identifying and managing the key source regions of non-point source pollution. But it neglects the influences on non-point source pollution from different terrain, soil, hydrological and meteorological conditions, so its application range is limited to some extent.

The unit areal load data can be obtained by adopting the continuous synchronous water quantity (Q) and water quality (C) within the selected catchment area, and the selected calculation area should be with a uniform type of land utilization. UAL (Unit Area Load) of different land utilization can be obtained by analyzing the data in the test area with different types of land utilization.

$$UAL = \frac{1}{AT} \int_{t1}^{t2} Q(t)C(t)dt \tag{2.11}$$

where, A—The area of the text catchment area, t_1—the start time of the surface runoff, t_2—the ending time of the surface runoff, and T—the duration of runoff outflow ($T = t_2 - t_1$). By definition, in order to obtain the representative unit areal load data, it is necessary to observe the rainfall-runoff generation-pollutant production processes of several rainfalls of different types.

If there exists point source pollution in the study catchment area, the calculation formula is correspondingly changed as follows:

$$UAL = \frac{1}{AT} \left[\int_{t1}^{t2} Q(t)C(t)dt - T \sum_{j=1}^{m} L_j \right] \tag{2.12}$$

where, L_j—the discharge load of the jth point source pollution, m—the number of point sources in the study region. The remaining symbols have the same meanings as before. It is assumed that the point sources are evenly discharged in the sampling period.

The unit areal load data obtained from the test can be used to estimate the non-point source pollution load within the study region, and this is the UAL (AKA Export

Coefficient) Model the American and Canadian scholars raised and applied while researching the land utilization-nutritive load-lake eutrophication relation in the early 1970s. its general expression is:

$$L = \sum_{i=1}^{m}(UAL)_i A_i \qquad (2.13)$$

where: L is the total annual export (kg a^{-1}) of a certain kind of pollutant of a certain kind of land utilization type; m is the number of types of land utilization; $(UAL)_i$ is the UAL or export coefficient (kg hm^{-2} a^{-1}) of the pollutant of the ith type of land utilization; And A_i is the area (hm^2) of the ith type of land utilization.

The UAL data of land utilization of different types are generally obtained from the field source tests, and the values are closely related to the soil conditions, rainfall intensity, rainfall duration, fertilizer amount, fertilizer time, fertilizer types and other agricultural activity conditions, and a lot of site test data are required to support the acquisition of the UAL data. If the type of land utilization is single in a small-basin test area (for example, over 90% of the land are in the same utilization form), the data obtained from the small-basin tests can be used to approximately calculate the UAL for such kind of land utilization.

When the UAL data obtained from the tests for special regions are applied in other regions, it is necessary to fully demonstrate the similarity between the underlying surface of the target region and the region where the test data are acquired, and these data can be only applied in the regions with similar soil conditions, rainfall intensity, rainfall duration, fertilizer amount, and fertilizer time. The UAL method doesn't consider about the chemical or biological change process of water quality indexes, strictly speaking, the method is only used to study the UAL of conservative substances, or calculate the pollution loss situations of TN and TP.

Improved Export Coefficient Method

The core assumption of the export coefficient model for non-point source pollution load is: it is assumed that all the factors have a universal expert coefficient considering the influences from the factors of arable areas, types and quantity of livestock, population and rainfall input. The model, while adopting different export coefficients to different types of land utilization, also adopts different export coefficients to different types of livestock, and the export coefficient of domestic pollution is mainly determined by the domestic sewage discharge and treatment situations.

The Improved Export Coefficient Method Which Considers about the Hydrological Factors and Basin Loss

The export coefficient model comprehensively analyzes the classification of types of land utilization and nutritive sources, but doesn't consider about the influences on the export coefficient of the model from the inter-annual changes of hydrological factors of the non-point source pollution, or the loss of basin pollutants in the transportation process. To overcome the disadvantage, the two parameters of rainfall influencing coefficient and basin loss coefficient are introduced, and the improved export coefficient model is as follows:

$$L = \lambda \left\{ \alpha \sum_{i=1}^{n} E_i \left[A_i(I_i) \right] + p \right\} \tag{2.14}$$

$$\lambda = \frac{1}{1 + aq^b} \tag{2.15}$$

$$\alpha = \frac{M_i}{\overline{M}} \tag{2.16}$$

where, α is the rainfall influence coefficient, λ is the basin loss coefficient, q is the annual basin runoff module, a, b are the parameters, M_i is the nutritive load in the ith year, and $\overline{M}$ is the perennial average nutritive load.

The traditional export coefficient model introduces the parameter of pollution load coefficient β to express the strength degrees of rainfall, runoff generation which transform the non-point source pollutants in the basin into the basin outlet pollution load. The studies show that there exists an extremely obvious positive correlation between the pollution load coefficient and the annual surface runoff module, so the improved model is shown in the formula below:

$$L = \beta \left(\sum_{i=1}^{n} E_i \left[A_i(I_i) \right] + p \right) \tag{2.17}$$

where: $\beta = ae^{bq_i}$ is the pollution load coefficient, q_i is the annual surface runoff module, $L/km^2/s$; and a, b are the parameters.

Rainfall Deduction Method

In a narrow sense, non-point source pollution is the rainfall runoff pollution, that is to say, the non-point source pollution of the surface water is the water body pollution caused by the pollutants in the atmosphere, on the ground and under the ground entering into rivers, lakes, seas and other water bodies under the leaching and scouring of the rainfall-runoff.

The rainfall deduction method assumes that all the pollution load of the basin outlet is contributed by the point source pollution when no surface runoff is generated in sunny or rainy days; only when the surface runoff is generated from huge rainfall or rainstorm, can non-point source pollution and point source pollution happen simultaneously. In general, the point source pollution discharge in the basin is relatively steady, and the annual point source pollution load can be considered as a constant, therefore, the pollution load caused by any rainstorm can be expressed into:

$$L = L_n + L_p = f(R) + C \tag{2.18}$$

where, L_n is the non-point source pollution load; L_p is the point source pollution load; L is the total annual load of the basin outlet cross section; R is the rainfall; $f(R)$ is the function relation between L_n and R; C is the constant and represents the constant discharged point source pollution load. So for any two rainstorms A and B, there is:

$$L_A - L_B = [f(R_A) + C] - [f(R_B) + C] = f(R_A) - f(R_B) = L_{n,A} - L_{n,B} \tag{2.19}$$

The physical significance of the aforesaid formula is that the deduction between the pollution loads (including point and non-point) generated by any two rainfalls (or any rainfalls in any two years) is equal to that between the non-point source pollution load caused by them.

If the function relations of $f(R)$ between L_n and R remain approximately unchanged between two rainfalls (or any rainfalls in any two years), the aforesaid formula is further rewritten into:

$$\Delta L = f(R_A) - f(R_B) = f(R_A - R_B) = f(\Delta R) \tag{2.20}$$

Its physical significance can be construed as: the deduction between the basin pollution loads generated by any two rainfalls is equal to that between the non-point source pollution load caused by them. The correlation between the rainfall deduction value and the pollution load deduction value (non-point source load) can be established on the basis of it, but it is unnecessarily to consider about the loads generated by point source pollution yearly.

The application steps of rainfall deduction method include:

1. The correlation between the basin rainfall and the measured pollutant load of the basin outlet cross section is established on the basis of measured hydrological and water quality data:

$$L = aR^2 + bR + c \tag{2.21}$$

where, L is the measured pollutant load of the basin outlet cross section, R is the average basin rainfall; and a, b, c are the fitting constants.

2. Calculate the rainfall deduction values every year and the measured pollutant load deduction values of the basin outlet cross sections every year, and make a correlated analysis:

$$\Delta L = d(\Delta R)^2 + e(\Delta R) + f \tag{2.22}$$

where, ΔL is the deduction value of measured pollutant loads of the basin outlet cross sections every year; ΔR is the rainfall deduction values every year; and d, e, f are the fitting constants.

3. Calculate and verify the basin point source pollution load. Substitute the rainfall deduction values every year into the aforesaid formula, and estimate the non-point

source pollution load of the basin outlet cross section L_n. The deduction between the measured pollutant value of the basin outlet cross section L and L_n is the basin point source pollution load L_p. If the calculated inter-annual variation of the basin point source pollution load is not large, it proves the correctness of the aforesaid assumption that the point source discharge within the basin is relatively steady.

4. Determine the correlation between the rainfall and the basin non-point source pollution load. Make a correlation analysis between the average rainfall and the non-point source pollution load of the basin outlet cross section L_n, and get the correlation formula between the rainfall and non-point source pollution load as follows:

$$L_n = \kappa R^\delta = L_n = \alpha R^2 + \beta R + \gamma \tag{2.23}$$

where, L_n is the non-point source pollution load; R is the average basin rainfall (mm); α, β, γ, κ and δ are the fitting constants. It can be seen from the formula that the formula of $f(R_A) - f(R_B) = f(R_A - R_B)$ can be completely established when and only when the fitting constant of δ is approximate to 1, or α is smaller, namely, when the outlet load of non-point source pollution and the average rainfall are in good linear relations, at the very moment, the rainfall deduction method can be applied to estimate the non-point pollution.

The rainfall deduction method is applicable to the actual situation where China has few non-point pollution monitoring data, and is a convenient and easy-to-use basin non-point source pollution load estimation method by using China's existing hydrological annals data and water quality monitoring data. The method can roughly distinguish the point source pollution load and the non-point source pollution load from the water quality monitoring data in a simple calculation process.

Concentration (Load) and Runoff Relation Model

The empirical statistical model is established in accordance with the relation between the pollutant load and the runoff quantity. For the same basin, there is always a strong correlation between the pollutant load and the runoff quantity, the pollutant quantity contained can be estimated in accordance with the runoff volume generated after rainfall, but such model is inapplicable to the situation where the types of the land utilization are changed, for example, it cannot be used to forecast the pollution load change after taking non-point source management measures.

The simplest model between concentration or load and runoff quantity—Rating Curve Method expresses the relation between the pollutant concentration C or the load L and the runoff quantity Q in the simplest form:

$$C = aQ^b \text{ or } L = QC = aQ^{b+1} \tag{2.24}$$

where, the parameters a, b are the empirical parameters. The aforesaid formula is applicable to not only the correlation between the fitting sand loss and the quantity Q, but also the other non-point source pollutants.

Furthermore, different fitting methods are available for the rivers with different pollution types. For a large, heavily polluted and evenly discharged river, the relation between its concentration and quantity can be generally expressed as:

$$C = a + \frac{b}{Q} \text{ or } L = QC = aQ + b \tag{2.25}$$

For a slightly polluted river with huge changes of hydrological processes, the relation between its concentration and quantity can be generally expressed as:

$$C = a + \frac{b}{Q} + cQ \text{ or } L = QC = aQ + b + cQ^2 \tag{2.26}$$

where, a, b and c are the empirical parameters; the parameter a is equivalent to the benchmark concentration, and $\frac{b}{Q}$ is the concentration contributed by point source pollution, and cQ is equivalent to the concentration contributed by the runoff increase. C can be the concentration of sand loss, N, P and other non-point source pollutants.

In order to distinguish the change of water quality-flow relation in different phases in the hydrological process, some researchers incorporated the changes of flow into the water quality and water quantity correlation model, and raised the following formula:

$$C = a + bQ + c\frac{dQ}{dt} \text{ or } L = QC = aQ + bQ^2 + cQ\frac{dQ}{dt} \tag{2.27}$$

Another type of concentration-runoff model adopts the actually measured data from different basins, and assumes that concentration (or pollution load) is a set of functions with dimensionless or normative hydrological and basin feature parameters, such as, the proportion of land utilization or soil types, rainfall, runoff quantity and other parameters. A typical case is the water quality model which was applied in the Tennessee Valley in the State of Tennessee, the water quality composition concentration is a power function which is expressed by runoff:

$$C = a\left(\frac{Q}{A}\right)^b \text{ or } L = QC = aQ\left(\frac{Q}{A}\right)^b = aA\left(\frac{Q}{A}\right)^{b+1} \tag{2.28}$$

where, A—the catchment area, the parameters a and b are changed in different catchment regions, and it can also be the function for land utilization, soil and other factors.

2.3.4.3   Basin Pollution Load Model Method

The basin pollution load model method comprehensively considers about the terrain change, differences in types of land utilization, differences in soil types, meteorological condition change, farm cultivation modes and other factors, have solid theoretical foundations in hydrological science and environmental science, and is widely applied.

The most commonly used non-point source pollution models mainly include SWAT (Soil and Water Assessment Tool), HSPF (Hydrological Simulation Program Fortran), AnnAGNPS (Annualized Agricultural Non-point Pollutant Loading Model), etc. and the other non-point source pollution models, such as, CREAMS (Chemicals, Runoff, and Erosion from Agricultural Management Systems), EPIC (Erosion-Productivity Impact Calculator), SWRRB (Simulator for Water Resources in Rural Basins), ANSWERS (Areal Non-point Source Watershed Environment Response Simulation), WEPP (Water Erosion Prediction Project), etc., to different degrees can be applied in the basin hydrological process, soil loss, nutritive (N, P) load simulation, BMPs (Best Management Practice) evaluation, etc.

Basin Load Model Establishment

The commonly used SWAT and HSPF are the semi-distributed and continuously simulated basin models, and are used to forecast the long-term influences on water, sand and pollutant export from the land management measures in the complex basins with various kinds of soil, land utilization and management conditions.

Data Demands

The export data required for basin models include a huge amount of spatial data and attribute data.

The spatial data mainly include digital elevation model (DEM), the actual water system distribution, land utilization and soil type distribution in the research region; the attribute data include meteorological data, hydrological data, water quality data, the data of reservoirs, lakes and gates and dams within the basin, pollution source investigation and statistical data, and agricultural management measure data, etc. Furthermore, the operation management data of reservoirs, gates and dams or irrigation stations are also required.

1. The following steps are required for the establishment of the basin model:
2. Establish the spatial database for attribute data;
3. Divide the catchment relations according to the DEM data;
4. Load the land utilization and soil classification data;
5. Divide the sub-basins and hydrological response units (HRU);
6. Calibrate and verify the model's hydrological, sand, water quality simulation and other important parameters.

The calibrated basin model is adopted to make a simulation calculation of the runoff generation and confluence and pollutant production processes, and to scientifically estimate the basin pollution load.

Model Selection

By referring to the Technical Guidelines of Non-Point Source Pollution Load Verification Based on Control Units, the corresponding suitable basin load models should be selected for different basin types to make load evaluation.

The hydrological simulation modules of most non-point source mathematical models adopt the soil conservation service curve number method (SCS-CN), and the soil erosion and sand loss are almost simulated by the empirical universal soil loss equation (USLE) and its various revised and improved versions.

Some of the pollutants of nitrogen and phosphorus and pesticides are simulated in simpler material balance calculation, and some simulate the nutritive circulation process (including the simulation of plant growth process) within the entire basin, in a more complex material balance calculation. The simulation abilities of different basin models and their comparison are shown in Table 2.27.

2.3.5 Water Environment Capacity Analysis

Water environment capacity is one of the grounds to formulate the local and specialized water discharge standards, the environment management departments use it to determine the permissible pollution load limit on the premise where the specific water body reaches the standards. The water environment capacity analysis is the important prerequisite and core for total pollutant distribution, the response relation between basin pollution load-water quality of the receiving water body can be established through appropriate basin model and receiving water body model, and then the environment capacity of the water body in each control unit can be determined.

Technology Process of Water Environment Capacity Analysis

The water environment capacity of each control unit and basin should be calculated respectively in different hydrological phases on the principle of "zoning, staging, and grading" and in accordance with the water quality targets of freshwater ecological function regions. Water environment capacity analysis should ensure that the river flow is continuous and the water qualities at the upstream and downstream freshwater ecological function regions meet the standards in accordance with the basin overall planning principle. Water environment capacity analysis should reasonably generalize the pollutant river (lake)-into spatial arrangement forms in the freshwater ecological function regions in accordance with the correlation of "freshwater ecological function region-control unit-estuary discharge outlet (including point source pollutant discharge outlets and non-point source confluence features)".

Table 2.27 Summary and comparison table of simulation abilities of different basin models

Model	Type		Complexity degree			Duration				Hydrological		Water quality						
	1	2	3	4	5	6	7	8	9	10	11	12	13	14	15	16	17	18
AGNPS	•	–	–	–	•	•	–	–	–	•	–	–	•	•	•	–	–	–
AnnAGNPS	–	•	–	–	•	–	•	–	–	•	–	–	•	•	•	–	–	–
BASINS	–	•	•	•	•	•	–	–	•	•	•	•	•	•	•	•	•	•
DRAINMOD	–	•	–	–	•	•	–	–	–	–	•	–	–	•	•	•	•	–
DWSM	–	–	–	–	•	•	–	–	–	•	–	–	•	•	•	–	–	–
EPIC	–	•	–	•	•	–	•	–	–	•	–	–	•	•	•	–	–	–
GISPLM	–	•	–	•	–	–	–	–	–	•	–	–	•	•	–	–	–	–
GSSHA	•	•	–	•	–	•	–	–	–	•	–	–	•	•	–	–	–	–
GWLF	–	•	–	•	–	–	•	•	–	–	•	–	•	•	–	–	–	–
HSPF	–	•	–	–	•	•	–	–	–	–	•	•	•	•	•	•	•	•
KINEROS2	–	•	–	•	•	•	–	–	–	•	•	–	•	•	•	•	•	–
LSPC	–	•	–	•	•	•	–	–	–	–	•	–	•	•	•	•	–	•
MIKE SHE	–	•	–	–	•	•	–	–	–	–	•	–	•	•	•	–	–	–
MUSIC	–	–	–	–	•	•	–	–	–	–	•	–	•	•	•	•	•	–
P8-UCM	–	–	•	•	–	•	–	–	–	–	•	–	•	•	•	•	–	•
PCSWMM	–	•	–	•	•	•	•	–	–	–	•	–	•	•	–	•	–	–
PGC-BMP	–	–	–	–	–	•	–	–	–	–	–	–	•	•	–	–	–	–
SHETRAN	–	•	–	–	•	•	•	–	–	–	•	–	•	•	•	•	–	–
SLAMM	–	–	–	–	–	–	–	–	–	–	–	–	•	•	•	•	–	–
SPARROW	–	•	–	–	•	–	–	–	•	–	•	–	•	•	•	•	–	–
STORM	–	•	•	–	•	–	•	–	–	•	–	–	•	•	•	•	–	–
SWAT	–	•	–	–	•	•	•	–	–	•	•	–	•	•	•	•	•	•
SWMM	–	•	–	–	•	•	•	–	–	•	•	–	•	•	•	•	•	•
Toolbox	–	–	–	–	–	•	–	–	–	–	•	–	•	•	–	–	–	•
WARMF	•	–	–	–	•	–	–	–	–	–	•	–	•	•	•	•	•	•
WEPP	–	•	–	–	•	–	–	•	•	–	–	–	–	•	–	–	–	–
WinHSPF	–	•	–	–	•	•	–	–	–	–	•	–	•	•	•	•	•	•
WMS	–	•	–	–	•	•	–	–	–	–	•	–	•	•	•	•	•	•
XP-SWMM	–	–	–	–	•	–	–	–	–	–	•	–	•	•	•	•	•	•

1 Grid, 2 River Path, 3 Export Coefficient, 4 Load Coefficient, 5 Physical based, 6 Within one day
7 Daily, 8 Monthly, 9 Yearly, 10 Surface Water, 11 Surface and Underground Water
12 User Defined, 13 Sediment, 14 Nutritive, 15 Toxic Substance/Pesticide, 16 Metal, 17 BOD, 18 Bacter

1. Select the appropriate water body models
   Appropriate water body models and calculation methods should be adopted for water environment capacity analysis in accordance with the richness and validity of the data acquired, the complexity degree of the target problems which need to be addressed and the types of study water bodies. For example, the eutrophic models should be adopted for the eutrophic water body, etc.

2. Define the environment problems and determine the pollution factors
   Make a preliminary analysis and evaluation of the hydrological and water quality features of the study water body, find out the current and future possible water quality problems, and determine the pollution factors to be calculated.

3. Define the water quality targets and determine the water quality requirements
   Water environment capacity analysis should be directed against the water areas with the planned water quality targets. The water areas without water quality protection targets should be determined in accordance with the freshwater ecological function regions, or by referring to the basin water function regions or water environment function regions. The control cross sections and water quality requirements of the hybrid regions are analyzed and determined in accordance with the features of the receiving water body, the water environment function region and surface water standards.

4. Determine the design hydrological conditions
   In accordance with the mastered hydrological data, the steady-state conditions, such as hydrologic method (30Q10), and biological method (30B3), might be adopted, or dynamic design hydrological conditions are directly adopted (such as, the hydrological processes of the 10 years, etc.). What calls for special attention is that it is much more suitable to independently calculate the steady-state design hydrological conditions.

5. Calculate the response relation between the pollution load and water quality
   In accordance with the give design hydrological conditions, the steady-state or dynamic water quality response coefficient of the unit discharge outlet load is calculated in each control cross section or the hybrid interface by using the related water environment model.

6. Determine the control unit background load
   The control unit background load mainly refers to the entrance load, and the non-point source pollution load also can be included. The upstream entrance load water quality has a great influence on the calculation of the permissible quantity of pollutants received, if the amount is too high, it is very likely that there is no permissible quantity of pollutants received for the downstream, but if it is too low, the downstream has an abundant permissible quantity of pollutants received, and this causes higher risks of the water environment bearing capacity of the receiving water body. It is suggested that the requirements for the inflow function region in the upper stream shall prevail or the degradation capacity in the water environment capacity (the pollutant concentration in the entrance water body is equal to the water quality target of the control unit) is considered as the calculation ground, if the permissible quantity of pollutants received of the control unit is

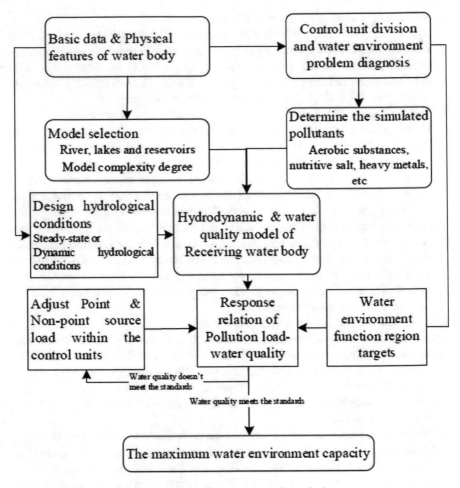

Fig. 2.42 Technology process of water environment capacity analysis

small, it is necessary to increase the water quality requirements of the upstream inflow.

7. Define and optimize the problems, and determine the maximum permissible quantity of pollutants received

    In accordance with the water quality targets and design hydrological conditions, the optimization problems to solve the maximum permissible quantity of pollutants received (linear programming or non-linear programming) are determined, the methods for the problems which are addressed in programming (simplex algorithm, genetic algorithm, particle swarm optimization algorithm, etc.), it is suggested to solve the non-linear programming problems more than 3 times to determine their optimization results due to their solution difficulties (Fig. 2.42).

2.3.5.1 Basic Data Investigation

Basic Data Investigation and Collection

1. Determine the types and spatial dimension range of the water body in the study water area.
2. The basic data which are used to calculate the pollutant receiving ability of river, lake (reservoir) water area should include: hydrological data, water quality date, river/lake (reservoir) discharge outlet data, river/lake (reservoir) side inlet and outlet data, the types of river channel cross channels, the underground terrain data of lakes (reservoirs), etc.
3. The hydrological data should include the flow quantity, the flow rate, the gradient, the water level of the monitoring cross section in each calculated stream segment, the water level, the storage capacity, the flow rate, and the inlet and outlet quantities of lakes (reservoirs), etc. the data should satisfy the design hydrological conditions and the requirements to calculate mathematical model parameters.
4. The water quality data should include the water quality states, the water function zoning and the water quality targets of the calculated water function regions in river segments, and lakes (reservoirs). The data should reflect the main pollutants in the calculated river segments, and lakes (reservoirs), and also meet the requirements of water quality parameters to calculate the pollutant receiving capacity in the water body.
5. The river/lake (reservoir) discharge outlet data should include the discharge outlet distribution, the discharge amount, the pollutant concentration, the discharge patterns, the discharge rules, the corresponding pollution sources for the river/lake (reservoir) discharge outlets and so on within the confluence range in the calculated water area.
6. The side inlet and outlet data should include the positions, water quantity, pollutant types and concentrations in the side inlets and outlets around the calculated segments, and lakes (reservoirs).
7. The river channel cross section data should include the horizontal and vertical section data of the calculated segments, and reflect its simple terrain states; and the lake (reservoir) underground terrain data should reflect its simple terrain states.
8. The basic data should come from the units with corresponding qualifications. When the related data cannot satisfy the calculation requirements, the data required should be obtained by expanding the investigation and collection range and field monitoring.

Determine the simulated pollutants

1. In accordance with the basin or region planning requirements, the pollutants determined by the planning management target should be considered as the ones for calculating the pollutant receiving capacities of the segments, and lake (reservoir) water areas.

2. In accordance with the pollutant and water area features within the calculated river and lake (reservoir) water areas, it is necessary to select the main pollutants which influence the water quality in the water function regions as the ones for calculating the pollutant receiving capacity of the water area.
3. In accordance with the water resource protection and management requirements, it is necessary to select the pollutants which have outstanding influences on the adjacent water areas as the ones for calculating the pollutant receiving capacity of the water area.

2.3.5.2 Selection of Numerical Models

The calculation technology of the permissible quantity of pollutant received is strongly targeted, and different calculation methods are available for different types of surface water bodies. The water quality model is the core of water environment capacity analysis, and it is the prerequisite for the quantified study, development and utilization of water environment capacity to express the migration and transformation rules of pollutants in the water bodies by applying reasonable water quality models.

Model Selection Principles

1. The selected models should combine with the requirements from the basin water quality administrators, and should provide supports to the basin administrators for implementing the total capacity scheme. At the same time, simple models should be adopted for the simple water quality problems within the basin, but complex ones for the complex ones.
2. The selected models should adapt to the acquired data state, and it should be much more cautious in selecting the models with higher data requirements, the use of complex models cannot recover the influences on the total capacity schemes from insufficient data, and the scientific model development strategy is to apply simpler models and then gradually complex ones according to the data situations.
3. The selected model should be a continuously improved open system. While satisfying the near-term demands of total capacity management, the model should have the ability to adapt to and improve the new medium-and-long-term demands, or can be continuously updated according to the understanding and data accumulation in the water environment process.

Model Selection Grounds/Processes

1. Determine the types of surface water bodies. In accordance with the water environment capacity analysis demand, firstly determine the types of surface water bodies, for example, the study water body is a river, lake, reservoir, or an offshore area.

2. Determine the spatial dimensions of water quality models. In accordance with the space and water quality distribution features of the simulated water area, determined the model spatial dimensions, such as one-dimensional model, two-dimensional model, and three-dimensional model.
3. Determine the temporal dimensions of water quality models. In accordance with the water environment problems, determine the model temporal dimensions, such as, the steady-state, dynamic, quasi-dynamic process simulation.
4. Determine the water quality simulation functions. In accordance with the actual demands, determine the water quality simulation indexes and components, such as hydrodynamic force simulation, sediments, nutrients, toxic substances, heavy metals, COD, DO (dissolved oxygen), bacteria, etc.
5. Determine the water quality model solution methods. In accordance with the actual demands, determine the water quality model solution methods. In accordance with the actual situations, the models which are applicable to the capacity model can the simple models (analytical formulas), the complex models (dynamic models), or the quasi dynamic models between them can be selected for capacity analysis.
6. Compare and select the water quality models. Compare and select the feasible water quality models, and select appropriate models to do simulation according to the water quality selection principles. The models which are suitable for the total capacity calculation should have the following features:

   (a) The selected models should be identical with the water quality target management requirements, namely, the models should have the ability to establish the accurate quantitative ability between the pollution load and the water quality target, and also reflect the temporal and spatial change features of water quality problems.
   (b) The selected models should coordinate with the scientific understanding of the water environment system, namely, the models should accurately reflect the basic features of main water environment processes, and the processes whose mechanisms (such as some aquatic ecological processes) cannot be determined is not incorporated into the simulation ranges.

7. Compare and selection of programming models. In accordance with the calculation results of the water quality models and the water quality target requirements, establish optimal target functions and constraints, in accordance with the forms of target functions and constraints, screen the feasible optimal models from the model base, and select the appropriate linear or non-linear programming models.
8. When there are two or over two types of surface water bodies in a basin, it is not suggested to use a mathematical model which is only applicable to one single type of water body, but use a mathematical model which are applicable to several types of surface water bodies, such as EFDC (The Environmental Fluid Dynamics Code), etc.

Classification and Application of Numerical Models

In accordance with the types of receiving water bodies, spatial forms, temporal and spatial heterogeneity of the hydrology and water quality, main water environment problems, detailed and reliable situations of data and other factors within the study area, the steady-state computer technology based on simple steady-state models or the dynamic computer technology based on multiple-dimensional complex models are selected. Under the dynamic hydrological conditions, it is necessary to select the water environment models with the joint hydrodynamic force and water quality simulation function. At present, there are many widely applied and maturely developed hydrodynamic and water quality models. In the daily water environment target management, appropriate models can be selected to calculate the water environment capacity in accordance with the management target demands.

River Models

1. The zero-dimensional river model. When the pollutants are uniformly mixed in the segments, the zero-dimensional models can be adopted to calculate the pollutant receiving capacity. Such kind of models are mainly applicable to the segments in water network regions. In accordance with the pollutant distribution situations, the evenly mixed segments with different concentrations should be divided and their pollutant receiving capacities are independently calculated.
2. The one-dimensional river model. When the pollutants are uniformly mixed in the horizontal section of the segments, the one-dimensional models can be adopted to calculate the pollutant receiving capacity. The models of such kind are mainly applicable to the small-medium-scale segments of $Q < 150 \, \text{M}^3/\text{s}$.
3. The two-dimensional river model. When the pollutants are non-uniformly mixed in the horizontal section of the segments, the two-dimensional models can be adopted to calculate the pollutant receiving capacity. The models of such kind are mainly applicable to the small-medium-scale segments of $Q \geq 150 \, \text{M}^3/\text{s}$. For the segments with continuous and constant pollutant discharge and rectangular horizontal section, the analytical method of the models can be used to calculate the pollutant receiving capacity.
4. The one-dimensional estuary model. The one-dimensional estuary models can be adopted to calculate the pollutant receiving capacity of the tidal reaches. The hydraulic parameters of the one-dimensional estuary models should be the average value of the tidal semi-cycle and the pollutant receiving capacity is calculated according to the steady flow situations.

Lake Models

1. The lake(reservoir) uniform mixture model. The uniform mixture models can be adopted to calculate the pollutant receiving capacity for the lakes(reservoirs) with uniformly mixed pollutants. The models of such kind are mainly applicable to small-medium-scale lakes(reservoirs).

2. The lake(reservoir) non-uniform mixture model. The non-uniform mixture models can be adopted to calculate the pollutant receiving capacity for the lakes (reservoirs) with non-uniformly mixed pollutants. The models of such kind are mainly applicable to large-scale lakes(reservoirs). In accordance with the lake(reservoir)-into discharge outlet distribution and pollutant spread features, different calculated water areas should be divided and the pollutant receiving capacities are separately calculated.
3. The lake(reservoir) eutrophication model. The Dillon Model should be adopted to calculate the pollutant receiving capacities of nitrogen and phosphorus for the eutrophication lakes(reservoirs). The Goda Model should be adopted to calculate the pollutant receiving capacities of nitrogen and phosphorus for the lake(reservoir) bay water areas with weak water flow exchange capacities.
4. The lake(reservoir) layering model. The layering models can be adopted to calculate the pollutant receiving capacities of the lakes(reservoirs) with water temperature layers. The pollutant receiving capacities of the lakes(reservoirs) should be respectively calculated according to the layering period and non-layering period.

Complex Models

The complex models should be adopted in the basins with complex water system and hydraulic connection, and also in the basins with serious non-point source pollution. It is necessary to collect the data required for developing the water quality models.

1. The one-dimensional unsteady-state water quality model should be adopted for the river basin capacity analysis, and the one-dimensional river network unsteady-state water quality model should be adopted for the plain network areas.
2. The two-dimensional unsteady-state water quality model should be adopted for the reservoir capacity analysis.
3. The one-dimensional or two-dimensional hydrodynamic model should be adopted for the river and lake capacity analysis to determine the flow field features under the design hydrological conditions.
4. The parameters of the water quality models should be calibrated and verified in accordance with the simultaneous monitoring data of the river water quality and quantity.

Table 2.28 lists 5 main recommended alternative water environment models of receiving water bodies, by comparing and analyzing their features in a horizontal manner, their advantages and disadvantages, data requirements and application situations can be discovered to support the water environment capacity analysis of the receiving water bodies and the model selection in the distribution work.

Table 2.28 Matrix of features of the recommended models

Alternative model		Bathtub	CE-QUAL-2K	CE-QUAL-W2	EFDC	WASP
Model dimension		0D	1D	Vertical 2D	3D	Quasi-3D
Modeling method		Single Equation	Finite difference	Finite difference	Finite difference	Finite difference
Simulated input conditions		Steady-state	Steady-state	Dynamic	Dynamic	Dynamic
Simulated index number		4	11	21	22	8
Simulated indexes	BOD/DO	✓		✓	✓	✓
	Nutritive salt	✓		✓	✓	✓
	Alga(e)	✓		✓	✓ (Class 4)	✓
	Others			Salinity, SS		
Data requirements		Low	Low	Medium	High	High
Hydrodynamic simulation		None	Simple	Relatively Simple	Complex	Relatively simple
Application region		Basin-shaped Lakes and Reservoirs	Channel Lakes and Reservoirs	Channel Lakes and Reservoirs	All kinds of Channel Lakes and Reservoirs	All kinds of Channel Lakes and Reservoirs

2.3.5.3 Determination of the Design Hydrological Conditions and Establishment of Load-Water Quality Response Relation

Design Hydrological Conditions

Steady-State Design Hydrological Conditions

1. River-type Basin

   (a) The average flow in the driest months with 90% of the assurance rate or the average flow in the driest months during the recent 10 years is considered as the design flow for the river water areas.

   (b) The minimum average monthly flow which is not zero is considered as the sample for the seasonal rivers and frozen rivers.

   (c) The corresponding flow at a low water level when the flow rate is zero at 90% of the assurance rate is considered as the design flow for the water network regions and tidal reaches with uncertain flow directions.

   (d) The minimum discharge flow or the channel ecological base flow is considered as the design flow for the segments with hydraulic engineering control.

   (e) The design flow of the offshore water areas should be calculated when the pollutant receiving capacity of the segments whose water function regions are divided by the bank side.

(f) The calculation of design hydrological conditions can be executed by referring to the regulations for Hydrologic Computation of Water Resources and Hydropower Projects.

2. **Lake-reservoir Basin**

(a) The minimum average monthly water level during the recent 10 years or water storage capacity corresponding to the average water level in the driest months with 90% of the assurance rate is considered as the design flow for the lakes (reservoirs). The corresponding water storage capacity of dead reservoir capacity can also be considered as the design flow for reservoirs.

(b) The design flow of the corresponding water areas should be adopted when the pollutant receiving capacity of lakes (reservoirs).

(c) The calculation of design hydrological conditions can be executed by referring to the regulations for Hydrologic Computation of Water Resources and Hydropower Projects.

3. **Complex Models**

(a) The measured perennial day-to-day flows are adopted for the design hydrological conditions in the phased water environment capacity analysis, and the hydrological series are generally not less than 3 years.

(b) The three wet, normal and dry hydrological years should be adopted for the water environment capacity analysis for basins with huge inter-annual hydrological situation changes.

A hydrological year is calculated according to related hydrological rules, the hydrological frequency of a wet hydrological year is 90%, that of a normal hydrological year is 50%, and that of a wet hydrological year is 10%.

- Risk analysis should be based on the protection targets of the aquatic ecological regions, consider about the pollutant features, and determine the permissible over-standard days and recurrence period. Under the technical conditions without risk analysis, the typical annual day-to-day flow can be considered as the design water condition.
- The special hydrological periods, such as the freezing period and ice flood season, and the cut-off period of seasonal rivers, should be separately considered about.
- The hydrological situations under the typical or optimized operation conditions of main water engineering should be comprehensively considered about during the water environment capacity analysis of the river and lake water bodies regulated by water engineering.

Dynamic Design Hydrological Conditions

When the environment risk conditions are determined, the dynamic water quality models can be adopted to make the environment capacity analysis and total quantity control programming. The dynamic design hydrological conditions, as a continuous hydrological process, are not definitely expressed as permissible average period and permissible recurrence period; but determine the loadable pollution load by adopting the programming methods in accordance with the permissible average period and permissible recurrence period. The dynamic design hydrological conditions express the water quality recurrence period in a higher precision and identify the emergent and toxic recurrence periods in a more reasonable manner. The dynamic design hydrological conditions can establish the random hydrological process for various boundaries not by adopting the measured hydrological series but based on the parameter analysis of the measured hydrological series. The dynamic design hydrological conditions can select design hydrological conditions by a more flexible way and express the recurrence period in a more precise manner (Design Risks).

Regarding the environment capacity analysis and total quantity control programming, the optimized solving process has higher time complexity under the dynamic simulation conditions, it is generally difficult to have a solution in the optimization methods, and the scheme comparison method is more adopted to determine the loadable pollution load.

Establishment of pollution load-water quality response relation

To calculate the water environment capacity by adopting pollution load-water quality response relation is to calculate the water quality concentration response situations of all the pollution sources on the control sections by virtue of the water quality model of the receiving water bodies, form the water quality response relation matrix between the pollutant sources and the control sections by integrating the influences on the control sections from the pollution sources, and calculate the permissible load of various pollution sources on the premise where the control sections meet the water quality targets. The method establishes the relation among the water quality control sections and the various correlated pollution sources (including the point source discharge outlets, non-point source branch estuaries) within the basin, and is a simplified method which can optimize calculation. It is necessary to select river and/or lake/reservoir water quality response coefficient matrix in accordance with the types of water bodies, and appropriate water quality models are selected for calculation.

The water quality response coefficient matrix is deduced by taking a two-dimensional water quality model as an example:

$$\frac{\partial h C_i}{\partial t} + \frac{\partial h u C_i}{\partial x} + \frac{\partial h v C_i}{\partial y} = \frac{\partial}{\partial x}(h K_x \frac{\partial C_i}{\partial x}) + \frac{\partial}{\partial y}(h K_y \frac{\partial C_i}{\partial y}) - h k_d C_i + S_m$$

(2.29)

where, C_i is the pollutant concentration of the ith calculation unit, h is the water depth, u and v are the flow velocities in the x and y directions, Kx and Ky are the vertical and horizontal turbulence diffusion coefficients, Kd is the comprehensive degradation coefficient, and Sm is the load if the mth pollution source.

When the hydrodynamic field conditions in the pollution source water quality response process are basically the same, the water quality response of the receiving water body to the discharge outlet should meet the linear superposition principle: based on the same hydrodynamic field conditions, the balance concentration field formed by the joint reaction of several pollution sources Wi(i = 1, 2, …, N) can be considered as the linear superposition that every individual pollution source independently influences the concentration field. That is, when Sij is the independent concentration field of the pollution source Wi at the ith pollutant discharge outlet from the jth water quality control point, then the formed concentration field S also meet the equation when N pollution sources exist simultaneously:

$$S(x, y, t) = \sum_{i=1}^{N} S_{ij} + Sb_j \qquad (2.30)$$

where, Sij is the water quality concentration response field of the water area control point j under the reaction of pollution source Wi, SBj is the background value concentration, S (x, y, t) is the water quality concentration response of the control point under the joint reaction of N pollution sources. Sij is defined as the response coefficient, and it represents the response relation of the water quality control section to a certain pollution source load.

For each and every pollution source, the concentration filed formed can be considered as the result of linear superposition under the reaction of multiple unit source strength value Ui (Different unit source strength can be set for different pollution sources).

$$S_{ij} = P_{ij} \times W_i \qquad (2.31)$$

where, Pij is the water quality response coefficient of the jth unit source strength.

Atmospheric sedimentation and endogenesis release are not incorporated into distribution, and are the important boundary conditions of water environment models of control units. Therefore, before calculating the load-water response relation of the lake/reservoir-type control units, it is necessary to study the temporal and spatial distribution of pollution load of atmospheric sedimentation and endogenesis release as a model input but without increase or decrease.

2.3.5.4 Water Environment Capacity Analysis of the Receiving Water Bodies

The maximum water environment capacity of a basin can be calculated by combining the water quality standards and water quality response coefficient matrix when the water quality target conditions are satisfied. The simulation calculation is conducted by applying the design hydrological conditions and upstream and downstream water quality targets to study the water environment capacity and make a rationality analysis and test.

Scene Analysis Method

A water environment Model is built to simulate the water quality process, and it is the foundation and ground for water environment capacity calculation. The water environment capacity model and the hydrological-water quality coupling model can be simply divided according to different modelling principles. The water environment capacity calculation methods can be divided into steady-state algorithm, dynamic algorithm and compound algorithm according to different models and calculation thoughts. The water environment capacity is directly calculated by using the model tool through the design of scene schemes.

Steady-State Algorithm

The steady-state algorithm is the traditional calculation method based on simple water environment capacity models in China's water environment quality management. The method often regards lakes and reservoirs as the zero-dimensional water bodies and adopts the zero-dimensional model; the analytic solution model or one-dimensional steady-state model is adopted in rivers, without considering about the time difference of water quality. Though the steady-state calculation method has a problem that its calculation result is with great uncertainty, it's simple and convenient, and applicable to the lakes/reservoirs and rivers with small temporal and spatial differences in hydrology and water quality and lack of basic data.

Dynamic Algorithm

The dynamic calculation method, through conducting a dynamic simulation of the hydrological and water quality processes of river or lake/reservoir water bodies, calculates the hydrological and water quality states under different input situations, establishes the load-water quality response relation and then calculates the water environment capacity according to the control targets.

Compound Algorithm

In accordance with the hydrological and water quality features and control targets of the control units in the basin, the hydrological, meteorological and pollution source data in typical periods re set or selected to compose the basic scene of water environment capacity calculation. The algorithm is adopted to simulate the temporal and spatial change processes of water body and water quality under the basic scene.

The trial-and-error method is adopted to increase or decrease the load successively on the basis of the current or design pollution load, to analyze and record the concrete index values of the corresponding control targets and form the quantified load-water quality response relation. According to the limit value of concrete indexes of the control targets, when all the conditions under which all the control targets are satisfied are discovered, the maximum pollution load the lakes/reservoirs can receive is their permissible amount of pollutants received.

By analyzing the short-term or local effects, and long-term and comprehensive effects of pollutants in lake/reservoir water bodies, it is necessary to consider about the influences from the two aspects on the calculation of the permissible amount of pollutants received of the lake/reservoir-type control units. Therefore, pollution source discharge strength should be restricted from the long-term and short-term temporal dimensions. The specific temporal dimensions can be analyzed and determined according to the actual situations of the control units and the pollutant features.

The compound algorithm integrates the advantages of the steady-state algorithm and the dynamic algorithm, firstly, it adopts steady-state algorithm to calculate the environment capacity of the target water bodies and works out the preliminary load distribution scheme, and then, it uses the dynamic algorithm to simulate and analyze the corresponding hydrodynamic and water quality processes, and determines the accurate water environment capacity in accordance with control targets.

Programming Analysis Method

By employing the load-water quality response coefficient matrix, considering the standard attainment of the water quality on the control sections as the constraint, and the total permissible load as the optimization target, the artifice intelligent method of Particle Swarm Optimization—Repulsive Particle Swarm Method (RPSM) or other optimizing mathematical methods, such as the programming solving method, are adopted to make calculation.

At present, the most generally commonly used programming method for calculating the permissible amount of conventional pollutants received is the simple linear constrained programming. Others include capacity programming of the multi-source hybrid regions, the capacity programming of the multi-source function regions, etc. The primary and common programming models are introduced below.

Linear Constrained Programming

The target function:

$$\max \sum_{j=1}^{m} X_j \quad , \quad j = 1, 2, \ldots \tag{2.32}$$

The constraint equation:

$$\max \sum_{j=1}^{m} a_{ij} X_j \leq C_i - C_{0i} \quad , \quad i = 1, 2, \ldots, m \tag{2.33}$$

$$X_j \geq 0 \tag{2.34}$$

where, the decision variable X_j is the discharge amount of the jth pollution source, a_{ij} is the response coefficient of the jth pollution source to the ith function region. C_i is the water quality control concentration of the ith function region, C_{0i} is the concentration contribution of all the background loads to the ith function region. It is assumed that the flows of all pollution sources remain unchanged, and the target function is the maximum permissible discharge amount for a certain conservative pollutant, and it is the simplest kind of linear programming problems. The maximum permissible amount of pollutants received of a basin can be obtained by using linear programming.

The aforesaid environment optimizing problems are very simple, but might be impractical, since the optimizing results are often unfeasible and cannot be accepted by the parties, such as the fairness, economy and efficiency problems. To make the solutions be practical, it is generally a better method to increase constraints, for example, the differential constraints are conducted in accordance with different features of pollution sources and different efficiencies of practical technologies. A distribution principle can be established by using the existing information (e.g.: the positions of the current discharge points, the amount of the current discharge points, the estimated discharge amount, the efficiency of processing technologies, etc.) and then the optimal solution of the total distributable load is raised under the distribution principle.

Constrained Nonlinear Programming

The target function is nonlinear: When there exists a linear response relation between the pollutant discharge amount and the water quality, but the target function is nonlinear, the optimal solution for the problems at the very moment can be discovered by the breadth and depth search technologies. The target function:

$$\max\,(\min) f(X_j)\quad,\quad j = 1, 2, \ldots, m \tag{2.35}$$

The concentration constraint equation:

$$\max \sum_{j=1}^{m} a_{ij} X_j \le C_i\quad,\quad i = 1, 2, \ldots, n \tag{2.36}$$

$$X_j \ge 0 \tag{2.37}$$

The upper and lower constraints of the decision variable:

$$X_{j,max} \ge X_j \ge X_{j,min} \tag{2.38}$$

Other constraints:

$$Y(x)_{j,max} \ge Y(x)_j \ge Y(x)_{j,min} \tag{2.39}$$

The constraints are nonlinear: When the pollutant discharge amount is related to the change of the sewage amount and there exists no linear response relation, the zero-dimensional water quality calculation can be considered as a sub-program of the programming method, and the optimal solution for the problems can be discovered by the breadth and depth search technologies. The method is only applicable to the simple water quality models with analytical solutions.

The target function:

$$\max \sum_{j=1}^{m} X_j \quad , \quad j = 1, 2, \ldots, m \tag{2.40}$$

The concentration constraint equation:

$$\max \sum_{j=1}^{m} f(X_j) \leq C_i \tag{2.41}$$

RPSM-Based Programming Model

The model is only applicable when the pollutant discharge amount is in a linear response relation with the water quality, but the target function and constraints are all nonlinear, at the very moment, the RPSM method can be used for solving.

The target function:

$$\max (\min) f(X_j) \quad , \quad j = 1, 2, \ldots, m \tag{2.42}$$

The concentration constraint equation:

$$\sum_{j=1}^{m} f(X_j) \leq C_i \tag{2.43}$$

The upper and lower constraints of the decision variable:

$$X_{j,max} \geq X_j \geq X_{j,min} \tag{2.44}$$

Other constraints:

$$Y(x)_{j,max} \geq Y(x)_j \geq Y(x)_{j,min} \tag{2.45}$$

The programming models are influenced by the function types and variable values of the target function and constraint, and the failure of iteration and termination of calculation often happen during the solving process, so the steady and mature models should be selected.

The programming models are more targeted, and the appropriate models which are selected according to the actual situations can effectively improve the calculation efficiency. The models which are easy to operate and convenient to use should be pre-emptively selected.

- When the target functions and all the constraints of the programming models are linear, the linear programming models should be pre-emptively selected.
- When the target functions and all the constraints of the programming models are non-linear, the dynamic programming models should be selected.

2.3.5.5   Rationality Analysis

The rationality analysis and test of the water area pollutant receiving ability should include the rationality analysis of the basic data, the rationality analysis of calculation condition simplicity and assumption, the rationality analysis and test of model selection and parameter determination, and also the rationality analysis and test of the water area pollutant receiving capacity calculation results.

Rationality Analysis of Basic Data

1. Hydrological Data: conduct representativeness, consistency and reliability analysis of the flow quantity, flow rate, and water level of rivers and lakes (reservoirs), and the analysis can be executed by referring to [3].
2. Water quality data: conduct representativeness, reliability and rationality analysis of the water quality monitoring sections, monitoring frequencies, period, pollution factors, water quality states and so on in accordance with the pollution source and discharge situations in the region.
3. Pollution Discharge Data: in accordance with the measured or investigation data of the pollution discharge outlets, conduct the rationality analysis of the sewage discharge amount and rules, and pollutant concentration and other data at the discharge outlets in the analogy method.
4. Land Pollution Source Data: based on the local economic and social development level, industrial structure, GDP, water intake quantity, industrial and agricultural water consumption quantity, domestic water consumption quantity, sewage processing level and other data, analyze the sewage discharge amount, the pollutants and discharge and so on and the corresponding rationality according to the water supplying, use, consumption and discharge relations.
5. River and lake (reservoir) feature data: compare the river and lake (reservoir) channel section, underground terrain, gradient and other data collected from investigation in different methods, and analyze the reliability and rationality.

Rationality Analysis of Calculation Condition Simplicity and Assumption

By comparison, analyze whether the boundary conditions, hydraulic features, discharge outlets of rivers, and lakes (reservoirs) are rational or not, satisfy the assumption conditions of the selected models, and the determined representative section can reflect the water quality of the aquatic function region or not.

Rationality Analysis and Test of Mathematical Model Selection and Parameter Determination

In accordance with the hydraulic features, boundary conditions and pollutant features of the water areas, analyze the rationality of the selected mathematical models and parameters and their application range. Analyze the rationality of the model parameters by comparing with the existing experimental results and study achievements; and also verify the model parameters and model calculation results according to the measured data.

Rationality Analysis and Test of the Water Environment Capacity Calculation Results

1. Analyze the rationality of the calculation results in accordance with the current pollutant discharge amount of the segments and by combining the water quality state.
2. Analyze the rationality of the calculation results by comparing with the water environment capacities of the upstream and downstream or the similar aquatic function regions.
3. Analyze the rationality of the calculation results by calculating the water environment capacity in different models and then making comparison.
4. Analyze and judge the rationality of the water environment capacity calculation result of a river, a water system or an entire basin in accordance with the local natural environment, hydrological features, pollutant discharge, water quality state, etc.

2.3.6 Pollution Load Distribution Based on Control Units

The pollution load distribution of water environment capacity based on control units, as an important link of the total pollutant quantity control, is the core for the water quality target management of control units, and also an important approach to protect the water environment quality and realize the basin sustainable development. The pollution load distribution target based on the basin water environment capacity should be determined from the perspective of the entire water quality aquatic ecological protection of the basin; meanwhile, it is necessary to consider about the upstream and downstream relation within the basin for determining the total pollutant load distribution, and the total permissible pollutant load target of control units is determined when the water quality target of the basin is totally met.

The distribution technology of the total pollutant quantity within the control units is strongly targeted, and the distribution methods are different for different types of surface water bodies and different basins. The pollution load distribution technologies are mainly determined by the types of pollutants, composition of pollution sources, load strength distribution, confluence relation, etc.

2.3.6.1 The Distribution Principles

The basin water environment capacity is a kind of public resource, and has the attributes of the public belongings of natural resources, and each and every stakeholder has the right to equally use the basin water environment capacity. When such kind of resource become a scarce resource, fair distribution seems crucial, and the resource utilization efficiency also becomes an issue which wins the social concern.

The Scientificalness Principle

The scientificalness principle means that the pollution load distribution process must be with sufficient scientific ground to make the distribution result satisfy the requirement of sustainable development. The scientificalness of distribution technologies quantifies the pollution source-load response relation and the distributable pollution load based on the calculation of water environment capacity by applying natural rules. By combining the actual situation of the pollution sources, the reasonable and well-grounded distribution schemes are formulated by comprehensively considering about the technical feasibility, economic factors, environmental benefits, stakeholders' opinions and other factors.

The Fairness Principle

The fairness principle means that various pollution sources should have relatively equal distribution right of the pollution sources in consideration of different factors. The pollution load distribution based the total basin water environment capacity is related to the vital interests of various pollution sources, and the distribution should fairly and rationally undertake responsibilities for different types of pollutant dischargers to meet the majority's interests. All the pollutant dischargers within the basin have the equal right to influence the water environment, that is to say, the pollution sources should undertake corresponding responsibilities according to the influence degrees on the water quality of the control sections, the pollution sources with greater pollution contribution have larger treatment and control responsibilities, namely, the pollution responsibilities accord with the influences. Furthermore, the water quality target of the total basin capacity should consider about the human's utilization demands for water resources and also protect the aquatic organisms.

The Efficiency Principle

The efficiency principle stresses the economic benefit maximization on the basis of technical feasibility, that is, the maximum benefit is obtained by the minimum investment or loss. The principle distributes the load reduction quantity of various pollution sources according to the different marginal treatment expenses of a certain pollutant, and its distribution target is to minimize the pollutant treatment expenses. The principle pursues to obtain the maximum environment benefits with the minimum economic investment within control unit ranges, and its primary representation is the minimum costs in the pollution load distribution methods. Therefore, the method of treatment expense minimization and the method of net marginal benefit maximization play important roles in realizing the economy-first principle. The Guidelines, starting from the b perspective of protecting and improving water quality, considers the degree of water environment improvement as the benefit pursued by the efficiency principle, the degree of load reduction as the investment or loss, and it is necessary to preemptively control the pollution sources which have huge load and significantly influence the water environment quality.

The Difference Principle

it is necessary to simultaneously consider about the principle to distinguish the basin and its internal control units for the pollution load distribution based on basin dimensions. Due to the differences in types of pollution sources, technical and economic features and influences from water quality, being fair is not equal to being average. The pollution load distribution based on the basin control units should consider about the requirements of differential protection on the basis of the fairness principle.

One, the economic development of different regions within a basin is influenced by the historical and resource environments, and certain development patterns and modes are formed, and certain function regions are planned, therefore, there exists regional differences for the function requirements of water bodies. Two, due to the differences in aquatic ecological system structure and function regions, the protection targets are different. Three, the different regional pollution degrees result in the different water qualities of different regions. Therefore, the total capacity distribution should properly consider about the regional differences according to the economic and environmental states of different regions, and make selection according to the differences in key points and requirements of environment management under their social and economic conditions.

2.3.6.2   Total Capacity Calculation and General Distribution Framework

General Rule

The total capacity is calculated in the following formula

$$TWCB = \Sigma MWCUi = \Sigma PLAm + \Sigma NLAn + MOS \qquad (2.46)$$

where, TWCB is the total water environment capacity of the basin; MWCUi is the maximum water environment capacity of each control unit; i is the serial number of the control unit; PLAm is the pollution load distribution amount of point sources; and NLAn is the pollution load distribution amount of non-point sources; and MOS is the margin of safety.

The total load distribution is in the two-layer distribution model, that is, "basin-control unit" distribution and "control unit-pollution source" distribution. The "basin-control unit" distribution can be called the total load distribution of a basin, and the control unit-pollution source" distribution the total load distribution of control units.

Total Basin Load Distribution Framework

The total basin load distribution process is influenced by the social, technology, management, resource and other technical and non-technical factors, it is always difficult to give consideration of both fairness and efficiency, and it is also hard to implement the distribution methods which blindly pursue fairness or efficiency. Therefore, it is essential to establish a distribution method which comprehensively consider about the social, environmental and economic development phases based on fairness and efficiency. Regarding the basin water target management requirements, based on the scientificalness, fairness, efficiency and difference principles, and by integrating China's total basin load management features, the double-layer and multi-task distribution optimal framework of total basin load is raised:

Double-Layer Distribution

The "basin-control unit-pollution source" double-layer distribution. The total basin load distribution is the medium product of total load distribution, and the distribution result must be under practical management by relying on the secondary distribution of control units, but the total basin load distribution based on total load distribution directly establishes a relation with water quality target attainment, which is the foundation for distribution of control units. The "basin-control unit" distribution focuses on solving the overall coordination (water quality coordination, regional coordination, etc.) of the total load distribution schemes, and the "control unit-pollution source" distribution focuses on solving the operability of the distribution results.

Multi-task Optimization

Basin-Control Unit Distribution

Since control units have no obvious social, economic or technical index features, the most rational method is the distribution method based on the total load which emphasizes the requirement of basin integrity of the water quality target attainment of the water quality control section. The water quality target attainment constraint which is the ground for the method based on the total load adopts the water quality response matrix of control sections and pollution sources, establishes the water quality target attainment constraints, and calculates the maximum permissible load of each control unit; or respects the current discharge pattern, considers the current or planned basin load reduction minimum as the target, and obtains the total permissible load of control units. The "Basin-control" total load distribution, on the premise of the scientificalness principle, should focus on reflecting the fairness and difference principles, and distribute the basin pollution load based on the match between the pollution responsibility and influence. It should focus on solving the overall coordination (water quality coordination, regional coordination, etc.) of the total load distribution schemes.

Control Unit-Pollution Source Distribution

The "control unit-pollution source" distribution process, stresses the fairness and efficiency principles. That is to say, the pollutant reduction amount of all the pollution sources is distributed according to the different marginal treatment expenses of a certain pollutant, and its distribution target is to minimize the pollutant treatment expense of each region, and its constraint is that the total load doesn't exceed the permissible load capacity after control units reduce MOS. After control units reduce MOS, based on the technical and economic features of pollution source treatment of the control units, the target function of the minimum pollution treatment expenses is established in accordance with the efficiency principle. Definitely, the pollution load which is related to the inside of control units should be distributed in some methods based on the fairness principle, such as, the Gini coefficient method, for evaluating the fairness (Fig. 2.43).

Fairness and efficiency are the difficulties in various resource distribution. There is only relative fairness, but no absolute fairness, or absolute high efficiency. The total basin load distribution should balance several targets and consider about several factors, and ensure the water quality target attainment and also have operability in the basin water quality management; therefore, it is the core and key for the distribution method to discover the balance points in fairness and efficiency and maintain the highly-efficient distribution mechanism on the basis of pursing the maximum social fairness.

The double-layer multi-task optimal distribution framework for the total basin load raised in the Guidelines considers about the fairness in distribution and also pays attention to efficiency. Reflecting the fairness in responsibility isn't the equal discharge quantity in a general sense or the simple fairness of the same reduction ratio.

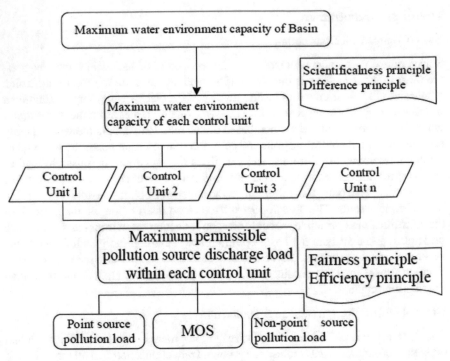

Fig. 2.43 Total basin load distribution technology framework

2.3.6.3 Basic Data Collection

The basic data of pollution load distribution are greatly similar with those of the basin water environment problem diagnosis and environment capacity analysis.

1. Based on the basic data collection of the basin water environment capacity analysis, focus on collecting the socio-economic data of the basin, the socio-economic statistical data should be divided into each and every control unit or the smallest administrative region.
2. The socio-economic development planning data of the basin, systematically collect the socio-economic development planning achievements of the basin and the administrative regions at various levels within the basin.
3. The basic information of point source pollution of control units and pollutant discharge amount and inlet quantity data.
4. The spatial distribution and total load of non-point source pollution of control units.
5. The pollutant reduction technology of pollution sources within the basin and social and technical index data.
6. The BMPs Control technology of non-point sources within the basin and social and technical index data.

The basic data should come from the units with corresponding qualifications. When the related data cannot satisfy the calculation requirements, the data required should be obtained from typical investigation.

2.3.6.4   Pollution Load Distribution Methods Based on Control Units

The water environment capacity is a very important kind of source, and especially precious in the areas with serious water pollution problems. The real core of control over the total water environment capacity is to scientifically and reasonably determine the allocation of the permissible pollutant source discharge, load within the basin, that is to say, to allocate the environment capacity resources to all the control areas according to pollutant discharge points, quantities and forms in different manners. The total load allocation is a process of multiple-scenario scheme optimization, compared with the total load allocation in a single water area, the basin scale involves a larger range and more pollutant sources, therefore, the capacity allocation is much more complex, and the possible scenario schemes are obviously on the rise, and it is a much more complex system optimization issue to choose which one scheme as the optimal scheme.

Allocation from Basin to Water Pollution Control Unit

Hierarchy Allocation Method

There is a huge amount of and a variety of pollutant sources, to give consideration to the identicalness, fairness and efficiency principles, in accordance with the sources and sinks relations of pollutants and by combining the confluence features of river systems, the pollutant load allocation hierarchies are determined and allocated hierarchy by hierarchy. The pollutant sources on the same hierarchy should be firstly divided into different kinds according to the differences of dominant factors, such as they can be divided into point source and non-point source; and then, they are allocated among different types of pollutant sources; the allocation of different pollutant sources of the same kinds is considered as the next allocation hierarchy and is conducted later. By such analogy, till the pollution load is allocated to each and every pollutant source or to the final hierarchies determined according to the demands. The allocation of the same kind of pollutant source should be dominated by the fairness principle; and that of different kinds by the efficiency principle.

The direct pollutant sources in the water within the basin can be decomposed into atmospheric sedimentation, endogenesis release, point source discharge, coastal non-point source, upstream inflow load, etc. As atmospheric sedimentation, endogenesis release and upstream inflow load are considered as background pollution loads, the water environment capacity should be firstly allocated between the point source discharge, coastal non-point source. The load allocation among different kinds of pollutant sources is dominated by the efficiency principle, therefore, the allocation ratio which has the greatest influence on the water quality should be selected to determine the permissible loads of the aforesaid two kinds of pollutant sources.

The main stream within the basin can accommodate several branches simultaneously, therefore, it is necessary to further allocate the pollution loads to each and every branch estuary, namely, the secondary allocation. For the basin generally, each and every branch estuary can be considered as a point source and belong to the same kind of pollutant source; such pollution load allocation is dominated by the fairness principle, therefore, the same fair allocation method should be chosen, such as, to determine the permissible load of each and every estuary.

The permissible pollution bearing capacity is obtained based on model calculation, and the pollution loads are allocated to each and every water pollution control unit or sub-basin outlet in the hierarchy allocation method, and the pollution load allocation scheme formed in the such way has abundant theoretical bases and implementing values. However, considering the existence of uncertainty and error, it is necessary to test the effect of the allocation scheme to ensure that the water environment quality of the water pollution control unit can reach the corresponding control targets. It is necessary to test the effect of the pollution load allocation scheme from several aspects, including the control target reachability, scheme operability, social and economic benefit analysis, etc.

The pollution load for each and every pollutant source is set according to allocation schemes, and the temporal and spatial changes of the water quality under the corresponding schemes are calculated in model simulation, and a comparison is made with the quantified control targets to test whether the water quality satisfies the expected target.

The Total Load Allocation Method Based on the Maximum Capacity in A Basin

The total water environment capacity is determined by the hydrological conditions, water quality target requirements, river pollution discharge space patterns, and composed of the permissible river load capacity of each and every river pollution discharge outlet (including the branch outlet), and the process of analyzing the water environment capacity actually implicitly has the process of allocating the total load in a basin, that is to say, the calculation of the total load in the basin is based on the total load allocation process which is oriented by target achievement.

Therefore, the analysis of the water environment capacity based on the water quality response coefficient is a kind of total load allocation method which is constrained by the water quality standard achievement, as it aims to maximize the total capacity in a basin, the method can be called the total load allocation method based on the maximum capacity in a basin.

1. Target Function

The maximum river permissible load amount in a basin:

$$\max (L) = \sum_{i=1}^{N} W_i \tag{2.47}$$

2. Constraints
   The water quality target constraint:

$$C_j = \sum_{i=1}^{N} \left(W_i P_{ij} \right) + SB_j \le ST_j, \ (i = 1, 2, \ldots, N; \ j = 1, 2, \ldots, M)$$

(2.48)

The pollutant source load constraint:

$$W_i \ge 0, (i = 1, 2, \ldots, N)$$

(2.49)

3. The solving methods of multi-objective optimization equation set:
   After observing the constraint equation above, there are three solving possibilities:

   (a) Positive definite constraint, the equation set whose constraints are equal to
       the number of the river pollution discharge (discharge) outlets can be directly
       solved to obtain the single solution of the maximum discharge capacity;
   (b) Indefinite constraint, the equation set whose constraints are fewer than the
       number of the river pollution discharge (discharge) outlets has infinite solu-
       tions which satisfies the constraint, cannot be directly solved to obtain the
       maximum discharge capacity, but can be solved in the linear programming,
       extremum search and artificial intelligence methods;
   (c) Over-determined constraint, the equation set whose constraints are more than
       the number of the river pollution discharge (discharge) outlets may have no
       solution, and only has solutions when some constraints are equality constraint,
       and can be solved in the linear programming, extremum search and artificial
       intelligence methods.

It is possible to calculate the Permissible Load of the receiving water area which
satisfies the water quality control target by combining the water quality control standard and the water quality response coefficient. It should be noted that the capacities
of large lakes and reservoirs should be calculated in accordance with the control
requirements for the mixing zone of the river pollution discharge outlets.

The total load allocation method based on the maximum total load in a basin
is based on the method that pollution influences the responsibility design, without
completely considering about the social and economic development stage of the water
pollution control unit and the aquatic ecosystem situation, therefore, the allocation
results might be inconsistent with the space layout difference required for the social
and economic development demands and the zoning protection targets of the aquatic
ecosystems in the basin.

The Total Load Allocation Method Based on Minimum Load Reduction in A Basin

The allocation results based on the total load allocation method based on the maximum total load might result in the surplus environment capacity in some water areas,
but the present or planned pollution discharge load in other water areas is far beyond

the water environment capacity, which finally lead to the deficiency in efficiency of the allocation schemes. Based on the overall economic consideration of the allocation schemes, the target function should consider about the current state of the water pollution control unit or the river load of the planned level years, and the function with the minimum load reduction amount is considered as the target function to calculate and allocate the maximum permissible load. Such the method is called the total load allocation method based on minimum load reduction in a basin.

1. Target Function

It is the total maximum river permissible load of a basin revised based on the aquatic ecosystem bearing ability of the water pollution control unit.

$$\min (X) = \sum_{i=1}^{n} X_i = \sum_{i=1}^{n} (WD_i - W_i), \quad X_i \leq 0 \tag{2.50}$$

$$W = \sum_{i=1}^{n} W_i \tag{2.51}$$

where, WD_i is the current or planned river load of the ith discharge outlet, and X_i is its current or planned river load reduction amount.

2. Constraint:

The water quality target constraint is the same as (2.48), and the pollutant source load constraint is the same as (2.49).

The Total Load Allocation Method Based on the Region Different Demands in A Basin

In order to revise the inconsistency between the allocation results obtained from the total load allocation method based on the maximum total load and the social and economic development demands and the protection targets of the aquatic ecosystems, it is necessary to introduce the different demands of water pollution control unit in the process of allocating the total load of a water pollution control unit in a basin and revise the target function of the total load allocation method based on the maximum total load, and it is called the total load allocation method based on the region different demands in a basin.

The researches on the bearing capability of the aquatic ecosystems in a basin pointed out the load bearing degree of the water pollution control unit, and the target function can be revised by using the aquatic eco-system load bearing degree (Rk = 0–1) of the water pollution control unit as the weight. That is to say, the water pollution control unit with lower load bearing degree is allocated with a total load which is lower than the allocation result obtained from the total load allocation method based on the maximum total load, and the surplus capacity automatically compensates the areas with higher load bearing degrees.

1. Target Function

It is the total maximum river permissible load of a basin revised based on the aquatic ecosystem bearing ability of the water pollution control units.

$$A_i = W_i \times F_i \tag{2.52}$$

$$\max (A) = \sum_{i=1}^{n} A_i \tag{2.53}$$

$$W = \sum_{i=1}^{n} W_i \tag{2.54}$$

where, F_i is the river load weight of the river pollutant source of W_i, $F_i = f(R_{ik})$; R_{ik} is the load bearing degree of k, the water pollution control unit where river pollutant source of W_i is located. In accordance with the principle where the lower the bearing degree of control unit is, the stricter the permissible load control is, and the weight distribution method should be revised.

2. Constraint

The water quality constraint is as the same as (2.48).

The pollution source load constraint

$$W_i \geq 0, (i = 1, 2, \ldots, N), \text{ when } R_{ik} = 0, W_i = 0 \tag{2.55}$$

definitely, the regional demand differences reflected by the aquatic ecological bearing capacity based on the basin control units is just a regulating method, and the regional different demands can be adjusted by adopting the current control unit states or socio-economic planning targets (e.g.: population, GPD, etc.).

Control Unit-Pollution Source Distribution

After the "basin control unit" load distribution, it is necessary to conduct the internal pollution source load distribution of the control units. The research scholars, based on the previous researches on total load control, raised and applied many pollution load distribution methods. The factors which influences the optimal scheme selection include: the positions and load sizes of pollution sources, types of controlled pollutants, the necessity to reduce the difficulty of pollution load, the operability in the execution process, the related planning coordination, etc. There are many methods for the permissible load distribution inside the control units, and they are applicable to different scene and targets, but in general, the commonly used distribution methods can be basically narrowed down to two types of the optimal distribution methods and the fair distribution methods.

The significant feature of the optimal distribution methods is that it has a single maximum (or minimum) target. The very target can be the total costs for pollutant removal, or the total removal amount of pollutants, or the adjusted amount of pollution load.

The fair distribution methods are to conduct the average distribution of the pollution load according to one certain attribute of the pollution sources. Generally speaking, fairness is specific to a certain quantitative index which can be the load contribution ratio, pollutant generation areas and so on. Fairness is a relative concept and different measurement standards and solving methods are available from different perspectives. The fairness principle should reduce disputes caused by distribution as much as possible in consideration of the regional population, economy, environment bearing capacity, current environment state, etc. At present, the fair distribution methods which have gained extensive attention are: the distribution method by rate of contribution, the equal percent removal method, the regional difference method, the Gini coefficient method, etc.

EPA (Environmental Protection Agency) raises 19 total load distribution methods in *Technical Support Document for Water Quality-based Toxics Control*, shown in Table 2.29.

The recommended pollution load distribution methods applied in the basic scenes in the Guidelines are as follows:

The Distribution Method by Rate of Contribution

The distribution method by rate of contribution is to distribute the permissible discharge amounts according to the contribution of pollution sources. The contribution can be measured by the pollution source discharged load, or by the influences on a certain water quality from the pollutants discharged by enterprises. Compare with the previous type of contribution, it is the negative contribution, if it is considered as a distribution principle, the greater the contribution is, the smaller the distributed discharge amount is, or in another way, the more pollutants the enterprises need treatment.

$$mi = F(a, mi0, l) = \frac{mi0}{1 + a_i^p l} \tag{2.56}$$

$$\sum aimi = \sum \frac{aimi0}{1 + a_i^p l} = Cp \ p \in (0, 1) \tag{2.57}$$

where, p is a constant, 0–1 is the reduction degree.
The Equal-Proportional Load Distribution Method

Based on the current pollution source discharge states, the equal-proportional distribution method distributes the pollution load among point sources and non-point sources and among individual point sources at an equal proportion according to their pollutant discharge amounts. It is a distribution method when the weight coefficient of each discharge outlet remain unchanged. And the calculation formula of weight coefficient of pollution source is as follows:

$$m_i = W_i / \sum W_i \tag{2.58}$$

Table 2.29 TMDL (total maximum daily load) pollution load distribution method suggested by EPA

SN	Distribution method
1	Equal percent removal
2	Equal effluent concentrations
3	Equal total mass discharge per day
4	Equal mass discharge per capita per day
5	Equal reduction of raw load (pounds per day)
6	Equal ambient mean annual quality (mg/l)
7	Equal cost per pound of pollutant removed
8	Equal treatment cost per unit of production
9	Equal mass discharged per unit of raw material used
10	Equal mass discharged per unit of production
11a	Percent removal proportional to raw load per day
11b	Larger facilities to achieve higher removal rates
12	Percent removal proportional to community effective income
13a	Effluent charges (dollars per pound, etc.)
13b	Effluent charge above some load limit
14	Seasonal limits based on cost-effectiveness analysis
15	Minimum total treatment cost
16	Best availability technology (BAT) (industry) plus some level for municipal Inputs
17	Simulative capacity divided to require an "equal effort among all dischargers"
18a	Municipal: treatment level proportional to plant size
18b	Industrial: equal percent between best practicable technology (BPT) and BAT
19	Industrial discharges given different treatment levels for different stream flows and seasons

where, m_i is the weight of the ith pollution source, and W_i is its pollution load capacity at the reservoir mouth.

It's assumed that the number of distributed pollution sources is n, and the pollutant discharge quantity of each pollution source is $W1, W2, …, Wn$, if the total permissible pollutant discharge quantity is Q, then the pollutant discharge quality distributed to each pollution source is:

$$Qi = \frac{W_i Q}{\sum_{i=1}^{n} W_i} \tag{2.59}$$

The equal-proportional distribution method actually reduces the contribution amount of the pollution sources to the reservoir mouth at an equal proportion, then establishes the load change relation between the "pollution source-reservoir mouth" loads, and then determines the actual reduction amount of the pollution sources. When the "pollution source-reservoir mouth" loads have a linear relation, the method reduces the pollution sources at an equal proportion.

The Marginal Benefit Maximization Method

The marginal benefit in the marginal benefit maximization method refers to the benefits by the current pollutant discharge level plus the unit pollutant discharge quantity. The pollution source benefit function is set as: $g(W, x)$, where, W is the pollution source discharge load, g is the function which is related to the pollutant discharge quantity of pollution sources, x represents other factors which influence the benefit function, and the function reflects the overall regional benefits in different discharge situations. And the unit marginal benefit is:

$$B(W) = \frac{\partial g(W, x)}{\partial W} \tag{2.60}$$

When the pollution load treatment expenses are considered about, the net marginal benefit is:

$$NB(W) = B(W) - C(W' - W) \tag{2.61}$$

In the actual distribution process, put each and every pollution source on a certain distribution level, on the equal-proportional distribution basis, calculate the corresponding net marginal benefit, pick up the two pollution source units with the maximum and minimum marginal benefits, then properly reduce the distribution quantity of the pollution source with the minimum net marginal benefit, and add the reduction quantity to the ones with the maximum net marginal benefit. Repeat the aforesaid processes again and again till the net marginal benefits of all the pollution sources are equal; if it is impossible to adjust the actual net marginal benefit to being equal, then adjust it to the overall maximum net marginal benefit.

The method is applicable to the secondary distribution of the pollution load with the region, and the making of marginal benefit function is the key and difficult points for the method.

The Fair Distribution Method Based on Gini Coefficient

The method was raised to quantitatively measure the income distribution difference degree by the Italian economist of Gini in 1912. Its application in the environment field can evaluate and correct the pollution load distribution work. It can be understood that the actual environment significance of the method is that it doesn't uniformly distribute the local proportions out of the total load in the pollution load distribution.

In the concrete load distribution, the overall regional Gini coefficient can be calculated based on the regional population, GDP, influencing weight of pollution sources, land area and other factors, and reduced to the lowest by adjusting the pollution load distribution schemes. The specific calculation formula of Gini coefficient is as follows:

$$Gini = 1 - \sum_{i=1}^{n}(X_i - X_{i-1})(Y_i + Y_{i-1}) \tag{2.62}$$

where: Xi is the cumulative percentile of population and other indexes; Yi is the cumulative percentile of pollutant discharge amount.

According to the economic standards to evaluate Gini coefficient, when Gini < 0.2, distribution is highly even; when 0.2 < Gini > 0.3, distribution is relatively even; when 0.3 < Gini < 0.4, distribution is relatively rational; when 0.4 < Gini < 0.5, distribution is greatly biased; and when Gini > 0.5, distribution is highly uneven. In the load distribution process, since there are many realistic factors which result in uneven distribution and objective non-uniformity (for example, the uneven population density might lead to the result that it is difficult to reduce simultaneously the Gini coefficient calculated respectively according to population and land area) existing in various environmental factors, in the evaluation of the environmental Gini coefficient, 0.5 is considered as the "warning line" for fair distribution, and the distribution is rational when the Gini coefficient of load distribution is controlled within 0.5, and vice versa.

The Efficiency-First Distribution Method Based on Multi-factor Gini Coefficient

The Gini Coefficient Method often needs to adjust the Gini coefficient it solves in the total water pollutant discharge quantity index distribution process, but only adjusting the single-index Gini coefficient with higher value reduces the fairness and rationality of the distribution schemes. On the premise where the fairness of total load distribution based on the other indexes doesn't become weaker, the efficiency-first method based on multi-factor Gini coefficient is used to optimize and adjust the distribution schemes

The optimizing rules for distribution schemes are formulated based on the following two points:

1. In optimizing and adjusting the Gini coefficient, ensure the order and positions of the units in all regions remain unchanged in the Lorenz curve based on the other single factors;
2. Ensure to decrease the other Gini coefficients as the target to make the reduction degree optimal and the reduction proportion in each region must be controlled within the stipulated range when the Gini coefficient of the single indexes tend to decrease (Fig. 2.44).

The pollution load distribution of each control unit considers the pollution load production quantity and area as the prior reference factors. The pollution load production quantity is obtained by calculating the pollution data in recent years and the pollution load is distributed based on the quantity, and the result can reflect the current actual pollution states in each region. Later, the pollution load is distributed by combining the land area, land area, as an invariable factor in the near future, can reflect the initialization degree of the pollution state of each region, in the initial distribution, larger pollution load is distributed according to larger regions, after the completion of the distribution, the distribution is further adjusted according to the actual situation to maximize distribution rationality. On the premise where the distribution which considers the pollution load production quantity and land area as

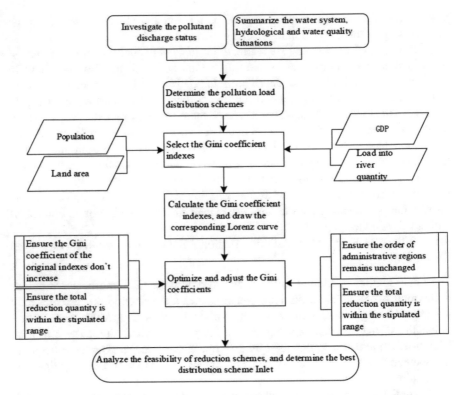

Fig. 2.44 Technology roadmap of multi-factor Gini coefficient water pollutant distribution

the main factors, distribution is conducted by comprehensively considering about population and GDP. The preliminary distribution principle is the factors with larger values are distributed with larger pollution load after comparing the factor values. Compare the result obtained and the distribution result with the comprehensive consideration about the pollution load production quantity and land area, then further weigh and compare them, redistribute the load, and the final result is the final distribution by combining the factors of pollution load production quantity, land area, population and GDP.

2.3.6.5   Determination of Margin of Safety (MOS)

The "basin-control unit" distribution is based on the water quality response coefficient matrix method, and adds the permissible loads distributed to the direct discharge outlets (point source) and branch estuary (the pollution sources in smaller basins, including point sources and non-point sources) together in accordance with the water quality target attainment of water function regions (aquatic ecological function region, water function region or water environment function region), the sum is the maximum water environment capacity of the water function region. The maximum water environment capacity has considered about the influences from the environment background value, but to ensure that the water body satisfies the water quality target, the uncertainty of environment planning should also be considered about; in the total load distribution schemes based on water environment capacity, a certain margin of safety (MOS) should be reserved in accordance with the features of water function regions and the uncertainty of data and methods in the total load calculation.

MOS is an indispensable important factor in the TMDL process executed by America. TMDL recommends the following 2 methods to calculate the MOS.

1. Implicit methods: several kinds of conservative assumptions adopted in the distribution scene design often directly contain MOS; if the water quality target is deduced in conservative assumption, the mathematical models are developed, the measures analyzed and the expected feasibility of actions repaired in conservative assumption, etc.
2. Explicit methods: set relatively conservative water quality target, or add safety coefficient in the pollution load estimation, or reserve some un-distributable pollution load as MOS.

The MOS design method about water quality target can be determined by adopting the tow rational implicit and explicit methods, but not limiting to the two methods.

The Explicit Calculation Methods

In accordance with the importance of the water area protection targets of control units and their sensitivity to hydrological and water environment, a part of total load of control units is reserved as MOS. The control units with higher water quality targets, larger changes of hydrological conditions and more complex pollution load have higher MOS proportion at 5–30%. For example, decrease the standard value of all the water quality targets by one percent (e.g.: 5%) or that of high-functioning regions by one percent (e.g.: 10%).

For example, among the TMDL cases in America, MOS is calculated in the form of TMDL proportion coefficient. In general, MOS is set to be 5–10% of the total maximum daily load, and its proportion is determined according to the types of pollutants, model precision and actual water environment management demands. Below are some cases with direct MOS distribution in proportion in TMDL projects, and also the proportions of MOS out of the total pollutant receiving capacity (Table 2.30).

The Implicit Calculation Methods

They are the methods to calculate and distribute the water environment capacity in accordance with various conservative design conditions.

MCS (Monte Carlo Simulation) Method

The MCS method is a widely applied method to analyze the model parameter uncertainty in recent years. The MCS method requires huge data and time quantity, can well reflect the posterior distribution of the model parameters, and calculate the model uncertainty.

The model uncertainty calculation is generally conducted in the following 6 steps in the MCS method:

1. Select the main parameters which are to be analyzed, and determine the parameter dereferencing range;
2. Analyze the initial Monte Carlo, and compare the simulation result with the monitoring data;
3. Analyze the parameter interval sensitivity;
4. Set the limit value;
5. Narrow down the parameter range;
6. Test and evaluate the analysis result.

MOS is determined on the basis of calculating the model parameter uncertainty in the MCS method.

FOEA (First-Order Error Analysis) Method

The FOEA method is a widely applied method to analyze the model parameters in recent years, and also one of the most important methods for definite MOS calculation in America's TMDL plan.

The prerequisite to calculate the model uncertainty in the FOES method is the model variables can satisfy the linear rules, that is to say, the variable median can represent the average value of the variables under the probability distribution. The FOEA

Table 2.30 Item list of direct MOS distribution in TMDL

TMDL item name	Control object	Core method	MOS proportion (%)
Tidal Potomac and Anacostia PCB TMDL	PCB	POTPCB Model	5
TMDL for Pathogens to Address 25 Lakes in the Northeast Water Region	Pathogen	Pollution Load Duration Curve Method	5
TMDL for the Wanaque Reservoir	Nutrients	LA-WATERS Model	7
Metals and pH TMDLs for the Tygart Valley River Watershed, West Virginia	Heavy Metal, pH	MDAS	5
TMDL development for Fort Cobb Lake	Nutrients	EFDC	5

method conducts the first-order expansion of the model in the Taylor polynomials, and the expansion formula is shown below:

$$Y = Y(X_e) + \sum_{i=1}^{p}(x_i - x_{ie})\frac{\partial Y}{\partial x_i}|_{X_e} \tag{2.63}$$

where: X_e is the dereferencing vector $(x_1, x_2, \ldots, x_n)$ of each variable at the expansion point; Y is the approximate value of the simulation result near X_e; and x_i and x_{ie} are the dereferencing value of the ith variable and its value at the expansion point.

According to the aforesaid equation, it can be deduced that the relation between the variance of Y and that of x_i is near the variable median point by using Taylor expansion, shown in the following formula:

$$\sigma_Y^2 = \sum_{i=1}^{p}\left[\frac{\partial Y}{\partial x_i}|_{X_e}\sigma_i^2\right] \tag{2.64}$$

According to the aforesaid formula, the coefficient of variation is used to establish the relation between the coefficient of variation of variable samples and that of the simulation result samples.

The five steps of the FOEA method in the MOS calculation process are shown below:

1. Determine the key variable of the models
   The hydrodynamic-water quality coupling models used in calculating the receiving water body's environment capacity are generally complex, and have lots of model input data and model parameters. In the model uncertainty research by using the FOEA method, it is extremely complex to analyze the influences from all the variable uncertainty, therefore, it is necessary to firstly find out the key variables which influence the model uncertainty. The key variables which influence the model calculation result uncertainty can be discovered through the study on the model mechanisms and parameter sensitivity.
2. Calculate the Partial Differential by Using the Finite Difference Method
   As it is impossible to solve the analytical expression form for the partial differential items $\partial Y/\partial x_i$ which is required to calculate in the Taylor Expansion Formula in many complex models, the method of numerical approximation is adopted and the finite difference method is used to calculate the partial differential.
3. Calculate the coefficient of variation (CV) of each key variable
   CV, AKA the variation of dispersion, is the ratio between the standard deviation and the average value and reflects the dispersion situation of variables.
4. Calculate the overall deviation and standard deviation of model output
   The FOEA method can determine the contribution degree of each and every variable to the deviation of the final model output results. In the model parameter uncertainty analysis, the FOEA equation can connect the overall deviation of the model output with the dereferencing value of each and every independent

parameter deviation, and the equation is shown below:

$$
\begin{aligned}
&(CV_Y)^2 \\
&= \sum_{i=1}^{P}(CV_{x_i})^2 \cdot S_i^2 \\
&= \sum_{i=1}^{P}(CV_{x_i})^2 \cdot \left[\frac{\Delta Y(x_i)/Y(x_i)}{\Delta x_i/x_i}\right]^2 \\
&= \sum_{i=1}^{P}(CV_{x_i})^2 \cdot \left[\frac{[Y(x_i+\Delta x)-Y(x_i-\Delta x)]/Y(x_i)}{(2\Delta x_i/x_i)}\right]
\end{aligned}
\tag{2.65}
$$

where: CVy is the CV of the output result of a certain simulation index of the model; CVxi is the CV of the dereferencing value of the ith model parameter; Si is the sensitivity of the ith parameter near the median point of the parameter samples; and $Y(x_i)$ is the output result of the ith model parameter at the dereferencing value of x_i.

5. Determine MOS based on the overall deviation or standard deviation result output by the models

   Seek for the maximum simulation result of Ymax generated under the situation of the dereferencing samples of key parameters, by limiting that the value of Ymax must be lower than the set water quality standard value, calculate difference value between the simulation result of y under the situation of actual parameter dereferencing and Ymax, in this way, it can be affirmed as the MOS based on the model uncertainty.

2.3.6.6 Effect Test

The concept of distribution is to find out the optimal scheme from lots of feasible distribution schemes. Under a distribution principle, there is one best load distribution scheme of the principle; under multiple distribution principles, there are several best load distribution schemes. Therefore, it is necessary to establish evaluation indexes for the results from the multiple distribution principles and evaluate the results in fair, efficient, economic and scientific manners, and determine the final decision distribution schemes.

The optimization problem of the multiple distribution principle is the multi-target optimization problem, and there exist certain conflicts among the multiple targets. There often exist infinite non-inferior solutions for a multi-task optimization, and the non-inferior solutions form the non-inferior solution set. The better schemes can be determined in the following forms:

1. The scheme comparison method. That is to say, the method firstly gets a certain quantity of non-inferior solutions, and then find the better schemes according to the parties' or decision makers' intentions who participate in distribution.
2. The interactive improvement method. The method firstly gets several non-inferior solutions, and find the better solutions step by step in the method of interaction among the analyzers, the distribution parties or the decision makers. The interactive improvement method is more like a negotiation game process for several

subjects of rights who participate in distribution. Regarding the total load distribution, a better scheme or several schemes are available through inter-regional interactive improvement, and then are decided by the competent departments in the basin. At present, China has few cases where interactive improvement is directly conducted in the basin distribution in the programming and calculation methods.

3. The optimization method. The decision makers are required to provide the relative importance degrees among the attention/evaluation targets in advance, and then the programming algorithm is based on it, transforms multi-target problems into single-target ones for solving, and find the optimal solution under such circumstance. The difficulty of the method is that how to distribute the true weighted information of the distribution participants and the decision makers.

The Guidelines mainly target to Methods (1) and (3), and put forward a quantitative evaluation index for the total load distribution which can better reflect the distribution weights and transforms multi-target problems into single-target ones, and the index can evaluate the different local optimal feasible solutions of the comparable schemes or non-linear programming on the same distribution schemes, and discover a better scheme; or directly work as a target function and reflect the distribution principles in the optimization process.

The weighted method can be adopted for the transformation into single-targets from multi-targets. The weighted method is actually the single targets formed by optimizing the positive linear combination of each target function. Its advantages are that it is concise and explicit and has low calculation complexity; and its disadvantage is that some non-inferior solutions might be omitted in the non-linear constraint problems, that is to say, the optimal solutions of single targets calculated in the positive weighted coefficient are definitely the non-inferior solutions, but some non-inferior solutions may not be expressed by a group of positive weighted coefficient.

TCRI can be adopted to evaluate the distribution results:

$$\text{TCRI} = \sum \alpha_k I_k \tag{2.66}$$

where: α_k is the weighting coefficient, $\sum \alpha_k = 1$; I_k is the single index value and is a dimensionless value between 0 and 1, the larger the value is, the more rational the distribution result is. The subscript represents the serial number of the factors which need considering about. The weighting coefficient can be determined in the general statistical methods, such as, the expert scoring method, the entropy method, the analytic hierarchy process (AHP), etc.

$$I_k = 1 - \sum (R_j - F_{kj})^2 \tag{2.67}$$

where: the subscript j represents the serial number of the pollution sources,
$F_{1j} = X_j / \max \sum_{j=1}^{m} X_j$—the maximum receiving pollutant load ratio,
$R_j = X_j / \sum_{j=1}^{m} X_j$—the distribution load ratio.

Table 2.31 The list for the sub-item indexes of the total load

Index		Content	Expression form	
The scientific index	I1	The environment capacity utilization ratio	The ratio of the total distribution load and the environment capacity of each discharge outlet in the distributed region Or in $1 - \sum (Rj - F1j)2$ $F1j$—The maximum pollution receiving load ratio	
The fair index	I2	The capacity utilization and water resource contribution ratio	In $1 - \sum (Rj - F2j)2$	$F2j$—The runoff contribution ratio
	I3	Population (Domestic pollution discharge right)	In $1 - \sum (Rj - F3j)2$	$F3j$—The population ratio
	I4	Farmlands (Domestic and production pollution discharge right)	In $1 - \sum (Rj - F4j)2$	$F4j$—The farmland area ratio
	I5	Urban Development (Standard accomplishment pollution discharge right	$1 - \sum (Rj - F5j)2$	$F5j$—GDP ratio
The economic index	I6	Lower treatment expenses	In $1 - \sum (Rj - F6j)2$	$F6j$—The current load quantity ratio

The meaning of economic index is the indirect minimum of environment investment, that is to say, the smaller the reduction or removal ratio the current pollution sources have, the smaller the investment is. Obviously, the method is applicable to the macro distribution, such as basin programming, and dominated by the total load distribution in large cities and branches. Regarding the smaller regions, the distribution focuses on minimizing the point source treatment expenses. The actual programming can increase or decrease the selected indexes according to the specific situations.

The pollution load distribution schemes based on the maximum water environment capacity of a basin from model calculation, and the pollution sources whose pollution loads are distributed to each and every control unit in accordance with the pollution load distribution principles and methods have the sufficient scientific theory grounds and implementation application values. However, due to the existence of uncertainty and errors, the effects of the distribution schemes should be tested to ensure that the water environment quality of the control units can reach the corresponding control targets. The method and thought to test the control target reachability of the pollution load distribution schemes is: set the pollution load of each and every pollution source according to the distribution scheme, simulate and Invert the corresponding temporal and spatial changes of water quality by using the model, and check whether the water quality satisfy the expected target by comparing with the quantitative control target (Table 2.31).

2.4 Conclusion and Suggestions

As an important constituent in objective management of river basin quality in China, objective management of water quality by control unit is a water quality management system based on controlling gross capacity of aquatic environment in the river basin, established in control unit and oriented to polluter in the aquatic environment management hierarchy system of "river basin-control unit-polluter". The control unit based objective management of water quality looks at the global situation based on controlling gross of river basin, targets to make the river basin's comprehensive water quality reach the mark, and overcomes limitation of water quality management of single region. Meanwhile, objective management of water quality by control unit is directly oriented to polluter, can provide concrete and operable suggestions and schemes for pollution control, and overcome river basin planning's drawbacks of inadequate consideration over details.

Dividing into control units can divide a complicated river basin into several mutually independent and interconnected units, facilitate calculation of aquatic environment capacity and fine distribution of gross pollution load. Solving the aquatic environment problem in the control units and handling well the relation between different control units can make the water quality objective in each control unit reach the mark and the general water quality objective in the river basin reach the mark, thereby protecting aquatic ecological function in the river basin. Meanwhile the control unit embodies characteristic of natural catchment and administrative management demand, thereby facilitating system management of river basin and carrying out scheme of water quality objective management in the river basin. The division of control units in the river basin should follow following principles: defining land based on water, independent water type, complete catchment basin, definite administrative responsibility, coordination with aquatic ecological functional area, etc.

Estimation over pollution load of river basin control unit is an important link in water quality improvement management by control unit, and can provide reference with regard to drainage load of various polluters for control of gross pollutant and formulation of exhausting reduction scheme. Appraisal of pollution load in the river basin needs to give comprehensive consideration to the influence of topographic changes, air conditions change, land use, soil types and agricultural cropping pattern, etc. in the river basin. Thus the river basin hydrology and pollution load model with firm theoretical basis of water science and environmental science has wider application.

Control unit based analysis on capacity of aquatic environment in the river basin is an important precondition and the core for distribution of gross pollutant. Based on the precondition of given water area scope, water quality objective, hydrological condition and drain position and discharge mode, the gross of certain pollutant able to be accommodated by the receiving water area is the capacity of aquatic environment. This definition here specially means gross pollutant that can be controlled and distributed. Proper river basin model and receiving water body model is used to build response relation between river basin's pollution load and receiving water body's water quality to define the environmental capacity of water body in each control unit.

Distribution of pollution load in river basin is based on a reasonable analysis on capacity of the aquatic environment in the river basin/control unit to present diversified distribution schemes according to scientificalness, equity, efficiency and difference of load distribution. The gross distribution should first consider selecting the distribution principle. If the aquatic environment has undercapacity or shortage of capacity, the principle of equity is especially important, while when the remnant environmental capacity is large, the efficiency principle can serve as the key. Generally speaking, a reasonable distribution way is mainly achieving equity with efficiency as the second priority. Based on the principle of equity and efficiency, this guide rule targets the requirements on management of river basin's water quality, combines the management characteristics of river basin gross in China, and put forward bi-layered multi-objective optimization distribution frame for gross capacity of the river basin.

To guarantee security of distribution achievement of aquatic environment's capacity, MOS is set to put forward conservative requirements on precondition of aquatic environment capacity analysis (water quality objective, hydrological condition, load estimate), selection of calculation model, and appraisal on effect of load control measures. As water quality objective deduction aquatic environment model and parameter selection, appraisal on effect of load control measures, etc. need support of substantive data and is relatively complicated, thus directly reserve MOS according to judgment of experts tends to be a common method.

As an important link for controlling gross pollutant in river basin, the control unit based pollution load distribution is the core for objective management of water quality by control unit of river basin, also an important way to protect aquatic environment quality and realize sustainable development of river basin. The distribution objective of gross pollution load based on capacity of aquatic environment in the river basin should be determined from the perspective of general aquatic ecological environmental protection of the river basin. Meanwhile, when determining distribution of gross pollution load, one should comprehensively consider the relation between upstream and downstream, left bank and right bank in the river basin. Objective gross of permissible load of pollutant in each control unit should be defined based on the general objective of the water quality of river basin reaching the mark.

References

1. GB 3838-2002 Environmental quality standards for surface water.
2. GB12941-91 Water quality standards for landscape entertainment.
3. SL278-2002 Hydrological calculation of water conservancy and hydropower project.
4. Abell, Frances, Michael Krams, John Ashburner, Richard Passingham, Karl Friston, Richard Frackowiak, Francesca Happé, Chris Frith, and Uta Frith. 1999. The neuroanatomy of autism: A voxel-based whole brain analysis of structural scans. *Neuroreport* 10 (8): 1647–1651.
5. Abell, Robin, Michele L. Thieme, Carmen Revenga, Mark Bryer, Maurice Kottelat, Nina Bogutskaya, Brian Coad, Nick Mandrak, Salvador Contreras Balderas, William Bussing, et al. 2008. Freshwater ecoregions of the world: A new map of biogeographic units for freshwater biodiversity conservation. *BioScience* 58 (5): 403–414.
6. Abell, Robin A. 2000. *Freshwater ecoregions of North America: A conservation assessment*, vol. 2. Island Press.
7. Albert, D.A., S.R. Denton, B.V. Barnes, and K.E. Simpson. 1986. *Regional landscape ecosystems of Michigan*. School of Natural Resources: University of Michigan.
8. Andersen, M.M., F.F. Riget, and H. Sparholt. 1984. A modification of the trent index for use in Denmark. *Water Research* 18 (2): 145–151.
9. Armitage, P.D., Dorian Moss, J.F. Wright, and M.T. Furse. 1983. The performance of a new biological water quality score system based on macroinvertebrates over a wide range of unpolluted running-water sites. *Water Research* 17 (3): 333–347.
10. Austrian Standards Institute. 1997. Entwurf, berechnung und bemessung in der geotechnik.
11. Bailey, Kenneth D. 1998. Social ecology and living systems theory. *Systems Research and Behavioral Science*: 421–422.
12. Bailey, R. G. Ecoregions of the United States.
13. Bailey, R.G. 2014. *Ecoregions: The ecosystem geography of the oceans and continents*. New York: Springer.
14. Bedford, Barbara L. 1996. The need to define hydrologic equivalence at the landscape scale for freshwater wetland mitigation. *Ecological Applications* 6 (1): 57–68.
15. Bedford, Barbara L., and Eric M. Preston. 1988. Developing the scientific basis for assessing cumulative effects of wetland loss and degradation on landscape functions: Status, perspectives, and prospects. *Environmental management* 12 (5): 751–771.
16. Biggs, Barry J.F. 1995. The contribution of flood disturbance, catchment geology and land use to the habitat template of periphyton in stream ecosystems. *Freshwater Biology* 33 (3): 419–438.
17. Bjerring, Rikke, Emily Gwyneth Bradshaw, Susanne Lildal Amsinck, Liselotte Sander Johansson, Bent Vad Odgaard, Anne Birgitte Nielsen, and Erik Jeppesen. 2008. Inferring recent

changes in the ecological state of 21 Danish candidate reference lakes (EU Water Framework Directive) using palaeolimnology. *Journal of Applied Ecology* 45 (6): 1566–1575.

18. Bojie, Fu. 2001. *Landscape Ecology Theory and Application*. Beijing: Science Press.
19. Bojie, Fu, Guohua Liu, Liding Chen, Keming Ma, and Junran Li. 2000. Schee of ecological regionalization in China. *Acta Ecologica Sinica* 21 (1): 1–6.
20. Bonada, Nuria, Narcís Prat, Vincent H. Resh, and Bernhard Statzner. 2006. Developments in aquatic insect biomonitoring: A comparative analysis of recent approaches. *Annual Review of Entomology* 51: 495–523.
21. Breine, Jan, Ilse Simoens, Peter Goethals, Paul Quataert, Dirk Ercken, Chris Liefferinge, and Claude Belpaire. 2004. A fish-based index of biotic integrity for upstream brooks in Flanders (Belgium). *Hydrobiologia* 522 (1–3): 133–148.
22. Brismar, Anna. 2002. River systems as providers of goods and services: A basis for comparing desired and undesired effects of large dam projects. *Environmental Management* 29 (5): 598–609.
23. Brookes, Andrew, and F. D. Shields. 1996. *River channel restoration: Guiding principles for sustainable projects*. Wiley.
24. Butler, David, and John Davies. 2000. *Urban drainage*. CRC Press.
25. Cao, Y., D.P. Larsen, R.M. Hughes, P.L. Angermeier, and T.M. Patton. 2002. Sampling effort affects multivariate comparisons of stream assemblages. *Journal of the North American Benthological Society* 21 (4): 701–714.
26. Chutter, F.M. 1972. An empirical biotic index of the quality of water in South African streams and rivers. *Water Research* 6 (1): 19–30.
27. Costanza, Robert, Ralph d'Arge, Rudolf De Groot, Stephen Farber, Monica Grasso, Bruce Hannon, Karin Limburg, Shahid Naeem, Jose Paruelo, Robert V. O'neill, et al. 1997. The value of the world's ecosystem services and natural capital. *Nature* 387 (6630): 253–260.
28. Costanza, Robert, Ralph d'Arge, Rudolf De Groot, Stephen Farber, Monica Grasso, Bruce Hannon, Karin Limburg, Shahid Naeem, Robert V. O'Neill, Jose Paruelo, et al. 1998. The value of ecosystem services: Putting the issues in perspective. *Ecological Economics* 25 (1): 67–72.
29. Crowley, John M. 1967. *Biogeography. The Canadian Geographer/Le Géographe canadien* 11 (4): 312–326.
30. Daily, Gretchen. 1997. *Nature's services: Societal dependence on natural ecosystems*. Island Press.
31. Darlington, P. J. 1957. *Zoogeography: The geographical distribution of animals*. Wiley.
32. David Allan, J., and María M. Castillo. 2007. *Stream ecology: structure and function of running waters*. Springer Science & Business Media.
33. Groot, De, S. Rudolf, Matthew A. Wilson, and M.J. Boumans. 2002. A typology for the classification, description and valuation of ecosystem functions, goods and services. *Ecological Economics* 41 (3): 393–408.
34. de Mérona, Bernard, Régis Vigouroux, and Véronique Horeau. 2003. Changes in food resources and their utilization by fish assemblages in a large tropical reservoir in South America (Petit-Saut Dam, French Guiana). *Acta Oecologica* 24 (3): 147–156.
35. Dinerstein, Eric., David M. Olson, Douglas J. Graham, Avis L. Webster, Steven A. Primm, and Marnie P. Bookbinder, George Ledec, and Kenneth R. Young. 1995. *A conservation assessment of the terrestrial ecoregions of Latin America and the Caribbean*. Number 333.79 BAN. Washington, DC: World Bank.
36. Engle, V.D., and J.K. Summers. 1999. Latitudinal gradients in benthic community composition in Western Atlantic estuaries. *Journal of Biogeography* 26 (5): 1007–1023.
37. European Commission. 2000. Directive 2000/60/EC of the European Parliament and of the Council establishing a framework for the Community action in the field of water policy.
38. Fiona, Wells, and Peter Newall. 1997. *An examination of an aquatic ecoregion protocol for Australia*. ANZECC Secretariat.
39. Fore, Leska S., James R. Karr, and Robert W. Wisseman. 1996. Assessing invertebrate responses to human activities: Evaluating alternative approaches. *Journal of the North American Benthological Society* 15 (2): 212–231.

40. Hawkes, Clifford L., David L. Miller, and William G. Layher. 1986. Fish ecoregions of Kansas: Stream fish assemblage patterns and associated environmental correlates. *Environmental Biology of Fishes* 17 (4): 267–279.
41. Hawkins, Charles P. 2006. Quantifying biological integrity by taxonomic completeness: Its utility in regional and global assessments. *Ecological Applications* 16 (4): 1277–1294.
42. Hellawell, John M., et al. 1978. *Biological surveillance of rivers; A biological monitoring handbook.*
43. Helliwell, D.R. 1969. Valuation of wildlife resources. *Regional Studies* 3 (1): 41–47.
44. Higgins, Paul A.T., and Stephen H. Schneider. 2005. Long-term potential ecosystem responses to greenhouse gas-induced thermohaline circulation collapse. *Global Change Biology* 11 (5): 699–709.
45. Host, George E., Philip L. Polzer, David J. Mladenoff, Mark A. White, and Thomas R. Crow. 1996. A quantitative approach to developing regional ecosystem classifications. *Ecological Applications* 6 (2): 608–618.
46. Houghton, David C. 2007. The effects of landscape-level disturbance on the composition of Minnesota caddisfly (Insecta: Trichoptera) trophic functional groups: evidence for ecosystem homogenization. *Environmental Monitoring and Assessment* 135 (1): 253–264.
47. Huang, Bingwei. 1959. Draft of China's comprehensive natural regionalization. *Chinese Science Bulletin* 18: 594–602.
48. Huang, Xiang, Yaning Chen, Jianxin Ma, and Yapeng Chen. 2010. Study on change in value of ecosystem service function of Tarim River. *Acta Ecologica Sinica* 30 (2): 67–75.
49. Hughes, Robert M., Thomas R. Whittier, Christina M. Rohm, and David P. Larsen. 1990. A regional framework for establishing recovery criteria. *Environmental Management* 14 (5): 673–683.
50. James, M. 1987. Ecoregions of the conterminous United States. *Annals of the Association of American geographers* 77 (1): 118–125.
51. James, M., James M. Omernik, and Robert G. Bailey. 1997. Distinguishing between watersheds and ecoregions. *JAWRA Journal of the American Water Resources Association* 33 (5): 935–949.
52. Jowett, Ian G., and Jody Richardson. 1990. Microhabitat preferences of benthic invertebrates in a New Zealand river and the development of in-stream flow-habitat models for Deleatidium spp. *New Zealand Journal of Marine and Freshwater Research* 24 (1): 19–30.
53. Karr, James R. 1981. Assessment of biotic integrity using fish communities. *Fisheries* 6 (6): 21–27.
54. Karr, James R. 1991. Biological integrity: A long-neglected aspect of water resource management. *Ecological Applications* 1 (1): 66–84.
55. Koeppen, W. 1931. *Grundriss der Klimakunde*. Walter de Gruyter.
56. Konarska, Keri M., Paul C. Sutton, and Michael Castellon. 2002. Evaluating scale dependence of ecosystem service valuation: A comparison of NOAA-AVHRR and landsat TM datasets. *Ecological Economics* 41 (3): 491–507.
57. Kong, Y., H. Jiang, X.Y. Zhang, J.X. Jin, Z.Y. Xiao, and M.M. Cheng. 2013. The comparison of ecological geographic regionalization in China based on Holdridge and CCA analysis. *Acta Ecologica Sinica* 33: 3825–3836. (in Chinese).
58. Kreuter, Urs P., Heather G. Harris, Marty D. Matlock, and Ronald E. Lacey. 2001. Change in ecosystem service values in the San Antonio area. *Texas. Ecological Economics* 39 (3): 333–346.
59. Lampert, Guy, and Amalia Short. 2004. *River styles, indicative geomorphic condition and geomorphic priorities for river conservation and rehabilitation in the Namoi catchment.* North-West NSW: Namoi CMA.
60. Lang, Claude, and Olivier Reymond. 1995. An improved index of environmental quality for Swiss rivers based on benthic invertebrates. *Aquatic Sciences-Research Across Boundaries* 57 (2): 172–180.
61. Li, X.M., R.B. Xiao, S.H. Yuan, J. An Chen, and J.X. Zhou. 2010. Urban total ecological footprint forecasting by using radial basis function neural network: A case study of Wuhan city. *China. Ecological Indicators* 10 (2): 241–248.

62. Li, Zhenyu, and Yan Xie. 2002. *Invasive alien Species in China*. China Forestry Publishing House.
63. Lohrer, Andrew M., Simon F. Thrush, and Max M. Gibbs. 2004. Bioturbators enhance ecosystem function through complex biogeochemical interactions. *Nature* 431 (7012): 1092.
64. Loomis, John, Paula Kent, Liz Strange, Kurt Fausch, and Alan Covich. 2000. Measuring the total economic value of restoring ecosystem services in an impaired river basin: Results from a contingent valuation survey. *Ecological Economics* 33 (1): 103–117.
65. Maxwell, James R., Clayton J. Edwards, Mark E. Jensen, Steven J. Paustian, Harry Parrott, and Donley M. Hill. 1995. *A hierarchical framework of aquatic ecological units in North America (Nearctic Zone)*. United States Department of Agriculture, Forest Service: Technical report.
66. McNab, W. Henry, and Peter E. Avers. 1994. *Ecological subregions of the United States*. U.S. Forest Service, ECOMAP Team, WO-WSA-5.
67. McNab, W. Henry, David T. Cleland, Jerry A. Freeouf, G. J. Nowacki, C. A. Carpenter, et al. 2007. Description of ecological subregions: Sections of the conterminous United States.
68. Millennium Ecosystem Assessment. 2005. *Ecosystems and human well-being: Wetlands and water*, vol. 5. Washington, DC: World Resources Institute.
69. Ministry of Environmental Protection of the country-region of the People's Republic of China. 2003–2012. *Annual statistic report on environment in place country-region of China*. China Environmental Science Press (in Chinese).
70. Moog, Otto, Astrid Schmidt-Kloiber, Thomas Ofenböck, and Jeroen Gerritsen. 2004. Does the ecoregion approach support the typological demands of the EU 'Water Framework Directive'? *Integrated Assessment of Running Waters in Europe*, 21–33
71. Odum, Eugene P. *Fundamentals of ecology*.
72. Odum, Eugene P., and Howard T. Odum. 1972. Natural areas as necessary components of man's total environment. In *Transactions of the North American wildlife and natural resources conference*.
73. Palone, Roxane S., and Albert H. Todd. 1997. Chesapeake bay riparian handbook: A guide for establishing and maintaining riparian forest buffers.
74. Reise, Karsten. 2012. *Tidal flat ecology: An experimental approach to species interactions*, vol. 54. Springer Science & Business Media.
75. Schaumburg, Jochen, Christine Schranz, Julia Foerster, Antje Gutowski, Gabriele Hofmann, Petra Meilinger, Susanne Schneider, and Ursula Schmedtje. 2004. Ecological classification of macrophytes and phytobenthos for rivers in Germany according to the Water Framework Directive. *Limnologica-Ecology and Management of Inland Waters* 34 (4): 283–301.
76. Seidl, Andrew F., and Andre Steffens Moraes. 2000. Global valuation of ecosystem services: Application to the Pantanal da Nhecolandia. *Brazil. Ecological Economics* 33 (1): 1–6.
77. Selig, U., A. Eggert, D. Schories, M. Schubert, C. Blümel, and H. Schubert. 2007. Ecological classification of macroalgae and angiosperm communities of inner coastal waters in the southern Baltic Sea. *Ecological Indicators* 7 (3): 665–678.
78. Shafer, Deborah J., Bryan Herczeg, Daniel W. Moulton, Andrew Sipocz, and Kenny Jaynes. 2002. *Regional guidebook for applying the hydrogeomorphic approach to assessing wetland functions of northwest Gulf of Mexico tidal fringe wetlands*. Engineer Research and Development Center Vicksburg MS Environmental Lab: Technical report.
79. Simpson, J.C., R.H. Norris, et al. 2000. Biological assessment of river quality: Development of AUSRIVAS models and outputs. In *Assessing the biological quality of fresh waters: RIVPACS and other techniques. Proceedings of an International Workshop held in Oxford, UK*, 16–18 September 1997, 125–142. Freshwater Biological Association (FBA).
80. Snelder, Ton H., and Barry J.F. Biggs. 2002. Multiscale river environment classification for water resources management. *JAWRA Journal of the American Water Resources Association* 38 (5): 1225–1239.
81. Stark, John D. 1993. Performance of the macroinvertebrate community index: Effects of sampling method, sample replication, water depth, current velocity, and substratum on index values. *New Zealand Journal of Marine and Freshwater Research* 27 (4): 463–478.

82. Suh, Sangwon. 2004. Functions, commodities and environmental impacts in an ecological-economic model. *Ecological Economics* 48 (4): 451–467.
83. Sutton, Paul C., and Robert Costanza. 2002. Global estimates of market and non-market values derived from nighttime satellite imagery, land cover, and ecosystem service valuation. *Ecological Economics* 41 (3): 509–527.
84. Turak, Ere N., Lloyd K. Flack, Richard H. Norris, Justen Simpson, and Natacha Waddell. 1999. Assessment of river condition at a large spatial scale using predictive models. *Freshwater Biology* 41 (2): 283–298.
85. United States Environment Protection Agency. Guide standard and protocol for testing micro-biological water purifiers. Technical report, 1987.
86. Wallace, Ken J. 2007. Classification of ecosystem services: Problems and solutions. *Biological Conservation* 139 (3): 235–246.
87. Warry, David N., and Marcus Hanau. 1993. The use of terrestrial ecoregions as a regional-scale screen for selecting representative reference sites for water quality monitoring. *Environmental Management* 17 (2): 267–276.
88. Whittier, Thomas R., Robert M. Hughes, and David P. Larsen. 1988. Correspondence between ecoregions and spatial patterns in stream ecosystems in Oregon. *Canadian Journal of Fisheries and Aquatic Sciences* 45 (7): 1264–1278.
89. Yan, Nailing, and Yu. Xiaogan. 2003. Objective, principle and system of ecological function zoning in China. *Resources and Environment in the Yangtze Basin* 12 (6): 579–585. (in Chinese).
90. Yang, Aimin, Kewang Tang, Hao Wang, and Jinhua Cheng. 2008. China ecological hydrology division. *Journal of Hydraulic Engineering* 39 (3): 332–338. (in Chinese).
91. Yuan, Lester L., Charles P. Hawkins, and John Van Sickle. 2008. Effects of regionalization decisions on an O/E index for the US national assessment. *Journal of the North American Benthological Society* 27 (4): 892–905.
92. Zhou, Baohua, and Binghui Zheng. 2008. Research on aquatic ecoregions for lakes and reservoirs in China. *Environmental Monitoring and Assessment* 147 (1): 339–350.

Chapter 3
River Health Assessment, Ecological Restoration and Management System

Liang Duan, Weijing Kong, Xin Gao, Juntao Fan, Francois Edwards, Richard Williams, Yonghui Song and Yuan Zhang

3.1 A Technical Guideline for Assessing Large River Health Assessment

by Juntao Fan, Francois Edwards, Weijing Kong, Yuan Zhang

Healthy river ecosystems maintain an integrity of their structural composition (including physical, chemical and biological composition), ecological processes and ecosystem functions. They have the ability to resist disturbance and restore their structure and function, and also provide ecological services in accordance with natural and human needs. River health assessment can indentify water bodies with poor health and their belonging watersheds, and provide possible causes affecting river ecological conditions. In Europe and the United States and other developed countries, river health assessment has turned out to be an effective method for environmental water management. The assessment results can be used to guide watershed management decision making, support allocation of management resources and choose priority areas for taking action, provide help for monitoring and assessing efficiency of human activities, and satisfy the current and future needs for water environmental management of our country.

3.1.1 Application Range

This norm specified the objectives, principles, working programs, criteria, methods of large river assessment and technical requirements of compiling relevant reports,

L. Duan (✉) · W. Kong · X. Gao · J. Fan · Y. Song · Y. Zhang
Chinese Research Academy of Environmental Sciences, Chaoyang, China
e-mail: duanliang@craes.org.cn

F. Edwards · R. Williams
Centre for Ecology & Hydrology, Wallingford, UK

© Springer International Publishing AG, part of Springer Nature 2018
Y. Song et al. (eds.), *Chinese Water Systems*, Terrestrial Environmental Sciences,
https://doi.org/10.1007/978-3-319-76469-6_3

tables and figures. The guideline applies to health assessment for mainstreams and large tributaries of Songhua River, Liao River, Hai River, Yellow River, Huai River, Yangtze River, Pearl River, and rivers in southeast, southwest and northwest China in the territory of the PRC.

3.1.2 General Principles

3.1.2.1 Objectives

River health assessment is a vital component of the strategy to protect and enhance the value of China's riverine ecosystems. The overarching objective is to provide reliable information concerning the health status of ecological assets that guides rational river management policy. Depending on river health status, the goal of management could be to achieve a target state of river health, maintain the current status, or improve on the current status.

A systematic, consistent, national approach to river health assessment will provide a periodic audit of the nation's river assets. These data will assist development of investment policy for river management by identifying priority rivers, priority issues, and priority management actions.

River health assessment provides reliable data to enable evaluation of the effectiveness of management actions in maintaining or improving river health.

Maintaining the integrity of a river's ecosystem is the foundation for maintaining a river's capacity to provide society with goods and services. River health information helps to determine the reasonable level of human utilization of a river's resources and services that is compatible with the desirable level of river health.

Finally, regular communication of river health information, using a simple report card that is understandable by the general public, will support the implementation of appropriate river management actions.

3.1.2.2 Main Steps

Large river assessment was carried out according to the steps below (Fig. 3.1).

3.1.3 River Classification

3.1.3.1 Objectives

River classification reflects the regional differences of physical geography, human activities, and composition of aquatic organisms. Regions for river health assessment are determined based on these above differences. Generally, three regions are

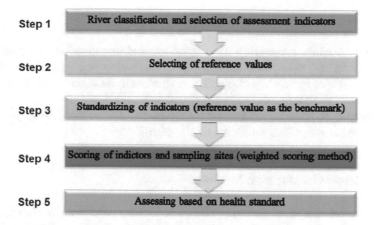

Fig. 3.1 Steps of large river assessment

included: namely the three basic catchment zones are highlands, midlands and low-lands. Alpine zone is a sub-zone of the highland zone, located above the tree line. A catchment will not necessarily have all zones present. Other catchment zones may be included if they are distinctly different from the three basic zones and the one sub-zone. The detailed characteristics of the zones will vary across China, but essentially they will divide catchments as follows:

Highlands (including Alpine):

High-altitude, high-gradient, cool/cold and moist climate, thin soils, main water and sediment source areas, mostly small streams, bedrock outcrops and coarse-grained bed material, shading from riparian trees, discontinuous and narrow flood-plains or no floodplains.

Midlands:

Mid-altitude, medium-gradient, sediment transfer zone, discontinuous and narrow to moderate-width floodplains.

Lowlands:

Low-altitude, low-gradient, sediment deposition zone, finer-grained bed material, wide floodplains.

The rivers that occur within each of the zones are expected to have a degree of similarity in their physical, water quality and biological character, so different indicators and reference values could apply to each zone. Also, these zones are used to aggregate data for the purpose of reporting results at the catchment and basin scales.

3.1.3.2   Working Programs for River Classification

Preparation: Mainly conducting data collection and investigation of aquatic eco-logical conditions, watershed subdivision, investigation and analysis of data in nature, society, economy and ecological environment, assessment on aquatic eco-environmental functions, etc.

Classification: Mainly conducting identification of river classifying indicators, quantitative classification, validation of classification results, etc.

3.1.3.3   River Classification Methods

Top-down: According to generic principles, identify dominant factors affecting regional differences of riverine ecology, and classify rivers step by step through spatial overlaying. Data used in the method are not high demand, so the method is suitable for regions lacking data and classification at macro-scale.

Bottom-up: On the basis of catchment units, delineate freshwater ecoregions by spatial cluster analysis combining with indicator species analysis, habitat analysis and assessment of riverine ecological functions. The method demands a mass of data as support, and it is more suitable for riverine ecological function regionalization at the small scale.

3.1.3.4   Investigation of Riverine Ecological Conditions

Collect and investigate historical and current aquatic ecological data, integrate and build dataset of riverine ecological conditions to provide basis for analyzing riverine ecological features at region scale and verifying regionalization results.

1. Aquatic organisms data collection
   Required data include phytoplankton, periphyton algae, zooplankton, macroin-vertebrate, fish, amphibians, rare and endangered species.
   Get data through historical data collection and field investigation. Historical data collection means collect historical and current species lists, species numbers, community structure, dominant species and distribution ranges of all species in published articles and books. Field investigation means set certain number of sampling sites for riverine ecological investigation at different times. Investigations should carried out at different seasons, and investigate aquatic species identity, numbers, biomass, community structure and distribution according to sampling methods.
2. Aquatic environmental data collection
   Environmental data include water physical and chemical indices such as pH, electric conductivity, ionic concentration, dissolved oxygen, etc., nutrient concentration indices such as total nitrogen and total phosphorus, organic pollution

Table 3.1 Candidate indicators

Indicator types	Indicators	Properties
Climate	Annual average precipitation (mm), annual average temperature (°C), accumulated temperature, etc.	Quantitative
Hydrology	Runoff coefficient and drought index	Quantitative
Geology	Lithology	Quantitative
Geomorphology	Geomorphic type, elevation, slope, and slope aspect	Qualitative, quantitative
Vegetation	Vegetation type, vegetation index, and coverage	Qualitative, quantitative
Soil	Soil type, soil composition, and soil texture	Qualitative, quantitative
Distribution pattern of aquatic organisms	Community features of fish, aquatic plant, amphibians, etc.	Qualitative, quantitative

indices such as chemical oxygen demand, ammonia nitrogen, etc., and heavy metal pollution indices such as Hg and Cu.

3. Data on riverine hydrology, morphology and physical habitats

   Hydrological data include river discharge, frequency, water level, water depth, duration and bankfull discharge, etc. Morphological data include slope, sinuosity, channel density, riparian width, riparian vegetation coverage, land-use, bank stability, etc.

3.1.3.5 Selection of Indicators for River Classification

Indicators include climate, hydrology, geology, geomorphology, vegetation, soil, etc. (Table 3.1).

Filtering indicators

1. Filter indicators suitable for regionalization from candidate indicators using statistical method combined with expert judgments. Principles for filtering indicators include:

   - **Dominance**: On the basis of comprehensive analysis of all factors, select the dominant indicators that affect regional differences of riverine ecological functions.
   - **Independence**: Indicators should reflect the interaction with objectives of regionalization independently of other indicators.
   - **Directness**: Select direct and single indicators, rather than complex and evaluative.

- **Spatial sensitivity**: Indicators should have enough variability, and clearly reflect spatial difference, to facilitate classification.
- **Time stability**: Select indicators that are relatively stable over time, rather than mutable.

2. Filtering indicators methods include qualitative and quantitative methods. Qualitative methods are mainly base on expert judgments and analysis to identify dominant indicators. Quantitative methods include sensitivity analysis of environmental indicators, principal component analysis, correlation analysis among indicators, spatial auto-correlation analysis of indicators and their correlation with riverine ecological factors.

3.1.3.6 River Classification Methods

Top-down method is adopted. Classify rivers base on spatial similarity of selected indicators to identify the ranges and boundaries of sub-regions. Technical route as shown in Fig. 3.2.

3.1.4 Filtering Indicators for River Health Assessment

3.1.4.1 Principles for Filtering Indicators

The core indicators suggested in this Regulation are advised on the basis of a number of principles, and these principles should also be used as the basis for selecting any supplementary indicators:

Comprehensive

To assess river health in a comprehensive way, a range of indicators should be selected in order to describe the river's physical, water quality and biological characteristics.

Ecological relevance

Indicators should have a known connection to ecological processes.

Scale independence

Indicators should be applicable and comparable over a range of river sizes from small creeks to large rivers.

Widely applicable

Indicators should be applicable to different river types across the range of geographic regions in China.

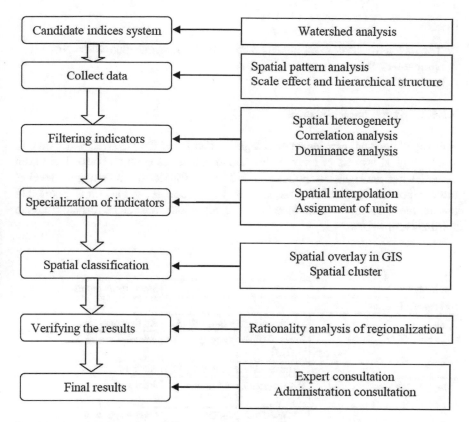

Fig. 3.2 Technical route for river classification

Sensitivity to stressors

Indicators should ideally have a locally demonstrated, or conventionally understood, sensitivity to (i) stressors, and (ii) rehabilitation due to management intervention.

Reliable

The response of indicators to stressors should be predictable and consistent over time and across a wide range of settings.

Interpretable, unambiguous and meaningful

The meaning of the indicators should be clearly understandable to, and be relevant to the main concerns of, those people with an interest in river health and river management. The indicators should guide managers with respect to the appropriate course of action to improve river health.

Practical

The selected indicators should be practical to measure, calculate and report, given the limitations of time, expertise, and funding available.

3.1.4.2   Suggested Indicators

This regulation suggests indicator categories that cover the physical, water quality and biological aspects of river health. These categories comprise a total of seven potential indicator sub-categories (Table 3.2). The indicator sub-category level is more important than the indicator category level, because river health is reported at the sub-indicator level. A combined river health index score can be formed from the sub-indicator scores.

Table 3.2   Suggested indicators

Indicators		Stress types represented
Biological elements		
Fish	Species diversity index (F_S)	All the stress types
	Index of biotic integrity (F_IBI)	All the stress types
	Shannon-Weiner index(F_SDI)	All the stress types
Macroin-vertebrate	Diversity index of families (M_S)	All the stress types
	BMWP scoring index (M_BMWP)	Organic pollution
Algae	Species diversity index (A_S)	All the stress types
	Berger-Parker index (A_BP)	All the stress types
Hydro-chemical elements		
Chemical oxygen demand (COD)		Organic pollution
Dissolved oxygen (DO)		Organic pollution
Electric conductivity (EC)		Eutrophication
Ammonia nitrogen (NH_3–N)		Eutrophication
Total phosphorus (TP)		Eutrophication
Physical habitat elements		
Water quantity (WQ)		Changes of hydrological conditions
Channel change (CC)		Changes of river continuity
Habitat diversity (MD)		Changes of habitats

3.1.5   Selection of Reference Conditions

The river health assessment requires assessing the status of surface water bodies by comparison with reference conditions, which can be defined as the set of conditions to be expected in the absence of or under minimal anthropogenic disturbances. For each indicator included in the assessment, there is a need to estimate the deviation from a reference condition (i.e., a control), expressed as a benchmark. Geographical reference sites (i.e., specific locations on a waterbody that is minimally impaired and is representative of the expected ecological integrity) can constitute the best choice for deriving reference conditions. When such sites do not exist for the river basin of interest, alternatives for setting the reference conditions can include: historical or modelled data, established criteria or standards, expert opinion or local knowledge.

3.1.6   Scoring for Indicators

Synthetically assess indicators giving priority to biological elements, and according to "trade down" principle (Fig. 3.3).

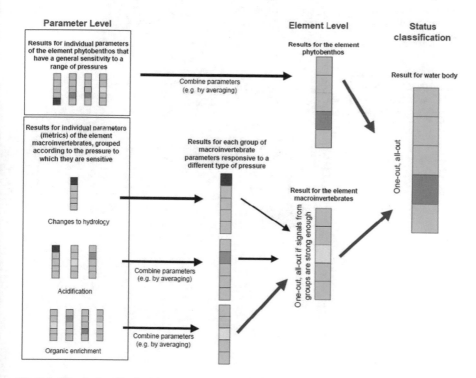

Fig. 3.3 "Trade down" principle when scoring for indicators

The input information is the indicator value, processed according to four thresholds (i.e., high/good, good/moderate, moderate/poor, poor/bad), and expressed as fraction of the Reference Condition (RC) value. The thresholds used in this application were set as default values at 0.8, 0.6, 0.4 and 0.2 for high/good, good/moderate, moderate/poor and poor/bad, respectively, on a scale from 0 to 1, where 1 is the best "condition".

3.1.7   Comprehensive Scoring for Sampling Sites

Steps of scoring for sampling sites (Table 3.3).

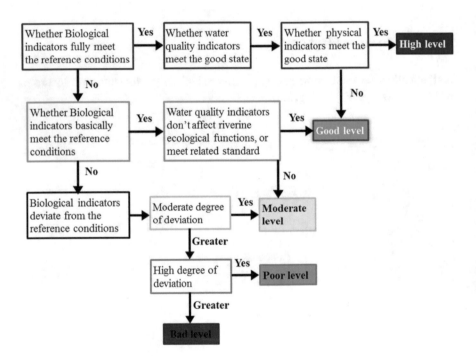

Table 3.3 Steps of scoring for sampling sites

Steps	Comprehensive methods	
	Biological elements	Other elements
Individual indicator score	Observed/Target	
Indicator group score	First, averaged among biological indicators belonging to the same impact typology; Second, results are aggregated to the same biological communities based on OOAO principle; Third, results are aggregated to the biological status based on OOAO principle	For each supportive LoE, one single indicator providing a negative response is enough to reject the inclusion into high and/or good status derived from the LoE biology
Site score	The sites score (ecological status) is determined by biological status; however, if biological status is high or good, the ecological status is determined by the supportive LoE status	
Final classifications	< 0.2 = bad; ≤ 0.4 = poor; ≤ 0.6 = moderate; > 0.6 = good; > 0.8 = high	

* "target" referred to RC values or thresholds for good/not good status; "observed" referred to observed values of sites

3.2   The System Report on Ecological Restoration Technology in Liao River Conservation Area

by Liang Duan, Weijing Kong, Yonghui Song

3.2.1   A Profile of Liao River Conservation Area

3.2.1.1   Natural Geographical Conditions

Geographical Location

Liao River is located in the northeast of China and Liao river basin is wider from east to west than from south to north, which presents the dendritic structure. Among the whole basin, the mountain accounted for 48.2% and the others are hills, plains and sand dunes. Liao River in south is in concern with the Heilongjiang River in north and they are the two largest water systems in northeast of China. If calculate in accordance with the length and watershed area of the main river, Liao River is the seventh largest river within the territory of China. The total length of the main river of is 1390 km and the basin area is 219000 km^2.

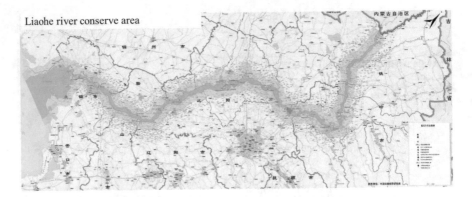

Fig. 3.4 The sketch map of Liao River conservation area

Liao River Conservation Area is built along the Liao river and the scope begins in Liao river interchange (Fude Dian in Changtu County of Tieling city), then goes into Liaodong gulf in Panjin city. The delineation of the boundary is as following: the area without the dam is delimited according to the flood level in 30 years. In the area of main tributaries, it is extended outward appropriately. The total area of Liao River Conservation Area is 1869.2 km^2 and the total length of main river is 538 km with a total of 33 tributaries and the watershed area is 37900 km^2, as shown in Fig. 3.4.

Topography-Geomorphology

The north of Liao River Conservation Area starts from Liao river interchange and Panjin estuary is located in the south. Topography is from north to south and tilt towards center by the east-west. And the flow direction is from the north to the south. In Liaoning and Shenyang area, it is about 60–40 m above the sea level; in Yingkou and Panshan area, it is about 7–4 m above the sea level; in the dam of Shifu Si area, it is about 40 m above the sea level, as shown in Fig. 3.5.

3.2.1.2   Economic and Social Conditions

Liao river area is rich in resources and is densely populated and urban centralized. Meanwhile it also has developed industry and convenient transportation. All of these make it is our country's important base of industries, equipment manufacturing industry, energy and commodity grain and plays a very important role in the northeast and even the whole country's economic.

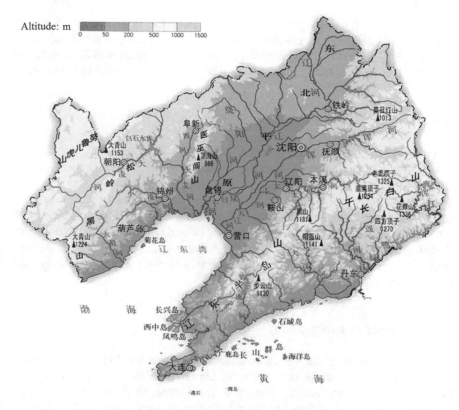

Fig. 3.5 The topographic and geomorphologic maps of Liao River basin

Administrative Region Division

Liao River Conservation Area from north to south is across four administrative city including Tieling city of Liaoning province, Shenyang city, Anshan city and Panjin city, 14 counties (districts), 68 townships (town) and involved 288 administrative villages in total. Among them, Shenyang has five administrative counties (district), 34 administrative township (town) and 143 administrative villages; Tieling city has 4 administrative counties, 16 administrative township (town) and 72 administrative villages; Anshan city has one administrative county, six administrative township (town) and 30 administrative villages; Panshan city has four administrative county (district), 12 administrative township, and 43 administrative villages.

Population Size

According to the investigation, Liao River Conservation Area has a population of about 298000, of which the population of Tieling city is 102000, the population of Shenyang is 118000, the population of Anshan is 43000 and the population of Panshan city is 35000. They are Han Chinese, Korean Chinese, Manchu and so on, of which Han Chinese accounts for 98% of the total population. And the population of conservation area is rather small, besides they are mainly in lateral of the dam.

The Current Situation and Structure of the Land Use

The total area of Liao River Conservation Area is 1869.2 km^2 and it overlaps the area of 829.2 km^2 with the national nature reserve—Shuangtaizi Estuary Natural Reserve. The main types of land use in Liao river channel in are farmland and vegetable greenhouses, river water, Oxbow Lake, natural wetlands, water conservancy facilities and residential areas, of which the proportion of farmland (paddy field, dry field and vegetable field), which accounts for about 41.22% and covers an area of 639.4 km^2. However, residential area is relatively small, which only account for 0.84%. Natural vegetation wetland accounts for 22.42%.The area of river water and shoaly land accounts for 28.70%.

The General Situation of Economy

Liao river flows through four main administrative city including Liaoning, Shenyang, Anshan and Panjin. And the mainstream of Liao river in Tieling city mainly flows through Changtu county, Kaiyuan city, Tieling county and Yinzhou district. Although agricultural proportion of Changtu County is relatively large and the output of grain is the highest per unit area, the income of per farmer is relatively small. However, Yinzhou district which is closer to Tieling city, the farmers' income is relatively high. Farmers' economic conditions in Shenyang are slightly better than farmers' in Tieling. The closer the farmers are to the city, the higher there income is, which is similar to Tieling. Liao river in Anshan district only flows through Taian County, so the output of per unit area yield in this district is lower than Liaoning's and Shenyang's. However, farmers' there overall net income is higher. Liao river in Panjin is called Shuangtaizi River and it flows through Panshan county, Dawa county, Shuangtaizi district and Xinglong district. The agriculture of Panjin district are mainly composed of rice and its output of per unit area is significantly higher than the previous three areas'. Besides, farmers' income there is the highest, which are influenced by the reeds and other sideline productions, as shown in Fig. 3.6.

The per capita net income of the farmers in the villages that Liao river flows through is lower than the other farmers', which shows the farmers' material life in the villages that Liao river flows through is not as good as the others' and economy that are only on the basis of the traditional agricultural need to be adjusted and cubic meters improved.

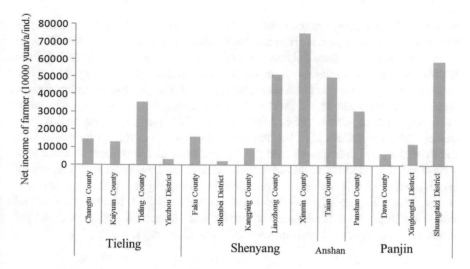

Fig. 3.6 The income of farmers in the cities along the Liao River Conversation Area

3.2.1.3   The Situation of Water Resources and Water Environment

The Situation of Water Resources

Liao river runoff mainly comes from rainfall, so the regional distribution and interannual changes of the runoff are consistent with the rainfall. Years of the average depth of the annual runoff declined from the southeast 300 mm to the northwest 50 mm. Interannual changes of the Liao river runoff is rather big. For example, the natural maximum runoff volume of the hydrological stations in Jiangkou, Tieling and Juliu River are 5.65 billion cubic meters, 9.47 billion cubic meters and 11.12 billion cubic meters respectively and their natural minimum runoff volume 262 million cubic meters, 707 million cubic meters and 805 million cubic meters respectively. And the ratio of the annual maximum runoff volume and annual minimum runoff volume are 21.6, 13.4 and 13.8 respectively. From years' average runoff's distribution, we can see that the annual runoff's distribution is uneven and the runoff volume in July and in August accounted for over 50% of the total annual runoff volume, which are larger than other months'.

Liao River Conservation Area has 38 reservoirs including seven large reservoirs, 10 medium-sized reservoirs and 21 small reservoirs and the total capacity of it is 2.785 billion cubic meters. All these reservoirs are mainly used to supplement the water environment of downstream river and ecological water's demand. What's more, these reservoirs guarantee the water usage of ecological environment in the dry seasons and guarantee water usage of the urban and rural residents, industry and agriculture.

The floods of Liao river basin are mainly because of rainstorm and because of the restriction of the rainstorm, 80–90% of floods occur in July and August during the flood season, especially from late July to the middle of August. During the non-flood season, the flood mainly occurs in the April or November. The flood period is divided as following: spring flood is from March to April; preflood period is from May to June; flood season is from July to September; the flood recession period is from October to November; freeze-up period is from December to February. The shape of the Liao river's flood includes unimodal type, bimodal type and multimodal type. Because the rainstorm lasted short and rainfall concentrated, the main tributaries such as Qing river, Chai river and for Pan river, Liu river and so on mainly flow through the mountains and hills. And because the speed of the confluence is fast, the flood rises and falls quickly and this kind of unimodal flood lasts no more than 7 days and main flood lasts within 3 days. Sometimes the rainstorm system occurs continuously and this makes the flood present bimodal type which generally lasts about 13 days and the interval of two peaks is 3–4 days. The time that multimodal floods last depends on the interval of flood peaks, the source of the flood and the rainstorm intensity. And floods usually last from 13–30 days.

The Situation of Water Environment

With the implementation of the strategy of rejuvenating old industrial bases in northeast China, the process of industrialization, urbanization and industrial clustering in Liao river basin is accelerated which severely polluted its water environment. By the 1990 s, Liao river had become one of the most polluted rivers and there was no live creature in it. And it also cannot be used for irrigation and drinking for people and livestock. And government departments at all levels have paid great attention to this kind of situation and they have taken the practical measures to improve the situation of Liao river basin according to Water Pollution Prevention and Control Law, the "ninth five-year plan" on the prevention of water pollution and the planning of 2010.

Compared with 1998', the deterioration degree of the river water quality in 2000 reduced. In 22 comparable sections, the proportion of sections that meets the standards of class III water quality increased from 4.5 to 13.6% and the proportion of sections that meets the standards of class V water quality reduced from 72.8 to 50%. But on the whole the water pollution is still serious. In 2000, of all the 27 sections that are monitored and met the standards of class III, IV and V water quality was 14.8%, 11.1% and 22.2% respectively, the proportion of class V water quality was 51.9% (Fig. 3.7).

By 2005, 87.1% of the sate-controlled sections did not meet the functional requirements and the key pollution indicators are COD, BOD5, ammonia nitrogen, potassium permanganate, volatile phenol, petroleum, and dissolved oxygen. And the water quality is poor in the dry season. During the Tenth Five-Year Plan period, the water of Liao river basin has no obvious changes and in 2005 the sections that belong to class V water quality account for 46.9% and decrease by 5% compared with the year of 2000. By 2009, the deterioration of Liao river basin water has been basically

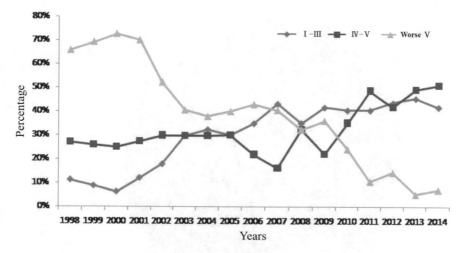

Fig. 3.7 The water quality change of Liao river year by year

restrained and the contamination risk of drinking water had been preventable and controllable. According to the standard of chemical oxygen demand (COD), Liao river had eliminated class V water quality and the goal of Tenth Five-Year Plan of Liao river's governance one year in advance. But the ammonia nitrogen of most parts of the river is still above of class V water standards and the pollution of tributary water is still very serious.

Liao River Conservation Area Administration was established in 2010 and it take responsibility to supervise the departments of the environmental protection, water conservancy, land and resources, traffic, agriculture, forestry, marine, fishery and so on and the construction of conservation area, which embodies the concept of integrated management and innovates management system.

From Fig. 3.8, we can see that between 2011 and 2014 the occurrence frequency of bad class V water body decreased from 11 to 1 and the occurrence frequency of bad class V water body decreased from 45 to 9. The occurrence frequency of class III water body and class IV water body increased from 9 to 22 and the occurrence frequency of class IV water body increased from 23 to 80 accordingly. And class IV water body became the main water body. According to the standards of water environment quality [4], Liao river was no longer heavily polluted at the end of 2012 and in 2013, 83% of tributaries met the standards of the water quality. We hope that in 2014 the sate-controlled sections achieve class IV water quality or above class IV water quality, especially the ammonia nitrogen meet the standards of water environment quality.

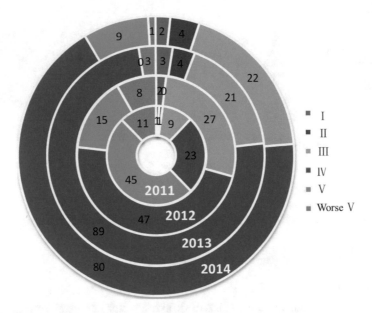

Fig. 3.8 The water quality change of Liao river year by year

3.2.2 The Current Situation of the Regional Ecological Environmental Quality

3.2.2.1 The General Situation of Liao River Ecological Environment

Since "the eleventh five-year plan", Liao river environment has got into the phase of the protection and restoration. Through the state and local governments' untiring efforts, the water quality of Liao river has got comprehensive improvement, ecological environment is gradually restored and ecological pattern nearly come into being. And the ecological corridor is built which is 538 km long and its area is 440 km² from Fude Dian to the mouth of Panjin. And the diversity of species in the conservation area has been significantly increased. For example, there have been 46 kinds of fishes, 6 kinds of amphibians and reptiles, 89 kinds of birds, 360 kinds of plants and 350 kinds of insects, which show the appearance of biologic chain in the conversation area. Meanwhile the vegetation coverage has been dramatically improved and the percentage of the vegetation coverage increased from 13% in 2010 to 81.3% in 2014. What's more, the percentage of the vegetation coverage in Binghe area even increased to more than 90%. In one word, ecosystem stability has been greatly restored. The kinds of sea calf gradually increased at the estuary of Liao river. And the knifefish began come back. The number of odontobutis obscurus and icefish significantly increased. An ecological corridor has come into being.

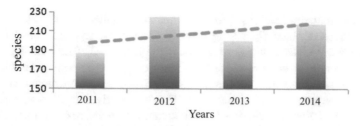

Fig. 3.9 The changes of plants from 2011 to 2014

3.2.2.2   The Biodiversity in Liao River Conservation Area

Vegetation

We choose to monitor plants in summer (July and August), because during this time the plants usually grow vigorously. We set 18 observation centers in the upstreams of the 11 rubber dams, Fude Dian in Tieling, Sanhe in Kangping, the wetland in Fan river, the reservoirs in Shifo Si, Daniu town in Taian county, Panshan, Liao river estuary and so on. And we use transects and quadrat sampling method to continuously monitor the species, number, height, coverage, density and frequency of ecological vegetation in the key observation centers (Fig. 3.9).

Monitoring results show that since the conversation area was built, the quantity of the vegetation increases in the trend of fluctuation. In 2012, the species of vegetation increase to 220 in a single observation. Although the species of vegetation decreased slightly in 2013, it increased again in 2014. The the species of vegetation that we observed increased to 360 in the 5 years. According to the disciplines of the ecosystem vegetation succession, the reason why the number of vegetation became the largest in 2012 is that after the ecological system was interfered, the number of composites and chenopodiaceae plants increased and later in the process of recovery vegetation gradually became dominant species and ecosystems kept stable. Based on historical data, the species of vegetation Liao river basin increased to over 1000 in the 70–80s of last century. Although Liao river vegetation recovers to a certain extent, there is still a long way to go to reconstruct the natural habitat.

Fish

According to the monitoring in 2010, we found that there were 24 kinds of fishes including crucian, minnow, topmouth culter, Chinese false gudgeon, abbottina liaoningensis, pseudorasbora parva, carp, chankaensis, rhodeus lighti, amur bitter-ling, huigobio chinssuensis, gobio rivuloides, Chinese hooksnout carp, megalobrama amblycephala, pseudogobio vaillanti, nemacheilus nudus, cobitis granocirendahl, catfish, ctenogobius girrinus, gobies, salanx ariakensis, medaka, haarder and chinese sleeper. In 2012, we found 20 kinds of fishes that belong to the 6 orders and 9 families and increased 5 kinds of fishes compared with the year of 2011. In 2013, we found 29 kinds of fished (including 1 hybrids) belonging to 6 orders and 9 families and increased 9 kinds of fishes compared with the year of 2012 including, Chinese false gudgeon, tinca, catfish, mirror carp, black carp, longnose gudgeon, snakehead, macropodus chinensis and rhodens sinensis. By 2014 there were more than 46 kinds of fished belonging to 6 orders and 10 families in Liao river. In 2014 we found more than 32 kinds of fishes (including 1 hybrids) in the monitoring centers and increased 3 kinds of fishes compared with the year of 2013 including gudgeon, pseudobagrus ussuriensis and goby. In 2012, there were 9 families and 20 kinds of fishes and increased 5 kinds of fishes compared with the year of 2011. And the appearance of odontobutis obscura and coilia ectenes indicated that the water quality in Liao river was further improved (Fig. 3.10).

According to the survey in 2009, there were 26 kinds of fishes in Liao river mainstream and after 6 years' governance and protection 20 kinds of fishes has returned. But the species of fishes in the Liao river still has a big gap with the data that Heilongjiang River fishery resources survey between 1979 and 1984 and articles said which written by Xie Yuhao [30] (Table 3.4).

Fig. 3.10 The changes of fishes from 2011 to 2014

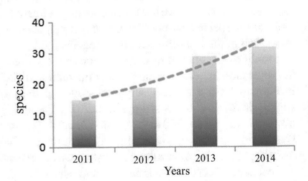

Table 3.4 Comparison between data in this survey and historical data

	Survey (1979–1984)	Literature in 1981	Survey in 1999
Species of fish	99	96	26
The number of family	23	23	8
Cyprinid	55 (55.6%)	53 (55.2%)	14 (53.8%)
Cobitidae fish	7 (7%)	8 (8.3%)	4 (15.3%)
Wei flying fish	4 (4%)	4 (4.2%)	1 (3.8%)
The rest family	33 (33.4%)	31 (32.3%)	7 (26.9%)
Typical freshwater fish	87	83	24
Anadromous fish	8	8	1
Brackish fishes	4	4	1
Inshore fishes	1	1	0

Fig. 3.11 The changes of birds from 2011 to 2014

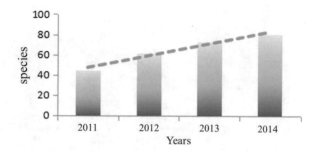

Birds

According to the survey in 2014, we mainly found 79 kinds of birds and the number of it doubled. And these birds were divided into 12 orders, 30 families and 49 genera. We found the First-Grade State Protection animal—oriental stork in Sutai river estuary and the wetland of Shifo Si, Second-Grade State Protection animal—Whooper Swan in the wetland of Mountain Qixing and Second-Grade State Protection animal— kestrels at the estuary of Fan River. Besides we also found Second-Grade State Protection animal—reef heron and many raptorial birds such as reef heron and falco amurebsis along the Liao river in 2014 (Fig. 3.11).

According to the data, there are 16 orders, 56 families and 340 species birds in Liao river basin in liaoning province, which account for 81.3% of all the known birds in Liaoning province. The area of Liao river Conversation Area is 1869.2 km². Now the birds we observed only account for 23.2% of the literature survey data. Although there are some errors in the observations, the habitats of the birds in Liao river Conversation Area remain to be further improved.

Mammals

In 2014 we found ten kinds of mammals in Liao river Conversation Area including the sewer rat, hare, yellow weasel, reed voles, the cat, leopard cat, harbor seal, cowfish, etc. From the perspective of the number, mice have the absolute advantages. Since the establishment of Liao river Conversation Area, the species of the mammals increased, but there is still a long way to go to reach the level of the 70–80s of the last century (Fig. 3.12).

According to the study on the survival environment of river creatures and the demand for land use at home and abroad, within the scope of the basin, the increasing usage of natural land is beneficial to maintain the biodiversity and ecological integrity. However, the increase of the construction land will reduce species and the number of it and damage the structure of river biological community. The loss of woodland and grassland in the basin will lead to the disappearance of the fish and macroinvertebrate. When the agricultural land is more than 30–50%, the habitat quality and biological integrity begin to decrease significantly. When the agricultural land is below 30–50%,

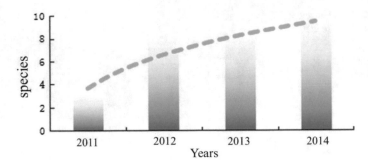

Fig. 3.12 The changes of mammals from 2011 to 2014

rivers can keep better biological integrity and habitat and once exceeding that figure, the river ecological integrity will change significantly. According to the foreign and domestic research results of the animal and plant habitat scale, when it is 50 m away from the river bank, we can better protect a few herbaceous plants such as cattail, reed and so on, invertebrates such as snails, mussels, and odontobutis, yellow catfish, silurus soldatovi and other precious fish and bufo gargarizans, rana nigromaculata, tree frog and other amphibians; when it is 100 m away from the bank, we can protect basket willow, honeysuckle, spiraea and other herbaceous plants and suffruticosa plants, provincial protected birds such as mallards and ruddy shelduck, reptiles such as snakes; when it is 200 m away from the bank, we can protect the plants such as iris lactea, carex enervis, setaria viridis and provincial protected birds such as Chinese merganser, morillon, peacebird; when it is 300 m away from the bank, we can protect the provincial and national protected animals such as tiger shrike, lanius schach, yellow weasel, sparrowhawk and hog badger; when it is 500 m away from the bank, we can protect provincial and national protected animals such as red crowned crane, black stork, grey crane, snowy owl, white bellied harrier, pied harrier, long eared owl, short eared owl, raccoon, leopard cat, polecat that are on behalf of the Liao river Basin. And it can also restore and protect the biological diversity in Liao river Conversation Area; when it is 1200 m away from the bank, we can create a natural landscape structure with the rich species and can protect national and provincial protected birds such as hawks, vultures, eagle owl, white crane, and red fox and so on.

Therefore, to accelerate the natural habitat restoration and improve the biodiversity and promote the stable development of ecosystem, we must further accelerate the construction of ecological restoration in Liao river Conversation Area, which will guarantee the health of the water environment and harmonious development of economy and society.

3.2.3 Ecological Restoration Technologies in Liao River Conversation Area

3.2.3.1 The Construction of the Artificial Wetland at the Tributary Inlet

The Construction and Technical Routes of the Artificial Wetland at the Tributary Inlet

The construction of the artificial wetland at the tributary inlet mainly includes the following aspects:

(1) The choice of the artificial wetland location. According to the contamination of the tributaries, we decide whether to build the artificial wetlands at the estuaries. If necessary, we then determine the placement and scope of wetlands according to the water quality and water quantity of the tributaries.
(2) The methods of constructing the artificial wetland. Considering the area of the estuaries, topography, hydrology, matrix, creatures and related factors, we get the related design parameters of the wetland construction and the construction arrangement. The area of artificial wetland of the estuaries depends on the pollution load and water volume in the dry season which lasts 5–7 days and 1 day in the flood period.
(3) Determine the water quality target. Aiming at meet the water quality standards of Liao river and considering functions of each estuary, we finally get the effluent quality of the artificial wetlands.
(4) The plant collocation in the artificial wetland. We should choose aquatic plants and helophytes. For the heavily polluted tributaries, we should choose reeds, cattails and the like that can help purify the environment. Meanwhile we can also make the most of the reeds, cattails and such economic crops to bring the economic benefits.

The Ecosystem Restoration of Estuarine Wetland

After the completion of the artificial wetland, we need to gradually restore estuary wetland ecosystem and specifically include the following aspects:

(1) The remediation of the underlying surface. After the buildup of the wetland, we begin the regulation of the underlying surfaces in wetlands and create the proper terrain for the plants.
(2) The recovery of hydrologic conditions. We restore the hydrology of estuarine wetlands mainly by raising water level with rubber dams. And when it is beyond the controlling of the rubber dams, we can use internal water-storing to regulate.
(3) The vegetation restoration. Vegetation is an important part of estuary wetland ecosystem and there are aquatic plants (water celery, etc.), marsh plants (sedges,

etc.) and hygrophytes (water smartweed, etc.) in 11 artificial wetlands. We should choose the plants that can purify the environment in the heavily polluted estuarine wetland, such as reed, etc. In the landscape estuary wetland, we choose plant landscape plants, such as water lilies, etc.

(4) Habitat restoration. Estuarine wetland habitats are mainly for birds. After the buildup of the artificial wetlands, the washland forms and then we can restore birds' habitats by natural closing.

(5) The hydrologic regulation and management. We should regularly monitor and control the water level changes in estuarine wetland. When the water level changes a lot, we need regulate it through the rubber dam. However, we can use internal adjustment to keep the water quantity.

(6) The ecological tourism management. We can develop ecological tourism around the landscape estuary wetland but we should prevent the pollution and protect wetland environment and functions.

3.2.3.2   The Restoration of the Pond Wetland

The Restoration and Technical Routes of the Pond Wetland

The construction of the pond wetland mainly includes the following aspects:

(1) The choice of pond wetland location. According to the distribution of sandpits, we choose concentration area of the sandpits and build pond wetland groups. The pond wetlands mainly concentrated in Mountain Mahu section of Qing river estuary.

(2) The method of constructing pond wetlands. According to the flow direction of Liao river, we regulate the underlying surfaces in the sandpit area. Considering the area, topography, hydrology, matrix and such related factors, we ascertain related design parameters of the wetland constructions and get construction layout.

(3) The plant collocation in pond wetland. We should choose indigenous plants and psammophytes that can purify the environment, such as campylotropis, sophora flavescens, scutellaria baicalensis and so on in pond wetlands and sandbars.

The Ecosystem Restoration of Pond Wetlands

The ecosystem restoration of pond wetlands includes the following aspects:

(1) The remediation of wetland underlying surface. After the connection of water system in the pond wetland, we need to regulate wetland underlying surface and form the different habitats for aquatic plants, marsh plants and hygrophytes.

(2) The vegetation restoration. The underlying surfaces of pond wetland are mainly made of the sand, so we had better choose psammophytes, such as sophora flavescens, erodium stephanianum, incarvillea sinensis and so on. In the heavily

polluted area, we should choose the plants that can purify the environment such as reeds, cattail and so on.

(3) The habitat restoration. The habitats that we restore are mainly for fish and amphibians. Combining the aquaculture, realize the integrity and biodiversity of of the food chain which is made of phytoplankton, zooplankton, small-sized fishes, big-sized fishes and so on.

(4) The hydrologic regulation and management. The controlling of water level in the pond wetland is very important. We can combine the rubber dam and water storage to control the water level.

3.2.3.3   The Natural Wetland Restoration of the Oxbow Lake

The Natural Wetland Restoration and Technical Routes of the Oxbow Lake

(1) The selection of the wetland location: According to the construction of rubber dams and the controlling plans of the water conservancy, we reduce the pollution of the natural wetlands in oxbow lake, enhance water self-purification ability and improve water quality, water conservation, flood detention and the diversity of landscape. Meanwhile, we choose the watercourses that have an advantage of area to carry out the natural wetlands restoration.

(2) The construction and recovery of wetland According to the original abandoned watercourses of the oxbow lake, we set the area of wetland central area and wetland activity area and carry out the proper slope restoration of the underlying surface in oxbow lake area. At the side of the new watercourses, we set water conservancy facilities to control the water conservancy according to the quantity of drainage and the demands of water in oxbow lake natural wetland. And we irrigate oxbow lake area by using rubber dams of water storage project to raise the water level and form stable ecological water and further form various habitats, including aquatic, marsh, wetland and meso conditions.

(3) The vegetation restoration: The introduction of aquatic plants and making them become the dominant species is the key to ecological restoration. We mainly choose the aboriginal plants of oxbow lake. According to the different flood depth and soil texture, we select different types of plants to form a stable plant habitat for wild animals, especially for fish and birds and make natural wetland ecological system of the oxbow lake has the function of self recovery. Technical routes are shown in Fig. 3.13.

The Ecosystem Restoration of Oxbow Lake Natural wetland

Aquatic ecosystem restoration is a whole process, so we need take the water, nutrient, soil, plants and all the main ecological factors into consideration. We reconstruct natural environment that are for the submerged plants, floating plants, emerging plants,

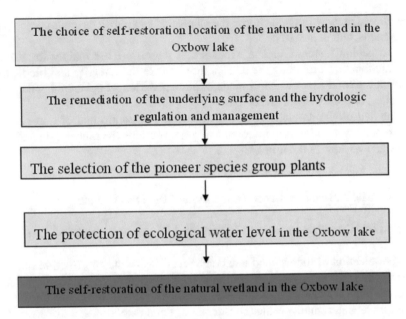

Fig. 3.13 The natural wetland restoration and technical routes of the oxbow lake

fish, birds and so on by modifying the underlying surfaces, maintaining ecological water level and introduce the appropriate pioneer species and dominant species. And then we make the oxbow lake ecological system into the stage of self recovery and natural succession and realize the integrity and biodiversity of the food chain which is made of big-sized fishes, small-sized fishes, zooplankton, and phytoplankton and so on, which completely restore the ecological functions and long-term self-sustaining of system structure in oxbow lake natural wetland.

Oxbow lake natural wetland system is connected with Liao river mainstream and forms the wetland network with estuarine wetland of the tributaries, pond wetland, and backwater natural wetland. And this kind of wetland ecosystem has the function of self recovery, which can strengthen the water self-purification ability, improve water quality, and develop its ecological functions of enhancing water self-purification ability and improving water quality, water conservation, flood detention and the diversity of landscape.

3.2.3.4 The Restoration of Backwater Natural Wetland

The Restoration and Technical Routes of Backwater Natural Wetland

We mainly adopts the methods of building artificial rubber dams, controlling wetland water level, returning farmland to forests and grasses and closing the wetland within the scope of the capacity of the reservoir. For the landslide area, we modify the

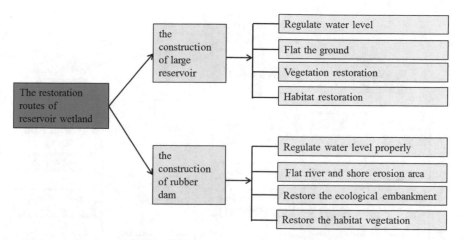

Fig. 3.14 The restoration and technical routes of backwater natural wetland

watercourse. In the sandpit area, we use the method of connecting water conservancy to keep river continuity.

Technical routes of backwater natural wetland are shown in Fig. 3.14.

The Ecosystem Restoration of Natural Wetland Group in Shifo Si Reservoir

Shifo Si natural wetland group is from ZuoXiao river artificial wetland to the submerged area of Shifo Si reservoir and its total area is about 25 km^2, including Chang river estuary natural wetland, Zuoxiao river artificial wetland, two oxbow lake wetlands, several pond wetland and Shifo Si reservoir wetland.

We plan to build estuarine artificial wetlands in Zuoxiao river artificial wetland and it covers an area of 1 km^2. And the type of estuarine artificial wetlands is subsurface-flow constructed wetland and the vegetations of it are mainly reeds and cattails. It is used to deal with the sewage that has been disposed in Shenyang city and help to purify the water.

Chang river estuary natural wetland belongs to the hybrid constructed wetland, which is made of oxbow lake wetland, Chang river estuary wetland and tidal flat wetlands. Its total area is about 6 km^2 and is mainly used for the establishment of Chang river estuary natural wetland. Through the natural recovery of the artificial watercourse, the natural wetland habitat comes into being, which can help purify the water and protect the biodiversity.

We mainly adopt the method of natural enclosing to restore the ecology in Oxbow lake wetland. And it can also forms a large number of biological habitats and provide the area which is independent of human society by controlling the water level, building sand heart, waters and other wetland types.

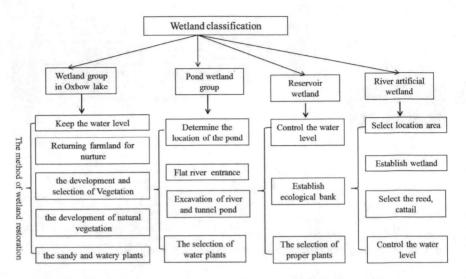

Fig. 3.15 The ecosystem restoration of natural wetland group in Shifo Si reservoir

We connect the Liao river with water surface of pond by the way of connecting water system, which forms its continuity. We mainly adopt natural enclosing to restore the vegetation. During dry season, it can Dian the water and in the rain season, it can form the bright water surface.

For the intercept of the dam, the water surface in the Shifo Si reservoir wetland is very big. We use the method of connecting water system to connect every wetland, especially with Liao river River. So when the water level increases, the water surface and some natural wetlands in submerged area are expanded. At the same time, we enclose the natural wetland to form multi types of wetland system.

Shifo Si reservoir controls the water level of Liao river from Tieling to Shifo Si and plays a very important role in the utilization of water resources, flood prevention and drought resisting. However, the ecological environment of the reservoir wetland is still deteriorating and people do not pay enough attention to the ecological environment protection. There are various types of wetlands in the reservoir wetland, so we should adopt different methods to restore the wetlands according to different types of wetlands. And technical routes are shown in Fig. 3.15.

3.2.3.5 The Ecological Restoration in Riparian Area

According to the principles and technologies of recovery and considering the river management, the restoration and technical routes in mainstream area are shown in Fig. 3.16.

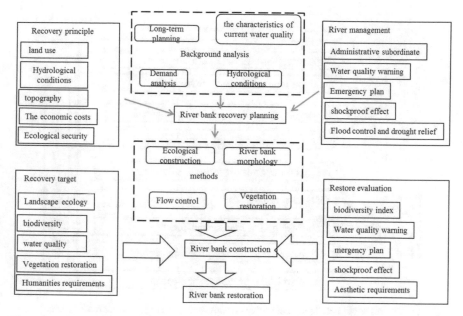

Fig. 3.16   The restoration and technical routes in mainstream area

3.2.3.6   The Vegetation Restoration of Main Streams

The vegetation restoration of Liao river includes the enclosure of the original plants and vegetation reconstruction. We can define the relevant areas and take the protective measures to make it natural restoration in the area where the habitat is not destroyed and there are not human activities. For the damaged vegetation areas that are caused by man-made interference, we should choose reconstruction. (Figure 3.17).

For the selection of specific plants, according to the changes of habitat conditions and the gradient of water level, it is suggested to adopt the following methods:

(1) The vegetations in the bright water surface. When the water level is over 1 m, we mainly choose submerged plants and floating plants including lotus, eichhornia crassipes, duckweed, azolla, pistia stratiotes, salvinia natans, etc.
(2) When the water level is between 0.5 and 1 m, we had better choose emergent plants, submerged plants and floating plants including cattail, cane shoots, wild celery, hornwort, watermifoil, hydrilla, eichhornia crassipes, duckweed, azolla, pistia stratiotes, salvinia natans, etc.
(3) When the water level is between 0 and 1 m, we had better choose helophytes including glycine ussuriensis, polygonum hydropiper, persicaria orientalis, rorippa islandica, adenocaulon himalaicum, alisma orientale, carex neurocarpa, chufa, water chestnut and so on.

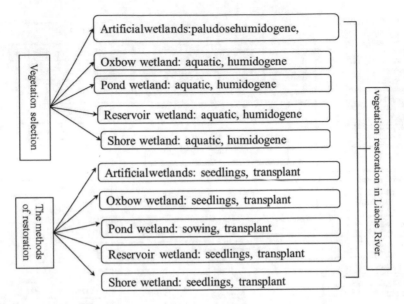

Fig. 3.17 The restoration and technical routes of the vegetation in Liao river mainstream

(4) Hygrophilous vegetation has no obvious water surface and the oil is rich in water. So in this case we mainly choose aboriginal plants including reed, polygonum hydropiper, persicaria orientalis, paris verticillata, glycine ussuriensis, chelidoniut rifolium lupinaster, impatiens noli-tangere, hypericum ascyron, viola acuminate, primula maximowiczii, mint, isodon japonica var.galaucocalyx, stachys chinensis, rubia sylvatica, garden balsam stem, dipsacus japonicus, schizopepon, convolvulus chinensis, leontopodium alpinum and so on.

(5) Sandy vegetation. After the buildup of above various types of wetlands, the water level raises and forms Shaxinzhou, Hexinzhou and so on. So in this case, we had better choose hygrophytes including campylotropis macrocarpa, sophora flavescens, geraniaceae, hibiscus trionum, stellera chamaejasme, isodon japonica var.galaucocalyx, scutellaria baicalensis, stachys chinensis, euphrasia, phtheirospermum japonicum, incarvillea, hieracium umbellatum, convolvulus chinensis, ajuga lupulina and so on.

(6) Along the embankment, we had better choose shrubs such as schisandrachinensis, glycine ussuriensis, delphinium grandiflorum, menispermum dauricum, hibiscus trionumwild, hippophae rhamnoides, iris tectorum and so on.

3.2.3.7   The Restoration of Liao River Natural Habitat

On the base of the analysis of the current situation of river natural habitat, the restoration of Liao river natural habitat focuses on the restoration of bird habitats. After

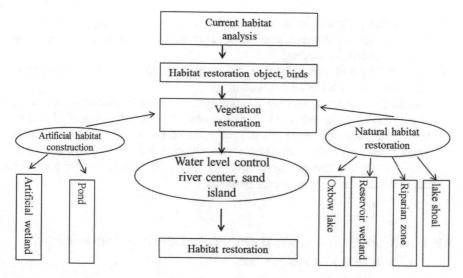

Fig. 3.18 The restoration and technical routes of the natural wetland in Liao river mainstream

analysis of bird species and their migration patterns and considering the construction of estuarine artificial wetland and pond wetland artificial wetland and the restoration of oxbow lake natural wetland and other natural wetlands, we can control the water level and create suitable environment for birds, restore birds' natural habitat and proposes the corresponding measures to maintain it. The specific technical routes are shown in Fig. 3.18.

3.2.3.8   The Agricultural Non-point Source Pollution Control of River Beaches

(1) The farmland non-point source pollution control: The existing farmland area of Liao river is 770.5 km^2. we plan to construct the corresponding optimal management measures by adjusting the agricultural structure, controlling growth of farmland area and encourage sprinkler irrigation, drip irrigation and other water-saving irrigation methods to improve the efficiency of irrigation water. At the same time we take corresponding policies to limit the application dose of fertilizer and pesticides, especially nitrogenous fertilizer and organochlorine pesticides and encourage sewage recycling and farmyard manure.

(2) The sewage control in residential area: There are 37 settlements in Liao river and they are mainly located in upper and middle part of the main stream. According to the successful experience of sewage treatment technology in local villages and towns, we plan to construct scattered sewage treatment facilities that are efficient and easily operated in villages and then collect the scattered residential sewage to dispose. We also gather the daily life garbage and transport outside.

(3) The tourism pollution control: We should strengthen the management and integrated control on the stream segments that are used to develop the tourism. (Such as Fan river estuary section).

(4) The breeding pollution control of livestock and poultry: Forbid breeding livestock and poultry within the watercourse.

(5) The dispose of the riparian zone: According to the ecological restoration planning of the riparian zone, we will restore the ecosystem of both sides of Liao river about 500 m and gradually form a natural river. We can utilize the construction of riparian zone to control non-point source pollution and purify the sewage from non-point sources and guarantee the water quality by intercepting and deal with the agricultural non-point source pollutants.

3.2.4   The Pollution Control and Water Quality Purification Technology of Liao River

3.2.4.1   The Pollution Control and Water Quality Purification Technology in the Headwater Region of Liao River

The headwater region of Liao river begins with the confluence of East Liao river and West Liao river and ends with the drop-down of Bazi River, Li River and Gong River (Fig. 3.19). The construction project of pollution control and water quality purification in the headwater region of Liao river Liao river is located in Fude Dian the drop-down section of San River and it is about 38 km long.

The Pollution Control at the Confluence of East Liao river and West Liao River (Fude Dian)

The confluence area of East Liao river and West Liao river (Fude Dian) emphasizes on the pollution control and the improvement of water environment. Considering the construction of wetland, the pollution control and project of vegetation restoration, the total area of it is about 4.5 km². And the area in East Liao river and West Liao river is 1.3 million square kilometers and wetland construction project covers an area of 2.2 km².

We improve the water quality of East Liao river and West Liao river and make water quality discharged of Fude Dian meet the standards of class V water quality, that is to say COD drop below 40 mg/L and ammonia nitrogen drop below 2 mg/L.

We build the submerged dam at the confluence of East Liao river and West Liao river-Fude Dian to form the backwater wetlands and control the water level and water quantity in the front of the dam, which will provide water for enclosing in the upstream area wetland.

Fig. 3.19 The pollution control and water quality purification technology in the headwater region of Liao river

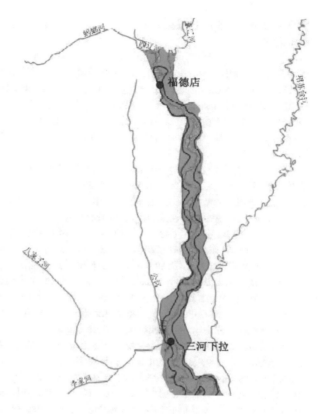

We clean up 104000 m³ of mud that deposit in the Liao river estuary. And the mud we clean is about 2.6 km and 0.4 m deep. We lay a foundation for improving water quality by dredging the pollutants.

The artificial wetland at the estuary of East Liao river and West Liao river is about 2–3 km long and covers an area of 4.5 km² and belongs to oxbow lake wetland. We mainly give priority to the native vegetations. At the same time we also choose some aquatic vegetation that have the function of physical block, reduce the resuspension of sediments and absorb the nutrients in the water and sediments. The retention time of waterpower in the wetland is about five days and the effluent COD drop to below 40 mg/L and the ammonia nitrogen drop to under 2 mg/L.

We renovate the ecology of Liao river main stream in the demonstration area and 6 km long bank slope of East Liao river and West Liao river and afforest the embankments by planting the aboriginal plants. The beachland ecological restoration in the demonstration area of confluence of East Liao river and West Liao river is based on natural enclosing and recover riparian wetland.

The Water Quality Purification from Fude Dian to the Drop-Down Section of San River

The water quality purification project from Fude Dian to the drop-down section of San River is based on the construction of riverside wetland and shoal wetland construction of riparian wetland and shoal wetland. And considering the riparian vegetation restoration and riverbank protection engineering construction, wetland project construction covers a total area of about 2.85 km^2 and vegetation buffer zone is about 22 km.

The non-point source pollution control and water quality purification engineering from Fude Dian to the drop-down section of San River include the ecological dredging in the key watercourses, beach wetland construction and pollution control engineering of riparian vegetation buffer zone.

We choose large and even bottomland from Fude Dian to the drop-down section of San River to build the shallow water wetland and 5 m long canal which make Liao river pour into it. However, the depth of it depends on the depth of the current watercourse and then forms the fluent watersystem in riverine wetland. After the completion of them, they can guarantee the smooth flow of the water in the canal in average water season and stop the water flow in the dry season. And this will further decrease the pollution and improve water quality.

We build a 22 km long vegetation buffer zone along the Liao river and use the vegetations of it to intercept and purify the pollution sources that pour in to the watercourses. Comprehensively considering various goals, the buffer zone along both sides of the river should be at least 10 m wide and choose shrubs and herbs that are tolerant to drought and moisture. Near the coast, we choose willows which are commonly seen there to construct the protection forest.

The Water Quality Purification in the Key Area of Drop-Down Section of San River

The water ecology demonstration area in the drop-down section of San River focuses on the estuary wetland pollution control and water ecology restoration and combine construction of rubber submerged dam, ecological governance of bank slope and so on. And the demonstration area covers a total area of about 3.25 km^2 and the area of wetland construction project is 1.2 km^2 and ecological restoration area covers 2.05 km^2.

The estuary wetland engineering in the drop-down section of San River adopt free flow surface constructed wetland and build buffer zone in the estuary wetland, which can purify the water and promote the water ecological restoration. The wetland vegetation is based on reeds and cattails. And the retention time of waterpower is five days and effluent COD drop to below 40 mg/L and ammonia nitrogen drop to below 4 mg/L.

We decide to build a submerged rubber dam in the Liao river main river channel to guarantee the water storage capacity and water level. And the dam is 2 m high

Fig. 3.20 The sketch map of the ecological restoration project of Shenkang highway bridges

and 60 m long. The backwater length of it is 7 km and the area of ecological water surface is about 500000 m^2 and the storage capacity is about 100 m^3.

3.2.4.2   The Pollution Control and Water Quality Purification Technology of Mountain Qixing

The Pollution Control of Shenkang Highway Bridges

The ecological restoration project of Shenkang highway bridges is located in the downstream of Shifo Si reservoir and is beneath Shenkang highway bridges (Fig. 3.20).

The Riverway Desilting and Ecological Riverway

(1) The ecological desilting of the riverway; We adopt mechanical dredging and human assistance to deal with the pollutants that deposited in the bottom of river.
(2) The modification of the riparian zone: We turn the riverbed that river flow through into the trapezoidal slope and the slope ratio is 1–2.5. And we make the sides of the riparian zone even and solid.
(3) The ecological revetment: We choose argillaceous revetment moving towards two sides. It is divided into two parts according to the constant level. When water level is above the constant level, we choose the lawn vegetation, hygrophytes

and so on. When water level is below the constant level, we choose submerged plants, such as the stem of cattail, etc.

Pond Wetlands

(1) The expansion of the original ponds: On the basis of keeping the landscape of original pond wetlands, we expand the original ponds and enlarge their water surface area. And we choose aboriginal plants in the wetland including aquatic vegetations, marsh vegetations, humidogene vegetations and mesophytes.
(2) Build the new ponds: On the basis of natural conditions of the riverbank, we make use of the big puddles that the sand excavation of left bank formed to build the new ponds and then make the water flow into the new ponds. And then we utilize the small tributaries of right bank to dig the ponds and make the water flow into the ponds forming the new pond wetland groups. We choose aboriginal plants in the wetland including aquatic vegetations, marsh vegetations, humidogene vegetations and mesophytes.
(3) The connected canals: We build the connected canals to make the river water flow between ponds and between ponds and watercourses. After the completion of them, they can guarantee the smooth flow of the water in the canal in average water season and stop the water flow in the dry season.

The bottom of the connected canals is made of pebble gravels. And the pebble gravels overspread the canals in a certain way and they are about 0.5 m thick. The project design is shown in Fig. 3.21.

The reconstruction of plant community: The construction project is located in Shenkang highway, so we had better choose aboriginal plants that have the ornamental value. Floating plants include water lily, nuphar pumila, Gorgon fruit, etc.; emergent plants include cattail, scirpus tabernaemontani, etc.; Hygrophytes include reeds, carex, etc.; submerged plants include waterweed, hornwort, etc.; silicicoles include seabuckthorn, calligonum mongolicum, etc. We choose shrubs or perennial herbs in the waterless area.

The Pollution Control of Pond Wetland Groups

According to the natural conditions from Shifo Si reservoir to Mountain Qixing, we choose the proper places to construct ponds or pond wetland groups. The construction of the pond wetland groups increase water volume, decreases the influence of non-point source pollution on water quality along the river and maintain water quality (Fig. 3.22).

Fig. 3.21   Gravel connected channel

Fig. 3.22   The sketch map of pond wetland location

The Water Quality Purification of Rubber Dam Backwater Wetland in Mountain Qixing

The rubber dam is near Mountain Qixing and it is 2.5 m high and 164 m long. Its backwater section is 7.9 km long backwater and the surface area is 1500 mu. Its storage capacity is 2 million cubic meters (Fig. 3.23).

The general layout of the engineering includes the establishment of the ecological rivers, the wetland natural restoration and returning farmland to forest land and grass land. For the serious collapsed areas and soil erosion areas, we main restore their edges and plant annual grasses to solid dam and establish ecological embankment. According to natural conditions, we choose the suitable places to establish pond wetlands, purify water quality and make the water quality of rubber dam backwater wetland meet the standards of class IV water quality. The restoration technology route of rubber dam backwater wetland is shown in Fig. 3.24.

Fig. 3.23 The engineering location map of the water quality purification of rubber dam backwater wetland in Mountain Qixing

Fig. 3.24 The restoration technology route of rubber dam backwater wetland

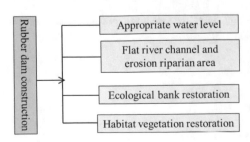

(1) Regulate the water level: We should regulate the water level to control the water level of artificial wetland. The rubber dam of Mountain Qixing is about 2.5 m high, so the water level that we regulate is generally no more than 2 m.

(2) Level the watercourse and the edges of the erosion area: We level the corresponding land and solid the cement earth near the dam. We conduct the wetland natural restoration and return the farmland to forestland and grass land. For the serious collapsed areas and soil erosion areas, we main restore their edges and plant annual grasses to solid dam.

(3) The restoration of ecological embankments: We turn the original flinty revetment into the ecological revetment and argillaceous revetment spreading towards two sides. It is divided into two parts according to the constant level. When water level is above the constant level, we choose the lawn vegetation, hygrophytes and so on. When water level is below the constant level, we choose submerged plants, such as the stem of cattail, etc.

(4) The restoration of habitat vegetations: We take measure of the natural restoration in the submerged area and return farmland to forest land and grass land. We plant emergent plants in the shoal water zone including reeds, cattail and so on. We choose shrubs or perennial herbs in the waterless area and silicicoles in the sandy soil area such as seabuckthorn, calligonum mongolicum, etc.

3.2.4.3   The Pollution Control and Water Quality Purification Technology of the New Lines

The Pollution Control of Pond Wetland Groups in Hada Railway Bridge

The project is set in the downstream of Hada Railway Bridge and located in DiaoBingshan city Tieling city, Liaoning province. The project location is shown in Fig. 3.25. According to the conditions of soil, topography and other natural factors and infrastructures, we decide to construct pond wetland groups on the left bank to regulate water resources, conserve the underground water, control water pollution and improve water eco-environment.

The construction project mainly includes the following aspects:

(1) Change and expand the existing pond wetlands. We should make full use of the topographic conditions of original natural ponds, shoaly land and so on to protect the original wetlands. For the small pond wetlands, we mainly restore the wetlands by the way of modification and expansion and plant aquatic plants, trees and flowers to form lotus ponds and reed ponds.

(2) Excavate new pond wetland. We can construct the artificial pond wetlands on the basis of big puddles that dredging and sand excavation left. This not only develops the local economy but also increases the area of water storage, regulates the flood peak and reduces the pressure of the downstream flood control. At the same time, we can also make use of the ponds that excavation formed to develop

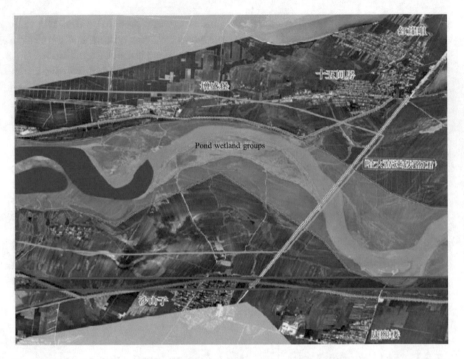

Fig. 3.25 The location map of pond wetland groups

ecological tourism and ecological livestock and poultry breeding and improve the water environment and increase people's income.

(3) Connect the river ponds. Ponds connect with each other through the connected canals. We can also plant shrubs (or perennial herbs) and hydrophilic plants such as calamus, reeds and so on to control the pollution, improve the water quality, prevent wind and stabilize sand and river regime.

The Pollution Control of Shelter Forest at the Bank of Hada Railway Bridge

The project is at the right bank of Hada Railway Bridge and 2 km long including green belts, slope and the platforms. Considering the flood control engineering construction, we afforest the right bank by plant cultivation and build pollution protection forest to improve the water quality. The construction of ecological environment along the river can also solid dams and embankments, prevent water loss and guarantee the safety of railway bridge.

The construction includes the following aspects:

(1) Renovate the existing embankments. It includes the levee regulation, the restoration of bank slope, bank protection and so on. We should make the bottomland

at the levee crown even. At the edge of the levees embankments, we choose the plants that have well developed roots and can solid earth and near the water system we grow aquatic plants.

(2) Plant trees near the levees and build shelter forests to control the pollution. Increase the vegetation coverage by planting trees and grass such as evergreen and deciduous plants, coniferous plants and broad-leaved plants, arbors and shrubs, flowering plants and foliage plants. Build shelter forests to control the pollution and the total area of it is about 50000 m^2 and the shelter forests also play a very important role in increasing the diversity of landscape and provide habitats and hidden places for birds and beasts.

Maintain the Water Quality of Ecological Water in Shashanzi

The project mainly aims for the bottomlands near the Shashanzi and constructs a man-made lake-Shashanzi ecological water storage lake between Road Guanli and the main levee (Fig. 3.26). And the water area is 1800 mu (1.2 million square kilometers). The maximum excavation depth of it is 4.45 m and the average excavation depth is 2 m. The water of it is 1.5 m deep. The boundary of the artificial lake is 50–80 m away from Road Guanli. The revetment of artificial lake is ecological embankment and its slope ration is 1–3.

The plant pollution control engineering includes aquatic plants area, bank protection forest belt and natural recovery area.

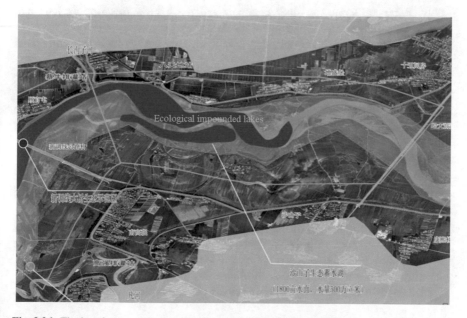

Fig. 3.26 The location map of ecological impounded lake in Shashanzi

We construct wetland aquatic plants area within 5 m scale near the lake bank and plant emergent plants and hygrophytes (reeds and cattails) in the low-lying wetland. We build the bank protection forest belt within 5–10 m scale near the lake bank and choose the land where the aquatic plant can not grow well as natural recovery area to make the wild plants grow naturally and restore the natural ecology. At the same time, we also plant psammophytes such as campylotropis macrocarpa, sophora flavescens, etc.

The Rubber Dam Water Quality Purification Under the New Line Highway Bridge

The rubber dam that is newly built under the new line highway bridge makes its upstream form a 6.9 km backwater section. And then we build the Shaxinzhou in the backwater wetlands to provide a good habitat for animals and plants, which plays a very important role in restoring biodiversity and protecting the riverine wetlands. This engineering aims at excavating artificial lake within the bottomland of the two 1.2 km sides of the new line highway bridge at the left bank of the rubber dam. The ecological engineering not only effectively restrains soil and water loss but also has the function of water quality purification (Fig. 3.27).

(1) The engineering design of water quality purification in the backwater wetlands is at the left bank of the upstream of rubber dam under the new line highway bridge and it mainly includes the following aspects:

   1. Renovate some the river channels and level the corresponding land and consolidate cement earth near the dam;
   2. Connect the water conservancy to keep the river continuity in the sandy pond area;
   3. Control wetland water level and connect the water system in the pond wetland to keep river continuity;
   4. Carry out the natural wetlands restoration in the flooded area and return farmland to forest land and grass land. Enclose the wetland within the capacity of the reservoir area;
   5. Plant hydrophilic plants such as calamus, reed and so on in the phytal zone, shoal area, wetland and water edge and plant shrubs (or perennial herbs) near the bank slope and bottomland to improve water quality, prevent the wind and stabilize sand and river regime.

(2) The engineering design of maintaining ecological water quality in the artificial lake. We construct artificial wetlands inside the bottomlands along the two side of the new line highway bridge at left bank of the rubber dam and dig two artificial lakes that connect each other closely. And each artificial lake has a crescent artificial island, which makes the ecological landscape more beautiful.

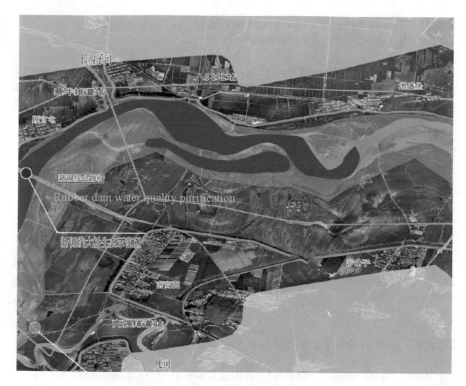

Fig. 3.27 The location map of the rubber dam water quality purification under the new line highway bridge

The Pollution Control in Changgouzi Estuary Wetlands

This project aims to purify the water quality in Changgouzi estuary area to alleviate the pollution load of Liao river and make the water quality that flow into Liao river meet the standards of class IV.

(1) The pollution intercept and purification engineering of riparian vegetations. The project makes full use of the river bank to build the 5 km long riparian vegetation buffer belt to intercept and purify the river pollutants. The buffer belt is 50 m wide and is made of shrubs and herbs. Near the shoreline, we plant willows that are locally commonly seen to build the shelter forest.

(2) The pollution control engineering in the artificial wetlands. According to the characteristics of the project, we build the surface flow constructed wetland in Changgouzi estuary area that is 100 m wide, 0.8 m deep and 4 kg long and its wastewater treatment capacity is about 40000 m³/d. After the completion of this artificial wetland, its effluent water quality will meet the standards of class IV.

3.2.4.4   The Pollution Control and Water Quality Purification Technology in Yubaotai

Of Liao river where river crossing under 0.8 km to yu bao platform bridge between 3.6 km about 13 km river to carry on the comprehensive improvement, in order to achieve the Liao river where—yu treasure of river water quality and river landscape obviously improved, and ensure the Liao river yu Po station water COD concentration is less than 30 mg/L, ammonia nitrogen concentration is less than 1.5 mg/L. Major construction projects including 13 km section of ecological river restoration project, where the river—yu bao section wetland network construction project, yu bao comprehensive resistance pollution control engineering, etc.

The Restoration of Ecological Watercourses

The restoration scope is between Juliu river rubber dam of the Liao river and Yubaotai Bridge and it is about 13 km long and 200 m wide and its channel gradient ration is 0.19 o/oo.

The Channel Remediation Design

(1) When the elevation is about 3.10–4.0 m, the bank is the argillaceous revetment. And then make its lower part solid and even and covered with planting soil to plant lawn vegetation. The xylophyta is commonly seen there.
(2) When the elevation is about 2.23–3.10 m and the water level is normal, the bank is still argillaceous revetment. And then make its lower part solid and even and covered with planting soil to plant hygrophytes and marsh plants to protect the bank.
(3) When the elevation is below 2.23 m, the bank is the ripped-rock revetment. In the bottom of the water, we plant submerged plants that can adapt to local conditions and grow rapidly such as waterweed, potamogeton crispus, and hornwort and so on. Then plant submerged plants along the banks of the river and it is about 1 m wide. Form the community long-term support mechanism through the asexual reproduction.

The Riparian Vegetation Buffer Belt

Plant shrubs and herbs that locally commonly seen to build the 20 m wide riparian vegetation buffer belt. Near the shoreline, we plant salix integra that are locally commonly seen to build the shelter forest.

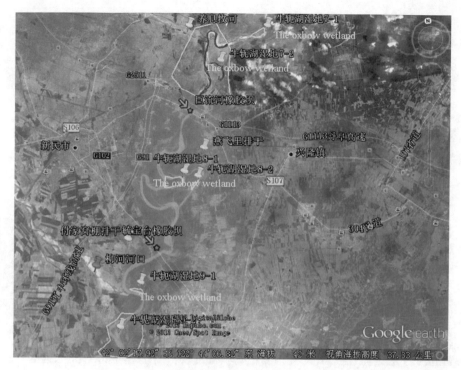

Fig. 3.28 The planning map of oxbow lake between Juliuhe rubber dam and Yubaotai rubber dam

The Purification of Oxbow Lake Wetland

The oxbow lake between Juliu river rubber dam and Yubaotai rubber dam is very typical (Fig. 3.28). The distance between two rubber dams is only 11 km and its water control conditions are good. So we plan to construct the large oxbow lake natural wetland which is about 36.93 km² through interfacial modification, vegetation restoration and construction of water conservancy. There forms many kinds of habitats including clear water area, deep water area, shallow water area, wetland, marsh wetland and so on. We can restore the aquatic animal groups such as fish and shrimp through the aquatic vegetation restoration and form the large oxbow lake natural wetland that has complete food chain, stable system and self-recovery ability.

3.2.4.5 The Pollution Control and Water Quality Purification Technology in Dazhangqiao Bridge Area

We mainly comprehensively renovate channel between Dazhangqiao Bridge and Hongmiaozi to improve its water quality and landscape. And it mainly includes river channel restoration project that is 10.6 km long, the water quality purification in Dazhangqiao bridge area and the pollution control in Hongmiaozi area.

The Ecological River Restoration

The restoration scope is the key river sections between Dazhangqiao Bridge and Hongmiaozi and it is about 10.6 km long. It mainly includes the cleaning of the river channels, construction of ecological revetments and riparian vegetation buffer belts.

The Remediation of River Channel

This key river mouth take Shi Yu slope protection technology, in the 1.06 km long on both sides of the slope protection renovation, slope ratio of 1:2.5. In flood elevation of 3.20 m above the line cutting stone material assembled, to ensure the stability of slope protection and safety.Upper with frame and wooden casing, within the framework of gravel and pebbles are embedded, took advantage of the gap between the gravel and pebbles slope protection plant. Slope surface of grass planting landscape in general.Flood level using gravel under revetment in fixed role at the same time using the attached to the gravel on the microbial decomposition of pollutants to water bodies to achieve the purpose of purifying water body. The measure convergence is hydrophilic, landscape, water purification function of ecological revetment structure.

The Riparian Vegetation Buffer Belt

Plant shrubs and herbs that locally commonly seen to build the 10 m wide riparian vegetation buffer belt. Near the shoreline, we plant willows that are locally commonly seen to build the shelter forest.

The Pollution Control in Dazhangqiao Bridge Area

The project is located in the downstream bottomland of Dazhangqiao Bridge and is about 60 m away from the artificial lake. It is between the right bank of Guanli Road and the main dam and its water area is 340000 m^2, the water perimeter is 2500 m, the altitude of lake bottom is 5.65 m, the maximum excavation depth is 2.95 m and the average excavation depth is 2.25 m. The water is 1.2 m deep and the boundary of the artificial lake is over 50 m away from Guanli Road to guarantee the safety of flood control dike. This project should include the water drainage engineering: we connect open channel with the artificial lake and build the rolled dam in the downstream of the artificial lake to adjust the lake water level and guarantee the safe operation of the rubber dam. Use ecological revetment and the excavation gradient of the lake is 1–3.

The Pollution Control in Hongmiaozi Area

The project focuses on the under-utilization of the water resource and ecological wetland restoration. It is 5.5 km long, 500 m wide on average and the total area of it is 335.25 m², of which the land area is 328.91 m² and water area is 6.34 m².

3.2.5 The Artificial Wetland Construction Technology of the Important Tributary Estuaries

3.2.5.1 The Ecological Comprehensive Demonstration Area in Qing River

Qing river is located in Kaiyuan city and its water column is controlled by the upstream Qing river reservoir. In the dry season, the water-carrying capacity is 10 m³/s and the depth of it is 1.5 m; in the rain season, the water-carrying capacity is 30 m³/s and the depth of it is 5 m. Qing river flow through Kaiyuan city and the industrial wastewater and domestic sewage that have been dealt with pour into it. So in the dry season, the water quality is rather poor and the main pollution factor is ammonia nitrogen; in the rain season, the water quality is below class V.

According to the water quality characteristics, regional conditions and its components, we launch the pollution control and water quality purification project at the estuary of Qing river on the basis of dredging the river channel. And we also construct the pollution control and water quality purification project which is 4.5 km long and covers 2 m² in the confluence area of Qing river and Liao river. The project asks us to plant the different kinds of plants in different areas such as fore-reservoir, ecological river channel, pond wetlands and estuarine wetland. On the basis of pollution control and expanding water environmental capacity, we further purify the water quality and comprehensively improve the estuary water environment and ensure that estuary water quality meet the standards of class IV surface water (Fig. 3.29).

3.2.5.2 The Comprehensive Ecological Demonstration Area in Fan River

The project aims to build the river system network, artificial wetland purification function areas, natural wetland restoration and reconstruction and water ecological restoration to improve the water environmental capacity of Fan River, strengthen the water self-purification ability and restore the water ecological restoration. The major construction projects include the following aspects: (1) broaden the water surface of Fan River and build the flap gate which is 1.5 km away from the entry port. Between the flap gate and the entry port section, we lay the gravel bed; (2) connect the original river channel to form the estuarine branch wetlands within the area of 5.8 m² at the

Fig. 3.29 The region of pollution control and water purification at the estuary of Qing River

entry port; (3) construct the estuarine branch wetlands; (4) restore and reconstruct the natural wetland; (5) build and restore the habitat in Hexinzhou (Fig. 3.30).

3.2.5.3 The Comprehensive Ecological Demonstration Area in Liu River

Liu River is located in Liuhegou town, Xinmin city in Shenyang city. It is 122 km long and about 300 m wide. It flows into Liao river in the Shaoguo village. Both sides of the river are arable land and the water quality is poor. COD is 40 mg/L and BOD is 14 mg/L.

According to the water environment and water ecology characteristics in Liu River estuary area, we dredge the river bank and set the pond protection project to eliminate endogenous pollution and enlarge the water surface. On the basis of improving the water environmental capacity, we construct shallow wetland water quality purification project, oxbow lake wetland ecological restoration project, build Liu River estuarine wetland group to control the pollution, purify the water quality and improve the regional environmental capacity and water self-purification ability.

Fig. 3.30 The region of pollution control and water purification at the estuary of Liu River

The major construction projects include the following aspects: (1) dredge the 3 km river segment at the estuary of Liu River, build the ecological river and revetment and afforest river embankment; (2) construct the estuarine pond wetlands, the river water diversion in pond wetlands and aquatic plant community; (3) The water quality purification and the ecological restoration at the estuary (Fig. 3.31).

3.2.5.4 Junction Entrance to Chang River

Wanquan River is located in Shenyang city and it is 32 km long. The river course is about 50 m wide and the river is 12–15 m wide. There are urban sewage treatment plants and sewage outfalls along Chang River. When Liao river River rises, the water volume near Chang River estuary and the upstream area become larger and a lot of trees and the crops along the riverbank are submerged. The water-carrying capacity

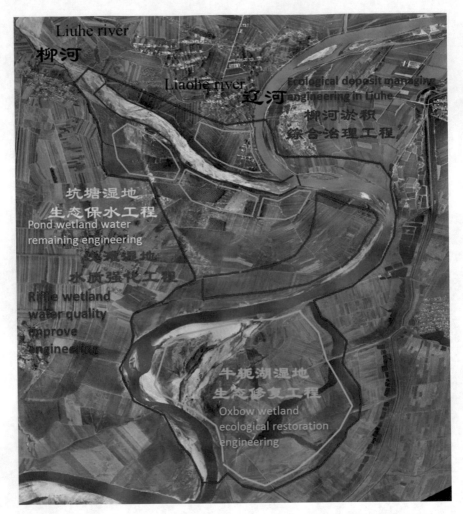

Fig. 3.31 The region of pollution control and water purification at the estuary of Liu River

is usually 3–8 m^3/s but most of the water belong to bad class V. The main exceeding standard factor is ammonia nitrogen (1–5 times) and COD is 20 mg/L.

The project comprehensively renovates Chang River sections that flow into Liao river and is 4 km long to improve the water quality and river landscape. It mainly includes the following aspects: (1) the construction of ecological fore-reservoir; (2) the dredging of 4 km river sediment; (3) the construction of 4 km ecological river course con; (4) the construction of artificial wetland that covers an area of 4.7 km^2 (Fig. 3.32)

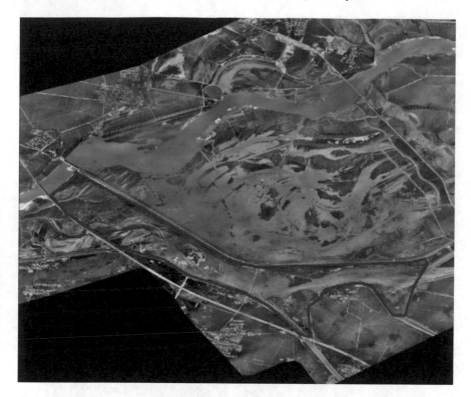

Fig. 3.32 Junction entrance to Chang River

3.2.5.5   Junction Entrance to Zhaosutai River

Zhaosutai River is located in Changtu county, Tieling city and there is water all the year round. In the dry season, the water-carrying capacity is 35 m³/s and the depth of it is 2 m; in the rain season, the water-carrying capacity is 300 m³/s and the depth of it is 5 m. The estuary of it is 70 m wide and terrain is flat. Zhaosutai River flows through Liaoning province and there are many villages along the river. The wastewater of Jilin industry pours into it, so the pollution is serious and in the dry season, the water quality belongs to class V. The main pollution factor is ammonia nitrogen and COD, which exceed 2–20 times and 2 times respectively.

The project comprehensively renovates Zhaosutai River sections that flow into Liao river and is 4 km long to improve the water quality and river landscape. It mainly includes the following aspects: (1) the construction of ecological fore-reservoir; (2) the dredging of 4 km river sediment; (3) the construction of 4 km ecological river course con; (4) the construction of artificial wetland that covers an area of 2 km² (Fig. 3.33).

Fig. 3.33 Junction entrance to Zhaosutai River

3.2.5.6 Junction Entrance to Liangzi River

Liangzi River is located in Qingyun town, Kaiyuan city and there is water all the year round. In the dry season, the water-carrying capacity is 3 m^3/s and the depth of it is 0.4 m; in the rain season, the water-carrying capacity is 10 m^3/s and the depth of it is 1 m. The estuary of it is 30 m wide and terrain is flat (Fig. 3.34). Liangzi River flows through Liaoning province and there are many villages along the river. The industrial wastewater pours into it, so the pollution is serious. There are many flotages on the river which emitted a disagreeable smell. The wastewater mainly comes from domestic sewage and industrial wastewater. The main pollution factor is ammonia nitrogen, which exceed 1 times.

Fig. 3.34 Junction entrance to Liangzi River

The project comprehensively renovates Liangzi River sections that flow into Liao river and is 2.5 km long to improve the water quality and river landscape. It mainly includes the following aspects: (1) the construction of ecological fore-reservoir; (2) the dredging of 2.5 km river sediment; (3) the construction of 2.5 km ecological river course con; (4) the construction of artificial wetland that covers an area of 1 m^2.

3.2.5.7   Junction Entrance to Chai River

Chai River is located in Longshan town, Tieling city and there is water all the year round. Its water column is controlled by the upstream Chai reservoir. In the dry season, the water-carrying capacity is 20 m^3/s and the depth of it is 2 m; in the rain season, the water-carrying capacity is 80 m^3/s and the depth of it is 4 m. The estuary of it is 300 m wide and terrain is flat. Chai River flows through villages. There is no industrial wastewater, so the water quality is good, which meet the standards of

Fig. 3.35 Junction entrance to Chai River

class V water. It mainly includes the construction of artificial wetlands, supporting infrastructure and flood diversion project (Fig. 3.35).

3.2.5.8 Junction Entrance to Wang River

Wang River is located in Zhenxi town, Tieling city and there is water all the year round. In the dry season, the water-carrying capacity is 0.5 m³/s and the depth of it is 0.5 m; in the rain season, the water-carrying capacity is 2 m³/s and the depth of it is 2 m. The estuary of it is 100 m wide and terrain is flat. There are 1500 people in Quanyangou countryside along the river. Wang estuary artificial wetland is composed of wetland and farmland. And the farmland has been retrieved by the reserve administrations. Farmland is mainly around the estuary. And the farther it is away from the town, the more suitable it can be used to construct the artificial wetland. It mainly includes the construction of grit pond, wetland subject project (xylophyta, herbs, wetland and so on) supporting infrastructure and flood diversion project (Fig. 3.36).

Fig. 3.36   Junction entrance to Wang River

3.2.5.9   Junction Entrance to Zuoxiao River

Zuoxiao River is located in the northern new district of Shenyang city and its total length is 14 km. The river course is about 70 m wide and the river is 30 m wide (Fig. 3.37). The river water is turbid and has peculiar smell. Zuoxiao River flow into the Lata Lake in the north of the northern new district of Shenyang city and finally flow into Liao river in the north of Lata Lake. The estuary is 150 m wide and farmland is around it. A lot of farmland and trees are submerged by the river. Because the sewage treatment plant in the northern new district of Shenyang city has stopped working, most urban domestic sewage pour into Zuoxiao River and pollute its water quality. The water-carrying capacity is usually 2–3 m^3/s but most of the water belong to bad class V. The main exceeding standard factor is ammonia nitrogen (1–12 times) and COD is 37 mg/L.

It mainly includes the river dredging and ecological river course project, wetland subject project (xylophyta, herbs, wetland and so on) supporting infrastructure and flood diversion project.

Fig. 3.37 Junction entrance to Zuoxiao River

3.2.5.10 Junction Entrance to Xiushui River

Xiushui River is located in Gongzhutun town Shenyang city. There are 600 people and about 200 households in the town. Domestic water is mainly from the well. When Liao river rises, the water column at the entrance of river is relatively large and a lot of fields and trees immersed in the water (Fig. 3.38). The average water-carrying capacity is 0.2 m³/s, COD is 37 mg/L and ammonia nitrogen is 2.36 mg/L.

It mainly includes the construction of grit pond, wetland subject project (xylo-phyta, herbs, wetland and so on), supporting infrastructure and flood diversion project, emergency handling project, etc.

Fig. 3.38 Junction entrance to Xiushui River

3.2.5.11   Junction Entrance to Yangximu River

Yangximu River is located in Beisan countryside, Shenyang city and farmland is along the river. When Liao river rises, the water storage is rather large and a lot of farmland is immersed in the water. The river estuary is about 150 m wide (Fig. 3.39). The water quality is poor, COD is 31 mg/L, ammonia nitrogen is 1.69 mg/L and BOD is 12 mg/L.

It mainly includes the wetland subject project (xylophyta, herbs, wetland and so on), supporting infrastructure and flood diversion project, emergency handling project, etc.

3.2.5.12   Junction Entrance to Yitong River

Yitong River flow through Panjin city and it is used to deal with the swage. Now we are planning to renovate the banks of the river and construct river parks. The water quality of Yitong River is poor. COD exceeds about 0.5 times, ammonia nitrogen exceeds 3–4 times and BOD exceeds 0.1–1.1 times. Yitong River estuary area is wasteland and tidal flats and under the natural conditions there is some vegetation. In addition, terrain there is high and affected by the floods, it is appropriate to construct the artificial wetland and the available land area is 0.60 km² (Fig. 3.40).

It mainly includes the wetland subject project (xylophyta, herbs, wetland and so on), supporting infrastructure and flood diversion project, emergency handling project, etc.

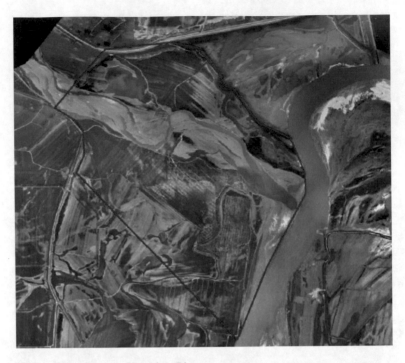

Fig. 3.39 Junction entrance to Yangximu River

Fig. 3.40 Junction entrance to Yitong River

3.2.6   The Construction Technology of Wetland Network in Liao River Conservation Area

By building tributary wetland, river wetland, pond wetland and estuary wetland and implementing the project of the riparian ecological restoration, we construct the river wetland ecosystem that has different sizes and self-recovery functions. The constructed ecosystem can cut into the river pollution load, strengthen water self-purification ability, and improve water quality. And at the same time, it can also conserve the water, flood detention, regulate climate, maintain biological diversity and landscape diversity and become the habitat for wild fauna and flora, especially for the fish and birds.

3.2.6.1   The Construction and Management of Pond Wetland

The Water Network Construction of Pond Wetland

The pond wetland groups project mainly includes the construction of river wetland water network, the reconstruction of aquatic plant community and the impedance control of island non-point source. The construction of wetland water network wetland is composed of pond excavation, construction of water system and the renovation of side. The pond excavation project can provide the habitats for birds by forming lake water and island in the hirst of Liao river mainstream. We reconstruct the aquatic plant community by alternating planting reeds and rushes and other aquatic plants in the wetland area and surface of the water to form aquatic plant zones, which can effectively remove the suspended solids, COD and ammonia nitrogen; we plant pterocarya stenoptera, shrub willow, salix integra and other aquatic plants to protect the banks and reduce non-point sources.

The Reconstruction of Aquatic Plant Communities

According to the differences of project construction, the reconstruction of aquatic plant communities can be divided into two kinds: one is above the Mountain Zhuer. Here the wetland community is based on the pond wetlands and mainly includes Xinglongtai wetland community, Qianxiatazi wetland community and Shengtaizi wetland community; the other one is under the Mountain Zhuer. It mainly includes Mountain Zhuer wetland community, Daliangangzi wetland community, Xingshutuozi wetland community, Xiawanzi wetland community, Hongmiaozi wetland community and Quanhe wetland community.

The reconstruction projects of aquatic plant community in pond wetland: we mainly choose emergent plants and floating plants that can adapt to the local habitats to reconstruct the aquatic plant community.

The reconstruction projects of aquatic plant community in lake surface and island wetland: we mainly choose emergent plants and submerged plants that can adapt to the local habitats to reconstruct the aquatic plant community.

The Impedance Control of Island Non-point Source

The impedance control of island non-point source mainly concentrates in line 1050 of Liao river conservation and green belt, slope and platform of Mid-lake Island. We choose plant grass and shrub vegetation within 30 m revetment to realize the pollution resistance control of agricultural non-point source. Afforestation plants mainly include pterocarya stenoptera, shrub willow, salix integra, ryegrass, festuca arundinacea, cynodon dactylon, vetiver, miao malan, cynodon dactylon, cynodon dactylon, etc.

3.2.6.2   The Construction and Management of Oxbow Lake Wetland

Igniting Flood and Enclosure

According to the condition and goal of enclosure, we determine the enclosure time: the enclosure time of trees is 6 years; the enclosure time of shrubs is 5 years; the enclosure time of shrub-grass is 4 years.

(1) The oxbow lake wetland construction in Juliu River
    Construction contents: The enclosure area of 1–1 and 1–2 oxbow lake natural vegetation is 25 km^2 and the fence of it is 25 km long. The area of artificial natural vegetation is 11 km^2 (Fig. 3.41).
    Specific measures: make use of the flood period and natural seed dispersal and artificial rearing seedlings to promote natural vegetation restoration. 1–1 oxbow lake wetland is near the city and its restoration can reduce the urban sewage pollution. After the purification of oxbow lake wetland, the water that flow into Liao river River can meet the standards of class IV water quality.
(2) The oxbow lake wetland construction in Lanqixiangong section
    Construction contents:The enclosure area of natural vegetation is 12.23 km^2 and the area of artificial natural vegetation is 10 km^2 (Fig. 3.42).
    Specific measures: We combine the artificial aids to promote the natural regeneration with enclosure restoration to protect and restore the vegetation along the edge of the lake, river and marsh of the wetlands.
(3) The oxbow lake wetland construction in Houwaiboshu area
    Construction contents: The enclosure area of natural vegetation is 23 km^2 and the fence of it is 61 km long. The area of artificial natural vegetation is 6 km^2. The dredging system of igniting flood irrigation is kilometers long (Fig. 3.43).

Fig. 3.41 The planning map of the oxbow lake wetland construction in Juliu River

Fig. 3.42 The planning map of the oxbow lake wetland construction in Lanqixiangong section

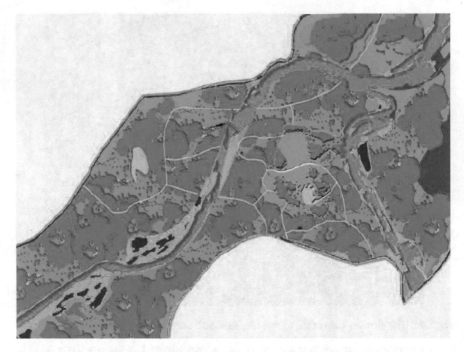

Fig. 3.43 The sketch map of vegetation restoration in the oxbow lake wetland

Specific measures: Make use of natural canal system and floodwater irrigation to increase the water area and take the measures of trenching and root cutting, root sprout and so on to recover the natural vegetation.

Plants can utilize the absorptive action, microbial action and physical action to improve water quality. After the vegetation restoration, the effluent water can meet the standards of class IV water quality, of which COD is 30 mg/L and ammonia nitrogen is 2 mg/L.

The Non-point Source Pollution Control in Oxbow Lake Wetland

(1) The oxbow lake wetland construction in Juliu River
    The oxbow lake in Juliu River is very typical. So we plan to construct the large oxbow lake natural wetland which is about 36 km^2 through interfacial modification, vegetation restoration and regulation of water conservancy. There forms many kinds of habitats including clear water area, deep water area, shallow water area, wetland, marsh wetland and so on. We can restore the aquatic animal groups such as fish and shrimp through the aquatic vegetation restoration and form the large oxbow lake natural wetland that has complete food chain, stable system and self-recovery ability.

(2) The oxbow lake wetland construction in Lanqixiangong section

Firstly, we should build the diversion dike at the estuary of oxbow lake wetland in Lanqixiangong section to gather the basin water to reach the wetland. At the same time, we should block one end of the river and guide the water flow straightly. Then the curved part gradually formed oxbow lake wetland and reduce the upstream pollution.

(3) The oxbow lake wetland construction in Houwaiboshu area

The mid-lake island in Houwaiboshu area makes use the existing highlands to excavate the drainage channels in the side of it and form artificial island which is surrounded by water. The inlet of the drainage channels should be as high as the original drainage channel to form the water surface easily. The outlet of the drainage channel should be 0.3 m higher than the original drainage channel and the outlet of the drainage channel is 15 m wide and the slope ratio of it is 1–2.5. The soil we excavatre the channel was heaped in the mid-lake to form the 3-level terrace: the upper part is level-one terrace, which covers an area of 5.27 mu; the level-two terrace is 25 m wide; the level-three terrace is 5 m wide, which is as high as the original terrace.

The Water Quality Purification and Habitat Restoration

Tieling: The oxbow lake wetland construction in Houwaiboshu area covers an area of 0.8 km^2. It includes the area that is in the north of Northwest Street of Houwaiboshu area, the natural oxbow lake in the west of the Longwang Si and the natural oxbow lake in north of Yangjiatangfang.

In conversation area, where the natural vegetation is sparse, we adopt the enclosure protection measure; where the human activity is frequent, we take the fence measure; we take the measure of enclosure and floodwater irrigation in the area that has the natural seeding ability and evenly distributed trees and shrubs to accelerate the vegetation restoration. The project covers an area of 29 km^2. It can block pollution and purify the water quality (Fig. 3.44).

In Houwaiboshu section, we maintain the stability of the basement, stabilize the wetland area and remold the landscape of wetland through the project measures. Dredge and expand the river course and construct the connected watersystem and form the connected oxbow lake wetland in1–4 oxbow lake section. In the wetland, we plant aquatic plants and give priority to reeds and cattails to reduce nitrogen and phosphorus and purify the water. In the high area, we choose to plant rock maple with golden leaf and prunus maackii.

In 1–5 oxbow lake wetlands, we take the abandoned channel as the center boundary of the wetland and appropriately modify the gradient to form the different habitats that have the different flooding depth after the oxbow lake was flooded. And plants there can also reduce the pollution and purify the water quality.

Fig. 3.44 The planning map of the oxbow lake wetland construction in Houwaiboshu area

Ecological Water Supplement and Dredging the River Course

The oxbow lake in Lanqixiangong section can form the beach face with the small gradient for the sediment accumulation, which is the pre-requisite of the wetland. On the basis of the natural situation of the original oxbow lake river course, we make use of the constructed river water storage projects and modify the interface to develop the lake wetland habitats that are suitable for aquatic, marsh and wet animals and plants.

We modify the underlying surface in Lanqixiangong section to enlarge the surface of the water and guide the water of Liao river pour into the oxbow lake. At the same time, the wetland can restore 0.3–0.5 million cubic ecological water and dilute the upstream pollution.

3.2.6.3 The Construction Management of Dam Backwater Natural Wetland

Backwater Wetland Construction

(1) The rubber dam backwater wetland construction of Hada high-speed highway bridge in Tieling
   The rubber dam is 3 m high and 80.6 m long. The backwater length of it is 9.21 km and water storage capacity is 1.38 million cubic meters. Its center is located in

42.436390° north latitude and 123.790916° east longitude. The farthest natural wetland is 20 km away from the dam and the average gradient is −1.0%. The nearest natural wetland is 7 km away from the dam and their elevation difference is 7 m. The artificial wetland at the estuary of Liangzi River is 4 km away and elevation difference is 2 m. Keep the water level of artificial wetland 0.5 m to form a large number of habitats and provide plenty of wetland habitats for animals and plants. In addition, according to the hydrologic data in Tongjiangkou area, the water level changes no more than 1 m, therefore we can choose to release flood water according to need during the flood period or control the water level of dam changes between 2 and 3 m. The water level control cannot influence the wetlands and enclosure recovery. The main water resource comes from the main stream.

(2) The rubber dam backwater wetland construction of Xinmin-Mountian Mahu Highway Bridge

The rubber dam is 2 m high and 137 m long. The backwater length of it is 5.5 km and water storage capacity is 350000 m³. Its center is located in 42.181818° north latitude and 123.485489° east longitude. When the water level of rubber dam increases 1 m, the water surface area will increase too. The connected water system can provide favorable conditions for wetland restoration and construction. In addition, according to hydrologic monitoring data in Mountain Mahu, the water level 30 year flood period is 1.7 m higher than the 5 year flood period. Therefore, according to the design, when the operation water level of the dam is 1 m, the 5 year flood period can be up to 2 m and the water level of 30 year flood period keep 1 m.

(3) The rubber dam backwater wetland construction in Mandu area

The rubber dam is 2 m high and 121 m long. The backwater length of it is 12.6 km and water storage capacity is 1.09 million cubic meters. Its center is located in 41.590081° north latitude and 122.686116° east longitude. The building of rubber dam can control water level, maximum elevation difference of wetland is 10 m and the average gradient is 0.9%. The operation water level of the dam is 2 m. During the flood period, the operation water level of the dam keeps 1.4 m. According to the design, we will further improve the dam and the operation water level of it can be up to 3 m.

(4) The rubber dam backwater wetland construction in Hongmiaozi area

The rubber dam is 2.5 m high and 111 m long. The backwater length of it is 8.4 km and water storage capacity is 1.66 million cubic meters. Its center is located in 41.441367° north latitude and 122.632326° east longitude. The maximum elevation difference of wetland is 10 m and the average gradient is 0.9%. The operation water level of the dam is 2 m. During the flood period, the operation water level of the dam keeps 1.4 m. According to the design, we will further improve the dam and the operation water level of it can be up to 3 m.

The Ecological Dredging of the River Sediment

(1) The rubber dam backwater wetland construction of Hada high-speed highway bridge in Tieling

In the dry season, we combine the mechanical way with artificial way to clean the surface with many pollutants and the length of it is 9.21 km. The average desilting depth is 0.5 m and the column of it is 322350 m^3. The bottom mud was transported to the wetland for the island restoration or shipped to the other place. These lay a foundation for improving the water quality.

(2) The rubber dam backwater wetland construction of Xinmin-Mountian Mahu Highway Bridge

Dredging method is same as above. The length of it is 5.5 km. The average desilting depth is 0.6 m and the column of it is 396000 m^3.

(3) The rubber dam backwater wetland construction in Mandu area

Dredging method is same as above. The length of it is 12.6 km. The average desilting depth is 0.5 m and the column of it is 630000 m^3.

(4) The rubber dam backwater wetland construction in Hongmiaozi area.

Dredging method is same as above. The length of it is 8.4 km. The average desilting depth is 0.5 m and the column of it is 420000 m^3.

In conclusion, we clean the four rubber dam backwater area. The total length of it is 35.71 km and the column of it is 1.76835 million cubic meters.

Level the Flood Land

(1) The rubber dam backwater wetland construction of Hada high-speed highway bridge in Tieling

In the rubber dam backwater area of Hada high-speed highway bridge in Tieling, we level the flood land along the river within the 80 m and the length of it is 5.2 km and the area of it is 416000 m^2. Within the line 1050, we construct the wetland system and the area of it is 416000 m^2. The wetland system can release the river pollution load and improve water quality.

(2) The rubber dam backwater wetland construction of Xinmin-Mountian Mahu Highway Bridge

We level the flood land along the river within the 60 m and the length of it is 3.5 km and the area of it is 210000 m^2. Within the line 1050, we construct the wetland system and the area of it is 210000 m^2.

(3) The rubber dam backwater wetland construction in Mandu area

We level the flood land along the river within the 70 m and the length of it is 5.6 km and the area of it is 392000 m^2. Within the line 1050, we construct the wetland system and the area of it is 392000 m^2.

(4) The rubber dam backwater wetland construction in Hongmiaozi area.

We level the flood land along the river within the 90 m and the length of it is 4.4 km and the area of it is 396000 m². Within the line 1050, we construct the wetland system and the area of it is 396000 m².

The Reconstruction of Aquatic Plant Communities

(1) The rubber dam backwater wetland construction of Hada high-speed highway bridge in Tieling

In this area, when the water level is over 1 m, we plant submerged plants and floating plants such as lotus, eichhornia crassipes, potamogeton crispus, ceratophyllum demersum and duckweed; when the water level is between 0.5 and 1 m, we plant emergent plants, submerged plants and floating plants such as phragmites australis, zizania caduciflora, potamogeton crispus, ceratophyllum demersum, vallisneria spiralis, ludwigia adscendens and duckweed; when the water level is between 0 and 0.5 m, we plant marsh plants such as wild soybeans; when soil water rate is high, we plant reeds and other plants. The restoration vegetation area is 416000 m². Then we further achieve the restoration and reconstruction of aquatic plant communities and the effective purification of water quality.

(2) The rubber dam backwater wetland construction of Xinmin-Mountian Mahu Highway Bridge

Planting plants and methods of it is same as above and the restoration vegetation area is 210000 m².

(3) The rubber dam backwater wetland construction in Mandu area

Planting plants and methods of it is same as above and the restoration vegetation area is 392000 m².

(4) The rubber dam backwater wetland construction in Hongmiaozi area

Planting plants and methods of it is same as above and the restoration vegetation area is 396000 m².

The Construction of Embankment Control Belt

(1) The rubber dam backwater wetland construction of Hada high-speed highway bridge in Tieling

We grow aboriginal plants along the 5.2 km long river bank to construct embankment control belt. We plant lawn vegetation such as vetiver, herba lycopi, winter jasmine and so on within the 15 m wide revetment and plant shrub willows and salix integra to construct the control forest within 60 m wide revetment. The area of embankment control belt bank is 390000 m².

(2) The rubber dam backwater wetland construction of Xinmin-Mountian Mahu Highway Bridge

We grow aboriginal plants along the 3.5 km long river bank to construct embankment control belt. We plant lawn vegetation such as vetiver, herba lycopi, winter jasmine and so on within the 10 m wide revetment and plant shrub willows and salix integra to construct the control forest within 50 m wide revetment. The area of embankment control belt bank is 210000 m^2.

(3) The rubber dam backwater wetland construction in Mandu area
     We grow aboriginal plants along the 5.6 km long river bank to construct embankment control belt. We plant lawn vegetation such as vetiver, herba lycopi, winter jasmine and so on within the 10 m wide revetment and plant shrub willows and salix integra to construct the control forest within 50 m wide revetment. The area of embankment control belt bank is 336000 m^2.

(4) The rubber dam backwater wetland construction in Hongmiaozi area
     We grow aboriginal plants along the 4.4 km long river bank to construct embankment control belt. We plant lawn vegetation such as vetiver, herba lycopi, winter jasmine and so on within the 20 m wide revetment and plant shrub willows and salix integra to construct the control forest within 80 m wide revetment. The area of embankment control belt bank is 440000 m^2.

3.2.7 Ecological Restoration Technology of Riparian Zone in Liao River Conservation Area

Considering gradient of the river bank, hydrological conditions, soil characteristics and environment characteristics around the river in natural enclosure district, we adopt artificial promoting technology which is composed of comprehensive ecological revetment technology and soil bioengineering. At the same time, we also strengthen the management technology of vegetation and constitute the key technology of natural enclosure artificial promoting technology of Riparian Zone in Liao River Conservation Area.

3.2.7.1 The Comprehensive Ecological Revetment Technology

We adopt the living branch cutting techniques to plant Salix integra or Tamarix chinensis Lour and plant Melilotus officinalis and Medicago to stabilize the earth's surface, improve soil structure, effectively control the soil erosion and intercept the pollutants in the bottomland prairie whose water level is above the constant water level. When the area's water level is close to the constant water level, we choose emergent plants such as typha orientalis, Phragmites australis and so on to control erosion and stabilize the side slope; when the area's water level is below the constant water level, we choose floating plants such as trapa according to the practical situation of embankment. In one word, the comprehensive ecological revetment technology can effectively reduce the surface soil erosion, improve the vegetation coverage and has landscape functions. Technical diagram is shown in Fig. 3.45.

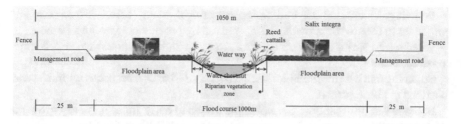

Fig. 3.45 The sketch map of comprehensive ecological revetment technology in natural enclosure

3.2.7.2   Soil Bioengineering

Soil bioengineering is a biological engineering which is based on reliable soil engineering and it use live plants and other auxiliary materials to build all kinds of slope structures to reduce water and soil erosion and improve the habitat. This kind of revetment technology uses a lot of ligneous plants that can grow the new roots rapidly and the most commonly used woody shrubs, aspens, and dogwoods and so on. Soil bioengineering is different from biological slope protection engineering technology and it has the features of large biomass, structure stability, low maintenance requirements, rapid habitat restoration, simple construction, low cost and so on.

Live branch cutting technique is a way of directly planting the live branches into the soil. Live branches get together with the soil particles to absorb more water in the soil; the roots and leaves of the branches play an important role in effectively improving the soil structure, increasing soil aggregate and preventing water and soil erosion. Taking live branches as the constructive species and combining them with revetment vegetation can quickly improve the habitat and build complete and nearly natural riparian vegetation buffer belt. And it has the features of small workload, low cost and wide adaptability. The concrete plans in natural enclosure district in Liao river conversation area is stated as following: When the area's water level is above the constant water level, we choose purple willow branches that are strong, tall and healthy. When the area's water level is below the constant water level, we choose floating plants such as typha orientalis, Phragmites australis and so on to control dredging and erosion. The new roots and leaves that live branches send forth can improve soil structure and effectively control the embankment water erosion. Taking live branches as the constructive species can quickly restore the river bank vegetation, improve the bank habitat and provide favorable conditions for the other local plants, forming a complete and natural riparian vegetation buffer belt.

3.2.7.3   The Cultivation and Tending Management Technology of Vegetation

Cutting: Choose 0.5 m long excellent osiers of which the diameter is 2–4 cm and then plant them between late April and early May.

Watering: Guarantee the bud water and critical water (between late May and mid-June) to ensure the growth needs. And timely provide the freezing water.

Weeding: In spring, before the osiers grow 1 m high, we should weed 2–3 times considering intertillage.

Stumping and keeping stubble: We keep osiers 2–3 or 3–4 branches within 3 years to extend the bush crown.

Remove the bifurcations: Between late May and early June, it is easy grow the new bifurcations, so at this time we should remove the bifurcations timely.

3.2.8 The Restoration Technologies of Mainstream Vegetation and Natural Habitats in Liao River Conservation Area

3.2.8.1 The Restoration Technologies of Mainstream Vegetation in Liao River Conservation Area

According to the geomorphologic features and climate characteristic of Liao river mainstream and the resource characteristics of freshwater fishes in Liao river, we can divide Liao River Conservation Area into two habitat sections: One is Changtu-Liaozhong section and the other is Taian-Panjin section. Their characteristics and recovery plant communities are stated as following:

Changtu-Liaozhong Section

Climate characteristics: Changtu-Liaozhong section is located in the middle and upper reaches of Liao river. The northernmost part of it is Kangping County and it is located in Horqin sandy land of Inner Mongolia. It belongs to the cold temperate semi-arid monsoon climate zone and the others belongs to the warm temperate semi-humid monsoon climate zone; annual average rainfall is 537.7–694.5 mm and annual average temperature is 6.7–8.1 °C. Most of them belong to north China flora. Tieling segments belong to Inner Mongolia flora, Changbai flora and north China flora. The plant and animal species there are complex and rich.

Main recovery plant communities: Chinese wildrye-reed meadow, leymus chinensis meadow steppe, hankow willow thickets along the river, willow thickets along the river, scirpus tabernaemontani and typha community, reed meadow wetland, the sandy willow community. The representative plants of this area include reeds, Chinese wildrye, salix, salix integra, salix kangensis, scirpus tabernaemontani and cattails, etc. Accompanying plants mainly include artemisia scoparia, calamagrostis epigeios, chloris virgata, setaria viridis, wild mugwort, cnidium monnieri, rumex acetosa, sedge, etc. And the wild soybeans are the important protective plant.

Taian–Panjin Section

Climate characteristics: Taian–Panjin section is located in the lower reaches of Liao river. It is based on Liao river estuarine wetland and belongs to the warm temperate semi-humid monsoon climate zone. The annual average rainfall is 623.9 800 mm and annual average temperature is 6.7 8.4 °C. Liao river estuarine wetland is low-lying and flat and the elevation of it is 1.3–4.0 m. The soils there are beach solonchak, coastal solonchak and meadow solonchak and marshy solonchak. It belongs to north China flora.

Main recovery plant communities: Floodplain small poplar + lobular poplar, leymus chinensis meadow, reed meadow, reed swamp, etc. The representative plants of this area include tamarisk, salix integra, reed, suaeda heteroptera, cattails, pennisetum, chinese wildrye and so on. And the wild soybeans are the important protective plant.

3.2.8.2 The Natural Habitat Restoration Technology of Liao River Mainstream

Implement the Ecological De-Farming Measures in Accordance with the Law

Liao river River has the typical characteristics of northern rivers and its main channel is generally 100–500 m wide. The flood draining mainly depends on the broad floodplain. However, the lack of management and the long-term fight for the land between people and water make Liao river almost completely reclamation, as mentioned above, which poses a serious threat to the safety of river flood control and ecological security. Based on the principles of respecting history, facing the reality and reasonable compensation and according to the article 23 and 39 of "land management law" "... Land use should be consistent with the planning of integrated harnessing, development and exploitation of rivers and lakes and be in accordance with the requirements of the rivers, flood storage and water transmission." "...... According to the general plans for the land utilization, we make a plan of returning land to farming to forestry and also to the lake." Since the 2010, Liao river has delineated 1050 mu main flood protection zone to implement natural enclosure, and created a total of 580 thousand mu farmland.

Compile "Sand Excavation Planning" and Comprehensively Complete the Rectification Work of Sand Field

Firstly, organize to compile "sand excavation planning" and get the provincial government' approval.

Secondly, from May to the end of December in 2015, we shut down 123 sand fields in the Liao river Conversation Area and end the history of disordered mining and digging.

Thirdly, from April to the end of June in 2011, we comprehensively renovate the closed sand fields combining with river ecological construction. We level off the sand fields that influence the river environment and flood safety; if the waste material, management houses and other piles of objects do not affect flood safety and do not pollute the river, we level off them. After leveling off the sand fields, we take measures of ecological protection and plant trees and grasses there. By the end of June 2011, the comprehensive improvement of sand fields in Liao river Conversation Area had been completed.

Carry Out the River Barrier Clearance

According to the "Regulations of the people's Republic of China on flood control law", "regulations of the people's Republic of China on the regulation of river management law" and "the regulations of Liaoning Province on the protection of the Liao river law" and other relevant laws and documents, Liao river reserve administration organize to complete the "Execute solution of the river barrier clearance in Liao river Conversation Area". And it states clearance principles as following: who put up barrier, who clear; flood safety should be put the first step and then pay attention to the river ecology and benefit local farmers; according to the flood control standard of the state, we expand or remove the bridge, approach road and ferry that seriously block the river within the prescribed period; dismantle the discarded production wells and its ancillary facilities. Formulate the clearance targets: before the flood season in 2012, achieve the full-close management in the main flood discharge areas and clear all the illegal small processing plants, management houses, breeding field and other water blocking buildings; at end of "Twelfth five-year", finish cleaning all sets of levee, trees, greenhouses, wells and affiliated facilities, adjusting flood land planting structures, expand or remove the bridge, approach road and ferry that seriously block the river and so on. In one word, we should fully restore flood diversion capacity of the river and flood control ability of the dike design.

3.2.9 The Agricultural Non-point Source Pollution Control of Liao River Floodplain

3.2.9.1 The Construction Technology of Ecological Barrier Zone

Liao River Conservation Area forms the ecological corridor from north to south in accordance with the vertical and horizontal space layout planning of the land usage, which also has been recognized by the people's government of Liaoning Province and can implement it. In 2011, in accordance with the research results of land use in Liao River Conservation Area, the people's government of Liaoning finance 3.6 billion yuan subsidies for the first time and return 61 million-mu flood land to the

辽河干流行洪保障区右岸阻隔带断面图

(单位: 米)

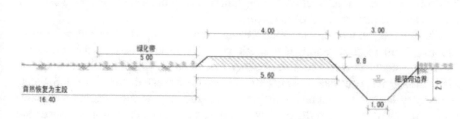

Fig. 3.46 The sectional drawing of the right bank of the ecological barrier zone in flood protection area of Liao river mainstream

river and the fully restore the vegetation. At the same time, combining the river barrier clearance with the relocation of residents, we achieve the flood control safety and ecological benefit.

The existing research results of river protection at home and abroad show that, within the scope of river basin, the increasing of the natural land is in favor of maintaining the biological diversity and ecological integrity. The increasing of the development and construction land will result in the decrease the species and quantity of the river creatures. The loss of forest land and grassland within the river basin can lead to the disappearance of fish and benthic invertebrates. When the agricultural land area ratio was 30–50%, the habitat quality and biological integrity begin to decline significantly; when it exceeds this percent, the ecological integrity of the river would change significantly.

The ecological barrier zone is composed of the side ditch and the green belt. Side ditch take the boundary of flood protection area as a starting point and excavate in the direction of the security zone. The upper part of the fracture surface is recommended to be 3 m wide and lower part of it is 1 m wide and 2 m deep, which forms 4 m wide and 0.8 m high tableland. And then construct 5 m wide green belt (Fig. 3.46).

The right bank of the ecological barrier zone in flood protection area of Liao river mainstream is 329.88 km. It is 187.80 km in Shenyang city, 63.20 km in Anshan city, 71.20 km in Tieling city and 7.70 km in Panjin city.

3.3   Guideline for Aquatic Ecosystem Protection: Goals Development

by Xin Gao, Francois Edwards, Weijing Kong, Yuan Zhang

The drainage basin, as a kind of complex ecological system, supports the development of human society with its rich resources. However, with the increase of population and the rapid development of the economy, drainage basins have suffered huge destruction, such as the deterioration of the water environment, soil erosion, biodiversity loss, the destruction of vegetation and so on, which seriously affected the sustainable development of the basin.

Throughout the world, watershed management has experienced a development process, from the utilization of a single resource to emphasizing the integrated ecosystem management. From the twenty first century, China has introduced the idea of integrated ecosystem management since the twenty-first century, and watershed ecosystem protection and management is gradually developing. Now the integrated management mode which takes the aquatic ecological function district as the unit has been recognized. On the base of systematically acquiring the characteristics of the aquatic ecosystem, we can determine the protection and management objectives within the units and formulate and carry out corresponding protection strategies. Therefore, setting the aquatic ecological protection goals is the prerequisite of developing the follow-up work.

The National Water Pollution Control and Treatment Science and Technology Major Project has proposed the integrated management system which is on the basis of the water ecological function area since the "11th Five-Year" and it also planed to realize the protection goal during the "12th Five-Year". To scientifically promote this project, we compiled the guideline for aquatic ecosystem protection goals the methods of determining the protection scope and the degree of protection, choosing the protection objectives, identifying the threat factors and evaluating the scheme formulations.

The guideline aims to provide a set of referable technical methods for the development of aquatic ecosystem protection goals and a reliable basis for aquatic ecosystem protection and restoration to achieve comprehensive management of the river basin.

3.3.1 Application Scope

The guideline stipulates the framework, procedures and methods of the aquatic ecosystem protection goals development.

This guideline is applicable to management units of different sizes in rivers, lakes, etc.

3.3.2 Basic Terms

The following terms and definitions apply to the specification of this paper.

3.3.2.1   The Aquatic Ecosystem Health

The aquatic ecosystem health refers to the integrity of the water ecosystem composition (physical composition, chemical composition and biological composition) and the integrity of ecological process (ecosystem function), which are embodied in the following aspects: (1) ecological system health means to maintain optimal operation ability under normal conditions; (2) resistance and recovery health namely resist the human threat and maintain optimal operation ability even under changing conditions; (3) healthy organizational capability means the ability of continuous evolution and development. A healthy aquatic ecosystem can not only maintain the stability of its structure and function, but also has the ability to resist disturbance and restore its own structure and function. Meanwhile it can also provide the ecological services that meet the needs of nature and humans.

3.3.2.2   The Aquatic Ecosystem Function Area

The water ecological function zone is used to keep the water ecosystem healthy and protect the integrity of ecosystem function [1] . It is based on regional differences of water ecosystem structure and function.

On the one hand, water ecological function areas reflect the spatial distribution characteristics of water ecosystem and its habitat and confirm the species and important habitats that need to be protected. On the other hand, it also reflects the spatial distribution characteristics of the aquatic ecosystem function and clarify the ecological function requirements of river basins and determine the eco-safety target, which facilitates the formulation of management goals and the implementation of the management plan.

3.3.2.3   Protection Targets and Priority Protection Targets

Protection targets reflect the characteristics of the ecosystem and should get more attention and protection from water managers. The protection target can be a certain species, biological community or typical ecosystem. Considering aquatic organisms, the protection targets include algae, benthic animals, fish, amphibians, large aquatic plants and other biological groups. The number of protected targets maybe very large and different protection targets may reflect the characteristics of the same water ecological system or need the same management measures to protect and restore.

The priority protection target is the protection target which is finally determined by combining the protection objectives and the nested analysis. The priority protection target is the ultimate target of protection and management, and it is the basis of subsequent protection objectives formulation of.

3.3.2.4   Protection Degree

Protection degree is used to measure the level of recovery that management hope protection targets return to after the implementation of protection measures. The protection degree usually reflects the index of a certain ecological attribute of the protection target and its expected recovery level is the state of the index in the ideal condition.

3.3.2.5   The Ideal State and Acceptable Range-Ability

Ideal state is a state that has no human interference. In fact, it is difficult to find a state without any human interference, so the minimum interference pressure condition can be considered as an ideal state. In one word, the ideal state namely is "no interference pressure condition" or "minimum interference pressure condition".

The acceptable range-ability is the range-ability of a key index or the key threat factors in the ideal state. When it is in the acceptable range, the water ecosystem is considered to be healthy.

3.3.2.6   The Important Ecological Attributes

Ecological attributes refers to the nature and relationship of a certain ecological characteristic of the protection target. Each protection target has a lot of ecological attributes, and the ecological attributes which determine the long-term survival of the protection target are called important ecological attributes.

3.3.2.7   The Key Index

The index is used to measure a certain ecological attribute. The important ecological attributes of the protection targets can be embodied by the different indexes, but the managers usually select some indexes that have clear meaning, a sensitive response and are easily used to reflect the changes of important ecological attributes, this kind of indexes are called key indexes.

3.3.2.8   Threat Factors and Key Threat Factors

Threat factor refers to the environmental characteristics that change the grades of the key indexes, including the direct threat factors and the indirect threat factors. The direct threat factor acts on the protection target and causes the fall of its key index grade. The indirect threat factors firstly act on other environmental characteristics or process and then lead to the decline of the key indexes.

There are many threat factors that can change the key index grade. The key threat factor is the threat factor with the highest grade by comparison to all threat factors, influence scope, controllability of the management.

3.3.2.9   Protection Objectives

Protection objectives specifically state the results that the managers hope to bring to the protection target. And it includes the recovery requirements of key indexes and the control requirements of key threat factors.

3.3.2.10   Protection Measures

Protection measures refer to the protective actions and methods that are formulated on the basis of protection objectives. They are mainly on how to improve the level of key indicators and how to eliminate the impact of the key threat factors.

3.3.3   The General Objectives and Framework for the Protection Objectives Development

3.3.3.1   The General Objectives

To keep water ecosystem healthy, we formulate a set of aquatic ecosystem protection goals which includes determining the protection scope, choosing the protection objectives, identifying the threat factors and evaluating the scheme formulations. All of this will provide the guidance for formulating the protection and restoration measures of water ecosystem and then serve on the comprehensive management of aquatic ecosystem health.

3.3.3.2   Technical Framework for the Development of Water Ecosystem Protection Goals

The technical framework for the development of water ecosystem protection goals is shown in Fig. 3.47, including 6 key steps [2]:

1. The purpose of determining the protection scope is to define the specific scope of the development of water ecosystem protection goals in combination with watershed management units (such as water ecological function zones) or planning conservation areas in different regions;

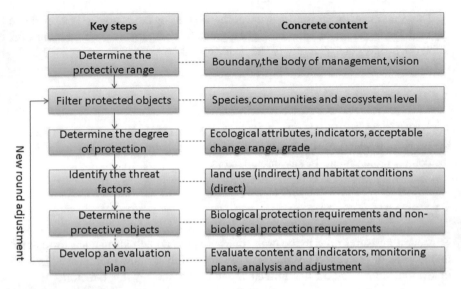

Fig. 3.47 Technical framework of the development of water ecosystem protection goals

2. The purpose of selecting protection targets is to identify the main protection target within the scope of protection. And according to the characteristic requirements of different levels, the main protection targets can be species, communities and ecosystems of different levels;

3. The purpose of determining the degree of protection is to know the status of the protection target and its health and determine the degree of protection and restoration that the protection targets should reach;

4. The purpose of identifying the threat factor is to recognize the abiotic factors that influence the protection targets at different scales including direct threat factors (e.g. hydrochemistry index) and indirect threat factors (e.g. land use) and then select the key threat factor;

5. The purpose of choosing the protection objectives is to develop a set of complete water ecosystem protection requirements on the basis of step 2, step 3 and step 4, including the biological requirements and non-biological protection requirements;

6. The purpose of establishing an evaluation scheme is to assess the expected protection objectives and then make adjustment to guarantee the work of protection and management can be carried out smoothly.

3.3.4 Determine the Protective Range

3.3.4.1 The Boundary Requirements of Protection

The scope of protection in this specification refers to the management area where the water management department carries out the healthy protection of the water ecosystem. The regional boundary can be defined by administrative boundaries or geographical and ecological boundaries. The division of administrative boundaries will destroy the integrity of the river system, and its biological and environmental information are not continuous. In order to ensure the integrity of water system and the continuity of biological and environmental information, we stress that the scope of protection should be defined by geographical and ecological boundaries in this specification [3–5].

Currently in our country, water ecological function areas, water function zone, ecological function areas and river basin or sub basin are delineated by geographical and ecological boundary. So we suggest to delineate the protection boundary and define the scope of protection in the same way.

3.3.4.2 The Method of Determining the Protective Range

Confirm the Management Body

The body of management is responsible for the implementation of relevant protection measures under the guidance of water ecosystem protection objectives. And it is also responsible for water ecosystem protection and comprehensive management within the scope of the protective range. The body of management may be located in the provincial, municipal, county and other relevant administrative departments and can also be located in the basin management department.

Establish the Protection Vision

The protection vision is the final state of protection that we ultimately achieve and it guides the development of the protection objectives. Different body of managements will set up their own protection vision (one or more) and the protection vision should be summarized, concise and predictable.

Delineate the Scope of Protection

The scope of protection is delimited by all the participators (managers, scientists and other related persons) on the basis of the protection vision that body of management set up and the basin boundary or related existing partition boundary based.

3.3.5 Selecting the Protection Targets

3.3.5.1 The Category and Selecting Process of Protection Targets

The Category of Protection Targets

The protection targets reflect the characteristics of water ecosystem and should be paid attention to and protected. From the point of view of the ecological system structure, it includes the different levels of species, community and ecosystem. The selecting of the protection targets has two important aspects: firstly, clarify the ideal state that the protection targets need to achieve and set clear protection objectives; secondly, clarify the specific actions or strategies we should take within the scope of protection according to the threats that protection targets face [4–6].

Species Level

Species level protection is to ensure that the population sizes of a certain species reach a certain level and each individual species can live normally.

It includes two categories of protection targets one is endangered species which lack resources, have small population size and need special protections due to all kinds of disturbance; another one is the species that have protection and management value on the basis of geographical distribution and biological characteristics, such as endemic species, umbrella species, indicator species etc.

Community Level

Community level protection is to ensure the integrity of the structure and function of biological assemblages. The protection targets we choose can reflect the characteristics of aquatic communities. Different aquatic organisms have different community indexes, so the community indexes we choose should be universal and representative such as species richness, diversity index, and integrity index.

Ecosystem Level

Ecosystem level protection is to ensure the integrity of the structure and function of the whole ecosystem. We mainly select some important and vulnerable ecological system as the protection targets. The type of the ecological system can reflect the different contents of streams, lakes, wetlands, stream nutrient, energy exchange, and typical habitat, such as spawning grounds and migration routes of fish. Ecosystem level protection is also the key to the protection of species and communities.

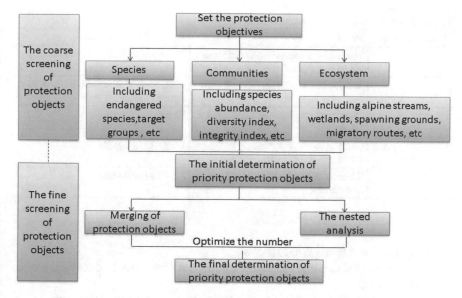

Fig. 3.48 Screening process of protection targets

The Screening Process of Protection Targets

Any living creature in the water ecosystem should be protected, but it is not realistic to protect all of the creatures, because that needs a huge investment. Therefore, we need to select the representative targets which can represent the biological diversity from the hundreds of thousands of organisms.

The screening process of the protection targets includes coarse screening and fine selection (Fig. 3.48). Coarse screening is to select representative targets from the numerous creatures. Fine screening is to further optimize the protection targets on the basis of coarse screening to achieve the final and feasible ideal protective effects.

3.3.5.2   The Coarse Screening of Protection targets

The Screening of Species Hierarchy Protection targets

The Species that Need Special Protection

Species that need special protection generally lack of resources and have small size for the human beings' disturbance. If we do not carry out effective protection in time, they will face extinction. It mainly includes two kinds of species: one is endangered species that the government departments and related protection organizations announced; the other one is in danger but not in the list of protection species [7–9].

Table 3.5 Relevant aquatic species protection list

Ordinal number	Name	Applicable instructions
1	The world conservation union's red list of endangered species	Amphibians, fish
2	Red list of Chinese species (volume 1–5)	Fish, large benthos, amphibians, large aquatic plants
3	China's endangered wild animals (fish)	Fish
4	Chinese biodiversity red directory—higher plant volume	Large aquatic plant
5	Chinese rare and endangered plant illustrations	Large aquatic plant

(1) Threatened and endangered species

   The purpose of screening: Protect the endangered species and reduce their chance of extinction.

   Target groups: Fish, amphibians, macrobenthos and large aquatic plants.

   Screening methods: Collect the existing aquatic organisms in the protected area. According to the aquatic species protection list (Table 3.5) that has been published and related protection laws and regulations of aquatic animals and plants, we choose endangered species as the protection targets within the protective range and arrange them from strong to weak ("very endangered" > "the endangered" > "vulnerable") according to the threatened level.

   If the protection targets are selected according to the protection list in the past 5 years, it can be used directly. If they are selected according to the protection list published more than 5 years ago, we should consider the use of threatened species and determine the threat level to the species.

(2) The threatened species

   The purpose of screening: Protect the endangered species and reduce their chance of extinction.

   Target groups: Fish, amphibians, macrobenthos and large aquatic plants.

   Screening methods: This method is applicable to the species that have long-term monitoring data (over 10 years) and a declining population.

   The monitoring data included the distribution of species, their range, habitat quality and the number of subspecies and mature individuals. The threatened level of aquatic species is classified into 3 levels: "extremely endangered", "endangered" and "vulnerable" and the specific standards seen in IUCN (3.1 editions). According to the evaluation results, we determine the protection targets and arrange them from strong to weak according to the threatened level.

Species with Protection Significance and Management Value

Some species that have protection significance and management value are protection targets. Species with protection significance include endemic species and species with management value include flagship species, umbrella species and indicator species [10–12].

(1) Endemic species

Endemic species is the species that are restricted to the area of protection.

Due to the limitation of the geographical scope, endemic species usually die out easily. Taking endemic species as the protection targets can provide a chance to protect these endangered species.

Target groups: Fish, amphibians, macrobenthos and large aquatic plants.

The methods of screening: We usually analyze the distribution of a certain species and habitat requirements according to the literature, expert experience and the actual investigation and determine whether the species is endemic within the area of protection.

(2) Flagship species

Flagship species are usually large and get public attention.

The purpose of protection: Flagship species can be used as an effective tool for public relations. But due to the limited scientific value, it should be used with caution.

Target groups: Fish and amphibian.

The methods of screening: The selection of flagship species is mainly based on the experience of experts and public awareness. In terms of biological attributes, the flagship species meet big size, long life and other ecological attributes (schedule A). In terms of management attributes, we should consider the corresponding characteristics of the management attributes in screening process of flagship species according to the different management purposes (Table 3.6).

(3) Umbrella species

Umbrella species are the species that if we take protection measures for, other species can be protected.

The purpose of protection: The habitat requirements of the umbrella species can cover the habitat requirements of other species, so the protection of the umbrella species can provide an umbrella protection for other species. Although there is no sufficient evidence to support this concept, the umbrella species can still be taken as a useful protection targets.

Target groups: Fish and macrobenthos.

The methods of screening: The premise of the screening of the umbrella species is abundant data. There are two methods for the screening of the umbrella species: the first one is to quantify the ecological attributes of umbrella species on the basis of data and make quantitative analyses to determine which species can be taken as the umbrella species; the second one is expert judgement of which species can be taken as the umbrella species and then verification according to the data.

Table 3.6 Management attributes in the screening process of flagship species

Area	Management purposes		
	Public awareness	Publicity/Fundraising	Create special protected areas
Be known to the public	•	•	
Be prefered to the public	•	•	
Cultural value	•		•
A wide range of living areas is needed		•	•
Has important ecological functions			•
Can indicate species abundance			•
Has its use value	•		
Has unique scientific value	•		

In combination with the application of the international biological protection field, we suggest to use the first method. And the quantification of ecological attributes of the umbrella species is realized by calculating the umbrella value.

The calculation of the umbrella value needs to consider 3 ecological attributes, that is, the degree of coexistence with other species (PCS), the degree of moderate distribution (MR) and the sensitivity to human disturbance (SC). And PCS pay attention to the species richness within the conservation area rather than the positions in which species appear. PCS can be quantified as 0–1, 0 presents the less species coexistence, 1 presents more species coexistence. And the calculation method is shown in the formula (3.1).

$$PCS = \sum_{i=1}^{l}[(S_i - 1)/S_{\max} - 1]/N_j \qquad (3.1)$$

In the formula (3.1), l is the number of sampling points; S_i presents the number of species appearing in sampling point i; $S_{\max}$ presents the total number of species; N_j presents the number of sampling points of species J.

MR requires the ideal umbrella species, which can not generally exist and can not be extremely rare, but it should be between the two. And the calculation method of MR is shown in the formulas (3.2) and (3.3).

$$MR = 1 - |0.5 - Q_j| \qquad (3.2)$$

$$Q_j = 1 - (N_{present}/N_{total}) \tag{3.3}$$

In the formulas (3.2) and (3.3), for each species J, $N_{present}$ is the number of emerging sampling points; N_{total} is the total number of sampling points.

SC needs the ideal umbrella species which can make appropriate response to different types of human disturbance. The quantitative calculation method is suitable for the biological groups that have quantitative characterization, such as the resistance value of macrobenthos. Qualitative method is suitable for the biological groups that are the lack of quantitative characterization, for example, the qualitative judgment of the characteristics of the fish's tolerance, which can be generally divided into 5 levels, namely extreme tolerance, moderate tolerance, general tolerance, sensitive, extreme sensitivity. Finally, according to the ecological attribute characteristic (schedule A), we determine the final umbrella species.

(4) Indicator species

The characteristics (presence/absence, density, dispersal, reproductive success, etc.) of indicator species can be used to assess habitat changes or the status of other species. In this specification, we emphasize the use of indicator species in habitat types and environmental stress intensity instead of emphasizing the indicator for species biodiversity.

The purpose of protection: The protection of indicator species can allow managers to understand the current the types of disturbance and the intensity of interference on the aquatic ecosystems and help the management department to develop a reasonable protection strategy.

Target groups: Fish, macrobenthos, algae, aquatic macrophyte.

The methods of screening: Indicator species reflect the habitat types, so we should firstly classify habitat types or use environmental data to identify the habitat types through cluster analysis and then determine the indicator species by the calculation of indicator value. When a species is distributed in the habitat type, the indicator value is the largest, and it can be used as the indicator species of the habitat type. The calculation method of indicated value is shown in the formulas (3.4), (3.5) and (3.6).

$$A_{ij} = Nindividuals_{ij}/Nindividuals_i \tag{3.4}$$

$$B_{ij} = Nsites_{ij}/Nsites_j \tag{3.5}$$

$$IndVal_{ij} = A_{ij} \times B_{ij} \times 100 \tag{3.6}$$

In the formula (3.4), $Nindividuals_{ij}$ presents the average individual number of species i in a single habitat type group j; $Nindividuals_i$ presents species i the sum of the average individuals in all habitat types.

In the formula (3.5), $Nsites_{ij}$ presents the number of sampling points of species i in the habitat type j; $Nsites_j$ was the sum of sampling points in the habitat type group j.

Indicator species reflect the gradient of environmental pressure and there are two kinds of screening: in the first method, we should firstly determine the pressure type within the scope of protection, including single pressure and multiple pressure. We can also use PCA to determine the type of single pressure. Then establish the relationship between different species and environmental pressure. In the second method, we need not clarify the type of pressure and we use the tolerance level of aquatic organisms to directly reflect the environmental pressure. This kind of pressure has no obvious directivity. Then we choose the species with low tolerance level as the index species.

The Screening of Protection Targets in Community Levels

The purpose of protection: Compared with the species protection, the scope of community protection on the water ecological system is wider. Community protection also combines the concept of ecological integrity and pays attention to the biological factors and non-biological factors.

Target groups: Fish, macrobentho, algae, aquatic macrophyte.

The methods of screening: We mainly use the method of expert recommendation to screen the community level and combining with the water ecosystem types and management purposes to choose the protection targets which reflect the community classification structure, function structure and the integrity of community. Appendix B provides the relevant community index of fish, macrobentho, algae, aquatic macrophyte for reference.

When selecting the suitable biological community index, we should select the small one whose selection index is easy to get. We should give priority to the index which has a quantity of ecological information, such as the index of biological integrity.

The purpose of protection:

Most aquatic species or communities are only distributed in a limited area. So when the managers carry out conservation management on a larger scale, they need to consider the protection targets of the ecosystem level. The distribution of aquatic organisms is closely related to the environment and ecological processes they are in. That is to say, the biological conditions and diversity in different water ecosystems is different. In addition, some of the water ecological system can also have the special needs, such as fish spawning and migration. The protection at the ecosystems level can protect the whole structure and function of ecosystem.

The types of water ecosystem: Because of the geographical and environmental difference, the indexes used for the classification of water ecosystem types can not be unified. Therefore, when determining the type of water ecosystem, we should mainly consider the local geographical and environmental characteristics and classification index to determine the appropriate type of water ecosystem. This specification provides an example of the type of water ecosystem classification, which is just for reference (Fig. 3.49).

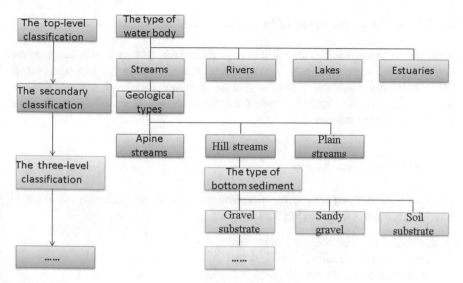

Fig. 3.49 Example of the type of water ecosystem classification

The methods of screening: When we select the water ecosystem as the object of protection targets, we should firstly determine the type of water ecosystem within the protection scope. There are two methods to divide the water ecosystem: one is the bottom-up classification method and the other is top-down classification method.

The steps of bottom-up method: ① knowing the environmental factors that influence the structure and function of water ecosystem (e.g. hydrological regime, the size of water body, connectivity, network position, substrate type, slope and elevation) and make these environmental factors digitize; ② determining the ecological categories of environmental factors; ③ assign each environmental factor by using GIS software; ④ According to the expert experience or ecological clustering analysis, we form the water ecosystem types. The details are seen in the "the water ecological functional partition specification of river basin".

The top-down method is to directly draw water ecosystem boundaries through the available digital data, paper maps and other appropriate data sources. When there is no suitable data, we can choose this method. Managers can use different strategies to build the top-down water ecosystem, such as natural geography, topography, vegetation and river network and so on. After determining the type of water ecosystem, we adopt the way of expert discussion to select the protection targets.

3.3.5.3   The Fine Screening of Protection Targets

After the course screening, the number of protected targets may be still large, so we need the fine screening. The process of fine screening includes the following steps: first, combine the protection targets species or community level; second, analyze the nested protection targets between the different levels; third, determine the final priority protection targets.

Merging of Protection Targets

In general, the merging of the protection targets of species or community level should have the following features (Fig. 3.50):

1. Growing in or inhabiting the same landscape;
2. Having the same ecological processes;
3. Facing similar environmental interference.

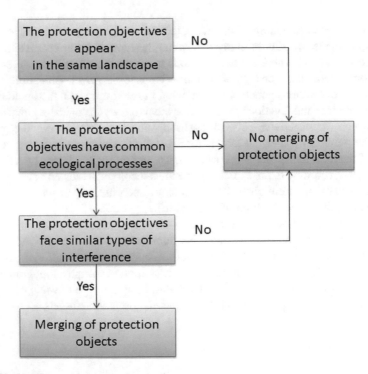

Fig. 3.50 Merging requirements of protection targets

The Nested Analysis of Protection Targets

Nested analysis is mainly to analyze the relationship between the ecosystem level and protection targets to further optimize and reduce the number of protection targets. For the protection targets of one ecosystem level, if the protection can meet the ecological attributes or protection requirements of some species, it indicates that there is nested relation between the different levels and then we retain the protection targets of this ecosystem level. Nested analysis mainly uses the method of expert judgement.

In general, when some species have the following features, there is no need to use the nested analysis:

1. Key species that affect the ecological processes;
2. The species whose different life stages exist in the different ecosystems;
3. The special species that disperse in the different ecosystems or use the different ecosystem resources. They are helpful in paying attention to the connectivity and environmental gradients of different ecosystems.

The Final Determination of Priority Protection Targets

After the merging of the protection targets and the nested analysis, we must determine the final protection targets. The final determination of priority protection targets should follow the following principles:

1. The final protection targets should be able to ensure the biodiversity within the scope of protection, and the protection of the rare and endangered species;
2. The number of the final protection targets should be no more than 8. If the number of the final protection targets is too large, we will have to take different protection strategies for each target, which will cause the management confusion. If the number of the final protection targets is too small, it is difficult to protect the biodiversity of freshwater organisms in one or more spatial scales;
3. The final protection targets should have the viability or resilience. If the protection target has been on the brink of collapse or required huge human intervention, it may not be the final protection target which can use the limited resources reasonably.
4. If other conditions are the same, we give priority to the protection targets that face the most serious threat;
5. It is reasonable to choose the final protection targets on the basis of the current protection.

3.3.6   Determine the Protective Degree

The main purpose of determining the protective degree is to estimate the effort that we need to make to maintain the protection targets and put forward the clear requirements of the continuation of the species, community and ecosystem. Therefore, the protective degree should be clear and reasonable.

The determination of protective degree includes two important steps: first, choose a key ecological attribute for each protection target and determine the key index which can reflect this ecological attribute; second, know the ideal state of the key index and establish the rating criteria.

3.3.6.1   The Important Ecological Attributes and Key Index of the Protection Targets

The Selection of Important Ecological Attributes

A protection target has the different attributes that reflect its characteristics. We should pay attention to the attributes which determine the long-term survival of the protection target. The degradation of these important ecological attributes will seriously threaten the sustainability of the protection targets in the next few decades.

Important ecological attributes are determined on the basis of expert discussion. The ecological attributes such as size, features, landscape pattern can help us to determine the important ecological attributes, as shown in Fig. 3.51, which provides more detailed and suitable information about ecological attributes.

The Selection of the Key Index

When selecting the key index, we should consider the following aspects:

1. We should use the simple and quantifiable indexes which are easy to monitor and have no technical limitation as far as possible;
2. Choose the sensitive index which is observable and proportional with the interference factors and the protection targets as far as possible.
3. Choose the representative index which can accurately reflect its ecological attributes as far as possible;
4. Try to choose an index to reflect its ecological attributes as far as possible. For example, for the community composition, we can choose the diversity index;
5. Choose the composite index when it needs to combine several indexes to reflect the ecological attribute. For example, it is recommended to use index of biological integrity to reflect the health status of community structure.
6. In many cases, the index can be ecological attribute itself, such as the population size which is the number of individuals.

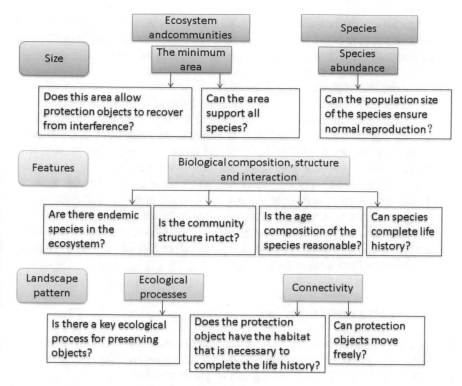

Fig. 3.51 The selection of important ecological attributes

3.3.6.2 The Ideal State of Key Indexes and Rating Criteria

The Determination of the Ideal State

In general, most ecological attributes always change, so the indexes should change correspondingly. Under the ideal condition, the protection targets can maintain the healthy development. If the index is beyond the range of change, it can be considered that the ecological attributes is degrading. The method of determining the index of the ideal state includes qualitative analysis and quantitative analysis.

The Method of Qualitative Analysis

When indexes lack research or need a lot of research effort or long-term research, we can use qualitative analysis. The method of qualitative analysis is based on the judgement of experts, the relevant historical data, or a similar ecological system to determine the acceptable rangeability of the index.

(1) The method of historical data

Generally, we determine the ideal state of the index by referring to the relevant literature, books and other types of information. For example, the building of a dam lowers the internal connectivity of the water ecosystem and reduces the distribution area of the protection targets. If we are sure that the species existed in many waters within the scope of protection, then we the distribution area of the species in history can be regarded as the ideal state.

(2) The method of experts' judgement

Generally, we determine the ideal state by experts' discussion. For example, if we take the population size as ecological attributes and take the number of species and the number of individuals as index, then according to the experts' experience, we can conclude that within the scope of protection there are at least 10 populations and each population has 200 individuals, which can guarantee the stable development of the population and meet or exceed the ideal state.

The above methods can be used alone, but it is recommended to comprehensively use them. At the beginning, the determination of the ideal state by the qualitative analysis need not be very accurate. With the development of protection work, we will gradually get more ecological information and then gradually improve the indexes of the ideal state.

The Method of Quantitative Analysis

The method of quantitative analysis is to determine the ideal state of the index by using statistics including the method of spatial reference condition, the method of prediction model and the method of pressure gradient and so on.

(1) The method of spatial reference condition

The method of spatial reference condition is to choose the waters that have a similar water ecosystem, less man-made interference and high quality habitat as the reference condition. And the rangeability of the index of some important ecological attribute is taken as an ideal state of this water area. The evaluation of reference conditions generally includes the method of water quality grade evaluation, the method of physical habitat evaluation and the method of comprehensive scoring. Finally, we select the waters that have higher score or higher level as the reference condition.

The method of water quality grade evaluation:

Use the method of single factor evaluation to evaluate the grade of the water quality and the lowest grade of water quality was taken as the final grade. Generally, we prefer to the waters with class II water quality as the reference condition. The assessment of water quality grade refers to "the quality standard of surface water environment" [4].

The method of physical habitat evaluation:

We select the waters that have higher scores as the reference conditions in consideration of hydrology, sediment composition, habitat complexity, riparian vegetation coverage and human disturbance etc. And the physical habitat evaluation standards of rivers and lakes can be referenced in appendix D and appendix C respectively and it can be adjusted according to the actual situation.

The method of comprehensive scoring

Combine the water quality grade assessment and physical habitat evaluation and select the waters that not only meet the requirements of water quality grade but also meet the requirements of the physical habitat scores as the reference conditions.

(2) The method of prediction model

The method of prediction model is applicable to the key indexes that reflect the distribution of species and the state of community structure and so on. According to the prediction, we get the ideal state of the index. At present, the method of using environmental data to predict the characteristics of biological community structure is more mature. We can select and adjust the models according to the teaching materials and the corresponding statistics.

In this specification, we take the multiple linear regression models as an example to predict the integrity of community and introduce the general process of determining the ideal state of the key indexes, including the following steps:

1. The choice of the reference conditions refer to the method of spatial reference condition;

2. The environmental indexes we choose can reflect the geography, topography, climate and other natural attribute and try not to use environmental indexes that reflect the characteristics of human interference to ensure the real situation of the predicted indexes;

3. In the process of constructing prediction equation, the predicted index is the dependent variable and the environmental index is the independent variable. We use statistical analysis software to construct the prediction model;

4. The acquisition of prediction results. We use the model to protect the key indexes and check the accuracy of prediction model within the scope of protection.

(3) The method of pressure gradient

The method of pressure gradient is to establish the corresponding relationship between key indexes and the main corresponding environmental pressure to obtain the ideal state of key indexes (Fig. 3.52). Firstly, screen out the most influential environmental pressures. Environmental pressure can be a specific environmental factor and can also be a few environmental factors. Secondly, construct the gradient response model between the environmental pressure and key indexes. Finally, use the method of quantile regression? to determine the ideal state of key indexes in a low pressure level (below 5 or 10% points of the ambient pressure).

The Establishment of Rating Criteria

The purpose of establishing rating criteria is to refine the acceptability of the key index when it is out of the ideal state. In this specification the key indexes are divided into 4 grades including the "bad", "general", "good" and "very good". The standards of 4 levels are shown in Table 3.7.

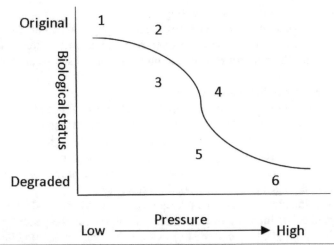

1	Original status.
2	Slight losses in population, slight changes in population density.
3	Some sensitive rare species disappear, the ecosystem function is complete.
4	Some sensitive species are replaced, Species distribution patterns are changed, the main function of the system still exists.
5	Enduring species are increasingly dominated, sensitive species become sparse, the main function of the system is changed.
6	The structure and function of the system have changed dramatically.

Fig. 3.52 The corresponding relationship of pressure between environmental pressure and indexes of ecological attribute

The level of "very good" refers to the range of the key indexes in the ideal state, so we only need to determine the other rating criteria. For the qualitative key indexes, we can determine the description of corresponding rating criteria on the basis of experts' judgment. For the quantitative key indexes, we can determine the rating criteria by the method of trisection or the method of quantile regression.

3.3.7 The Identification of Key Threaten Factors

Generally, the water ecosystem has been degraded or needs a series of protective actions to eliminate the serious threat. So firstly we need to find the threat factors that influence the protection targets and determine the order of the threat factors to focus on the limited resources to carry out the protection.

Table 3.7   The description of rating criteria for key indexes

Key indexes level	Bad	General	Good	Very good
Level description	It is not possible to restore the protected object and prevent it from being eradicated under such conditions for a period of time	The indicators are out of the range of acceptable variation and human intervention is needed. If you do not intervene, the object of protection is severely degraded	The indicators are in the range of acceptable variation and human intervention is needed	The indicators are in an ideal state, and there is little need for human intervention

3.3.7.1   The Category of Threat Factors

Threat factor refers to the physical and chemical factors that affect the health of aquatic ecosystem. It can be divided into direct threat factors and indirect threat factors.

The Direct Threat Factors

The direct threat factors refer to physical and chemical factors that can directly affect the protection targets and lead to ecological crisis. Chemical factors generally refer to the water quality parameters such as dissolved oxygen, ammonia nitrogen, and electric conductivity and so on. Physical factors generally refer to the physical parameters which reflect the habitat quality, such as the type of sediment, velocity of flow, vegetation coverage of the riparian zone and so on.

The Indirect Threat factors

The indirect threat factors refer to environmental factors that can indirectly affect the protection targets and lead to ecological crisis. Because of the nested relationship of spatial scale, the indirect threat factors generally exist in the larger scale, such as the characteristics of watershed landscape, economic development, and population. Considering the correlation among these characteristics and the operability of management, in this specification the indirect threat factors are summarize as the land cover type within the watershed scale.

Land cover types include natural type and human disturbance type. The former refers to the land cover without the human intervention or human development, such as forests, grasslands and so on; the latter is different from the original land cover, such as towns and farmland and so on. The decrease of natural type or the increase of human disturbance can affect the protection targets by changing the water physico-chemical factors. So the changes of land cover type can be regarded as the indirect threat factors.

3.3.7.2   The Methods of Identifying the Key Threat Factors

The key threat factors refer to the most influential factors that affect one of the most important ecological attributes. The identification of key threat factors includes two steps: the first one is to identify all the threat factors that lead to ecological crisis and then rank these threat factors to determine the key threat factors.

The Determination of Threat Factors

If one of the key indexes of the protection targets change, then we determine that there is a threat factor.

The methods of identifying the threat factors include the statistical analysis and expert judgment. The method of statistical analysis includes regression analysis, ranking analysis, correlation analysis. Under the condition of sufficient data, we identify the threat factors. The method of experts' judgment is mainly aimed at the qualitative description of the threat factors, such as the river connectivity features that affect fish migration.

The Determination of Key Threat Factors

The Judgement of Threat Level of the Threat Factors

According to the influence of threat factors on the protection targets such as influence scope, severity and the recovery degree, we rank the threat level of each threat factors. The influence scope of the threat factors refers to the geographical scope that may affect the protection targets in the next 10 years according to the present situation. The influence scope include 4 evaluation grades, including "very wide", "wide", "medium" and "not wide", each of which has 4 points, 3 points, 2 points and 1 point respectively (Table 3.8).

The severity of the threat factors refers to the degree of damage to the protection targets in the next 10 years according to the present situation. The severity of the protection targets has 4 evaluation levels, including "very serious", "serious", "medium" and "not serious", each of which has 4 points, 3 points, 2 points and 1 point respectively (Table 3.9) [13, 14].

Table 3.8 The ranking criteria of the influence scope

Level of influence	Very wide	Wide	Medium	Not wide
Level description	Threat factors are widely available within the scope of protection, and impact on protection targets of all districts	Threat factors are widely available within the scope of protection, and impact on protection targets of many districts	Threat factors are partly available within the scope of protection, and impact on protection targets of some districts	Threat factors are partly available within the scope of protection, and impact on protection targets of limited districts
Score	4	3	2	1

Table 3.9 The ranking criteria of the severity

Level of severity	Very serious	Serious	Medium	Not serious
Level description	Threat factors may destroy or completely destroy protected targets in certain areas of the protection	Threat factors may lead to serious degradation of protected targets in certain areas of the protected area	Threat factors may lead to moderate degradation of protected targets in certain areas of the protected area	Threat factors may lead to light damage of protected targets in certain areas of the protected area
Score	4	3	2	1

Table 3.10 The ranking criteria of recovery degree in management

Level of recovery degree	High recovery degree	Middle recovery degree	Low recovery degree	Unrecoverable
Level description	The impact of the threat factors is recoverable, with a low cost	The impact of the threat factors is recoverable, with an acceptable cost	The impact of the threat factors is recoverable, with a high cost	The impact of the threat factors is unrecoverable
Score	4	3	2	1

The degree of recovery after management refers to the extent which the impact of the threat factor can be restored. The recovery degree of management includes 4 evaluation levels, including "high", "medium", "low" and "not recovery", each of the level respectively has 4 points, 3 points, 2 points and 1 point (Table 3.10).

The ranking of the threat level of the threat factors has four levels including "very high", "high", "medium" and "low" from the aspects of influence scope, severity and the recovery degree of management and the score of each level is 10–12, 7–9, 4–6 and 1–3 respectively (Table 3.11).

Table 3.11 The ranking criteria of the threat level of the threat factors

Ranking level	Very high	High	Medium	Low
Ranking scores	10–12	7–9	4–6	1–3

Table 3.12 The identification of key threat factors

The index of ecological attributes	Threat factors	Threat level	Type of threat factors	Key threat factors
1	Threat factor 1	Very high	Direct	Yes
	Threat factor 2	Low		
	Threat factor 3	Low		
	Threat factor 4	High	Indirect	yes
	Threat factor 5	Medium		
	…	…	…	…

The Identification of Key Threat Factors

The threat factors that influence key indexes are often multiple. According to the above steps, we can know the threat level of each threat factors. And the threat factors with the highest ranking are chosen as the key threat factors (Table 3.12). The key threat factors include direct threat factors and indirect threat factors.

3.3.8 Set the Protection Objectives

Protection objectives of the aquatic ecosystems specifically state the protection results of the water ecosystem that the managers hope to bring to the protection target and put forward the biological and non-biological protection requirements.

3.3.8.1 The Principles of the Development of Protection Objectives

The development of protection objectives should follow the following principles:

1. The principle of specificity: The results that we expect should be described in the clear terms so that people have a common understanding;

Table 3.13 The corresponding protection requirements of different grades

Index Level	Bad	Common	Good	Very good
Grades				
Protection requirements				

2. The principle of measurability: The results that we expect should refer to the relevant metrics (such as numbers, percentages, scores) to ensure that the results are measurable;
3. The principle of implementability: Under the conditions of current social and political background, time range and resource allocation, we should ensure that the expected results can be achieved;
4. The principle of timeliness: Clarify the time when the expected results are achieved.

3.3.8.2 The Biological Protection Requirements

According to the rating criteria of indexes in Sect. 3.3.6.2, we determine the grades of the key indexes of different waters and put forward the corresponding protection requirements (Table 3.13). When the key indexes are good and very good, the protection requirements should maintain the current grades of the key indexes; when the key indexes are general and bad, the protection requirements should improve it to the good grade.

3.3.8.3 The Abiological Protection Requirements

First, we need to identify the acceptable range of key threat factors and then put forward the corresponding protection requirements.

The Acceptable Range of Key Threat Factors

Analyze the acceptable range of key threat factors and find the level of key indexes when key threat factors are in the ideal state.

If the key threat factors are described in the qualitative way, the range is to completely eliminate the threat factors. If the key threat factors are described in the quantitative way, we analyze the relationship between the key threat factors, and the key indexes by building the relationship (Fig. 3.53). A is the linear model relationship, C and B are nonlinear model relationship, and the relationship between D, E and F is the threshold model relationship. For the different models, we can determine

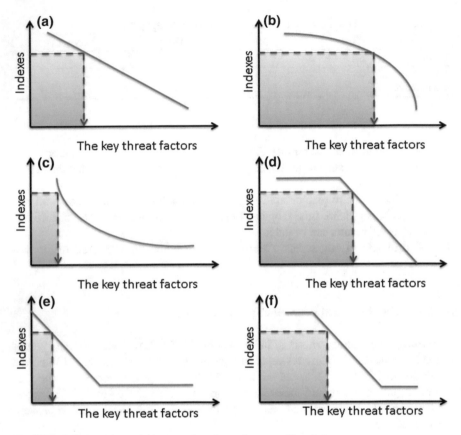

Fig. 3.53 Relationship model between key threat factors and key indexes

the acceptable range of key threat factors according to the range of key indexes in the ideal state.

The Protection Requirements of Key Threat Factors

The protection requirements of key threat factors should consider the current management ability and elimination ability to eliminate threat as far as possible or return to the acceptable range.

3.3.9 Design the Evaluation Scheme

The management department develops protective measures and takes actions according to the protection objectives. But we also need to develop a set of assessment programs to guarantee it. The evaluation scheme helps to determine whether we can get the expected effects and adjust the protection objectives. The management department can adjust the protection objectives according to the evaluation results and form the new protection measures.

3.3.9.1 The Content of Evaluation Scheme

The evaluation scheme includes two parts: the effective evaluation and the situation assessment. The purpose of effective evaluation is to evaluate whether the protection measures can get expected results according to the protection objectives. The purpose of situation is to assess whether the protection targets and threat factors change without the protection measures.

3.3.9.2 The Evaluation Indexes

The Index of Effective Evaluation

Generally, it is easy to select the effective evaluation indexes. Because the protection measure is to eliminate the key threat factors or improve the degradation of the key indexes, the key threat factors the key threat factors or the key indexes can be used as the index of effective evaluation.

The Index of the Situation Evaluation

The index of the situation evaluation can be determined from non-critical threat factors or the indexes that reflect other ecological attributes.

The Priority of Evaluation Index

In the process of implementing the protection measures, it is impossible to monitor all the evaluation indexes simultaneously, so we need to determine the priority of different indexes. Firstly, we should consider the index of effective evaluation and then consider the index of the situation evaluation.

3.3.9.3 The Monitoring Scheme

The determination of evaluation indexes lets us know the content of the monitoring scheme, but we also need to refine other monitoring information to ensure reasonable assessment results. The monitoring information that is needed to refine includes the following aspects:

1. Time: Determine the monitoring frequency of indexes and monitoring time of the year.
2. Location: Determine the geographical location.
3. Persons: Determine the relevant skilled persons and the main leaders.
4. Cost and the source of funds: Determine the resources needed for including the required time and funds, and funds of each monitoring indexes.
5. Index status: If we know the index status, we should record them in the monitoring plan in detail. The first monitoring results of the index are generally called background data. It is very important to collect the background data as early as possible, because the other monitoring data in the later period will be compared with the background data.
6. The summary reports: Convert the monitoring data into the information that can guide the protection management regularly. Therefore, it is necessary to write a summary report at regular intervals.
7. The state of execution: The monitoring progress (monitoring/without monitoring) should be marked clearly, which will help us to adjust the later monitoring scheme.

3.3.9.4   The Analysis and Adjustment

The purpose of the analysis is to convert the original monitoring data into the useful information, which makes the following problems clearer:

1. Are things going in the right direction?
2. Have the protective measures produced the desired effects?
3. Has the status of the protection targets improved?
4. Have the key threat factors been relieved?

The analysis process should ensure that the monitoring data can be used to assess the protection actions and completed work instead of evaluate whether the results of the monitoring have statistical significance.

The purpose of adjustment is to change and improve the protection management, which includes updating monitoring indexes and monitoring plans and change the protection objectives and protection measures, etc. This may be the partial adjustment of the protection management, and may also be the repeated process of setting the protection objectives.

3.4   The Evaluation of Wetland Ecosystem Restoration in Liao River Conservation Area

by Liang Duan, Weijing Kong, Yonghui Song

3.4.1 The Health Evaluation Index System of Wetland Ecosystem

3.4.1.1 The Selection and Description of the Indexes

The Ecological Characteristic Index

1. The vegetation along the river banks and the edges of river beds: We combine the qualitative analysis and quantitative analysis to study the disturbed state and coverage changes of vegetation.
2. The river channel erosion and deposition: We combine the qualitative analysis and quantitative analysis to reflect the stability condition of the river channel and the deposition degree of sediment.
3. Water quality: We evaluate the water quality of the river channels and wetlands according to the standards of class III water quality that *the quality standards for surface water environment* [4] formulate.
4. The guarantee and replenishment of water sources For the marsh wetlands, we calculate mainly according to the ratio of the replenishment ratio of water sources.
5. The diversity of species: It refers to the percentage that the number of plant species in marsh wetlands account for among all the species in wetlands.
6. The size or specifications of animal individuals: We get the average size of the fish in the same wet area and then compare them with the original data to get the change rate (the data we compare should be in the same period).
7. The size of plant individuals: We get the average height of dominant plants and then compare it with the historical data that we recorded to get the change rate.
8. The biomass: We calculate the average annual biomass of plants and get the change degree of this historical period.
9. The degradation of wetland area: We can use the percentage of the degraded wetland area in the existing wetland areas to reflect it and can also use salinization and desertification of wetland and vegetation degradation area to measure it.
10. The threatened status of wetland: On the basis of a variety of human disturbances such as, overly fishing, mowing and reclamation and other threat factors, we combine the qualitative analysis and quantitative analysis.

The Function Conformability Index

1. The flood control: Because the area of the natural wetland reduced, the flood control capacity decreased. Therefore, we need use artificial engineering to make up, such as building embankment, reservoir, flood detention area and so on. So we use the increasing rate of flood control surcharge to express.
2. Hydrological regulation: It provides the water for agricultural irrigation, industry and so on. And we use the change rate of water supply to represent.

3. Erosion control: It prevents the soil from being eroded by wind and water. And we use the change rate of wind erosion and water erosion to represent.
4. The disposal and purification of wastes: It includes the disposal of wastes, pollution control, the removal of toxicity and other aspects. And we use the change rate of the disposal of wastes and purification rate to represent.
5. The habitats For the habitats of wildlife and brooding habitats, we use destruction degree or degradation rate to represent.
6. The food production: It includes fishes, fruits and so on. And we use the change rate of the annual yield to represent.
7. The raw materials: It includes wood, fuel, feed, etc. And we use the change of the quality to measure and combine qualitative analysis and quantitative analysis.
8. The entertainments: It includes wetland tourism, fishing and other outdoor recreational activities. And we use the increasing or decreasing of entertainment days to represent.

The Social and Political Environment

1. Peripheral population quality: It is represented by the percentage that the people who get the junior high school education accounted for.
2. Human activity intensity: It is represented by the population density.
3. Population health status: It is calculated by morbidity and mortality.
4. The index of material life: It is calculated by the level of per capita income and its unit is yuan/a.
5. The use intensity of pesticides: It is calculated by the annual usage quantity of pesticides per square hectometer and its unit is kilogram per square hectometer, here we can use the utilization of pesticides to represent.
6. The use intensity of fertilizer: It is calculated by the annual usage quantity of fertilizer per square hectometer and its unit is kilogram per square hectometer, here we can use the utilization of fertilizer to represent.
7. The industrial and domestic sewage: It is represented by the treatment rate of sewage and waste water.
8. The awareness of wetland protection: It is calculated by the proportion of the people who have the awareness of wetland protection account for.
9. The implement of relevant policies and regulations: It is calculated by the proportion of the people who have accepted the relevant policies and regulations account for.
10. The level of wetland management: We combine the qualitative analysis and quantitative analysis and measure the overall level of wetland management team.

3.4.1.2 The Method of Evaluation

We adopt the fuzzy comprehensive evaluation method.

3.4.1.3   The Criteria of Evaluation Indicators

Evaluation indicators are divided into five levels including very healthy level, healthy level, rather healthy level, general morbidity and morbid level (Tables 3.14, 3.15 and 3.16).

3.4.1.4   The Index Weigh

We adopt the expert consultation.

3.4.2   The Evaluation of Wetland Ecosystem Health in Liao River Conservation Area

3.4.2.1   The Evaluation Model of Wetland Ecosystem Health

The Establishment of Index System [15–20]

(1) The composition of the candidate index system
On the principles of integrity, representativeness, operability, feasibility and qualitative and quantitative analysis, we select 11 characteristic indexes as the candidate indexes of the evaluation of wetland ecosystem health which can fully reflect the water environment quality, aquatic biological characteristics and habitat environment quality in Liao River Conversation Area. And among them, there are 8 indexes reflecting the water environmental quality status, including water temperature, PH, chemical oxygen demand (COD_{Mn}), the concentration of ammonia nitrogen (NO_3–N), the concentration of nitrate nitrogen (NO_3–N), the concentration of nitrite nitrogen (NO_2–N), conductivity (EC), the concentration of total phosphorus (TP); there is one index reflecting aquatic characteristics and it is the concentration of chlorophyll a (Chl-a); there are two indexes reflecting the habitat quality, including the concentration of dissolved oxygen (DO) and oxidation reduction potential (ORP).
(2) The selection method of candidate indicators
We use principal component analysis (PCA) to extract principal components from the candidate indexes. According to the principal of varimax, we select the candidate indexes that whose loading value is about 0.4; And then we do the normal distribution test on the rest candidate indexes, if the indexes conform to the normal distribution, we adopt the Pearson correlation analysis, if the indexes do not conform to the normal distribution, we adopt the Spearman rank correlation analysis. And at last, we determine the degree of correlation between the indexes according to the significance level. Combined with the actual importance degreeof the indicators, we select the relatively independent and important

Table 3.14 The ecological characteristic index

Index	Grade				
	Good health	Healthy	Less healthy	Morbidity	Illness
Vegetation along the river edge	Undisturbed original or local land, vegetation coverage, >80%	Slightly disturbed with some exotic species, vegetation coverage, 60–80%	Medium coverage with the mixed and originally introduced species, vegetation coverage, 40–60%	Strongly disturbed with more exotic species, vegetation coverage, 20–40%	Bare or sporadic land, vegetation coverage, <20%
The channel scour/silting	stable without scour/silting	Sporadic scour	Medium scour	Obvious scour	Extensive and intense scour
Water quality	I	II	III	IV	V
Water guarantee rate	>70%	60–70%	50–60%	40–50%	<40%
Species diversity	>40%	30–40%	20–30%	10–20%	<10%
Animal Individual scale	The individual is significantly enlarged without deformity, the change rate 20%	The individual is rather enlarged, the change rate 0–20%	The individual does not change obviously, the change rate 0–20%	The individual becomes small obviously, the decrease rate >20%	The individual becomes small obviously with deformity
The individual plant scale	The individual height increased and stem diameter increased, the change rate, >10%	The individual height relatively increased and stem diameter increased without obvious change, change rate, >10%	The individual does not change obviously, the decrease rate 0–10%	The individual becomes small obviously, the decrease rate >10%	The individual has significantly changed or mutation
Biomass	Biomass increased, the change rate >10%	Biomass increased without obvious change, the change rate 0–10%	Biomass does not change obviously, the change rate 0–10%	Biomass decreased, the change rate >10%	Biomass decreased significantly the change rate 50%
The degradation rate of wetland	<5%	5–15%	15–25%	25–35%	>35%

(continued)

Table 3.14   (continued)

Index	Grade				
The threat to wetlands	No overfishing, mowing, picking up eggs, reclamation and so on	Proper fishing, mowing, without picking up eggs, reclamation and so on	Overfishing, mowing, without reclamation	Over fishing, mowing, picking up eggs, reclamation and so on	Over fishing, mowing, picking up eggs, reclamation and so on

indicators as evaluation indicators; and the above analysis process is completed by the statistical software, SPSS.

(3) The determination of the index weight

According to the characteristic value, variance contribution rate, accumulated variance contribution rate and initial load value of principal components of each corresponding indexes, we calculate the weight of every index. The formula of calculating the weight is showed as the following:

$$b_j = \sum_{i=1}^{n} \left(f_{ij}/\sqrt{\lambda_j} \times \theta_j / \sum_{j=1}^{l} \theta_j \right) \tag{3.7}$$

In the formula, b_j stands for weight; f_{ij} stands for the initial load value of the j principal components that the i index corresponds to; λ_j stands for the characteristic value that the j principal component corresponds to; θ_j stands for the variance contribution rate that the j principal component corresponds to.

The Evaluation Method of Wetland Ecosystem Health

(1) The evaluation method

The comprehensive index method is a common multi-index test. Through the comparison of the survey data with the reference value, we get the quantization value and then through the weighted synthesis, we can obtain the comprehensive index value of the wetland ecosystem health. According to the classification of the total index, we determine the health level of the wetland ecosystem [21–23]. The calculation formula of the comprehensive index method is shown as following:

$$CI = \sum_{i=1}^{n} W_i I_i \tag{3.8}$$

Table 3.15 The function conformability index

Index	Grade				
	Good health	Healthy	Less healthy	Morbidity	Illness
The control of flood	High control ability and no additional engineering cost	For the reservoir embankment, it has strong control ability	Strong control ability due to reservoir embankment and detention basins	Without the obvious control ability and additional engineering cost is high	Without the control of flood
Hydrological adjusting	Water supply and replenishment capability is being improved	The building of the reservoir embankment improves the replenishment capability	The building of artificial facilities improves the replenishment capability	Water supply and replenishment capability decrease	Without the water supply and replenishment capability
The control of erosion	No soil erosion, the change rate of erosion ≤ 0	The soil erosion is not obvious, the change rate of erosion $<2\%$	The soil erosion occurs in some area, the change rate of erosion $2-5\%$	The soil erosion is serious, the change rate of erosion $5-10\%$	The control of erosion decreases obviously, the change rate of erosion $>10\%$
The purification ability	The purification capacity increases, the change rate ≥ 0	Without he obvious purification ability, the rate of decrease $<5\%$	The purification capacity decreases, the rate of decrease $5-10\%$	The purification capacity decreases, the rate of decrease $10-20\%$	The purification capacity decreases obviously, the rate of decrease $>20\%$
Habitat	The destruction or degradation rate $<2\%$	The destruction or degradation rate $2-5\%$	The destruction or degradation rate $5-8\%$	The destruction or degradation rate $8-12\%$	The destruction or degradation rate $>12\%$
Food production	The annual harvest yield increases, the increase rate $>5\%$	The annual harvest yield increases, the change rate $2-5\%$	The annual harvest yield is stable, the change rate 0	The annual harvest yield decreases, the change rate $0-5\%$	The annual harvest yield decreases, the change rate $>r5\%$
The raw materials	The quality remained stable and the plant height and stem diameter did not change obviously	The quality in some area decreases but it does not pose a threat	The quality in some area decreases but it is in control	The quality decrease obviously and it has bad effect on the yield	The quality decrease obviously and the plant height and stem diameter change obviously, bringing profound influence

(continued)

Table 3.15 (continued)

Index	Grade				
Entertainment	The value of landscape and aesthetics is high and the number of entertainment activities increase	With the long entertainment days	With the entertainment days in specific period	With the fewer entertainment days. The value of landscape and aesthetics is not high	Without the entertainment days and value of landscape and aesthetics

Table 3.16 The social and political environment indicators

Index	Grade				
	Good health	Healthy	Less healthy	Morbidity	Illness
Quality of surrounding population	>15%	15–10%	10–5%	5–2%	<2%
Human activity intensity	<10%	10–20%	20–30%	30–40%	>40%
The health status of population	<2‰	2–5‰	5–8‰	8–10‰	10‰
Material life index	>4000	4000–3000	3000–2000	2000–1000	<1000
Pesticide utilization rate	>50%	50–40%	40–30%	30–20%	<10%
Fertilizer utilization rate	>50%	50–40%	40–30%	30–20%	<10%
Sewage treatment rate	>80%	80–70%	70–60%	60–50%	<50%
The consciousness of wetland conservation	>1‰	1–5‰	0.5–0.2‰	0.2–0.1‰	<0.1‰
Implementation of policies and regulations	Policies and regulations are fully and actively implemented	Some policies and regulations are seriously implemented	Some policies and regulations are implemented	Policies and regulations are not seriously implemented	Policies and regulations are not completely implemented
Management level	Reasonable management mechanism, high quality of personnel and scientific allocation of personnel	Quite reasonable management mechanism and high quality of personnel	Corresponding management mechanism, but managers are lack of the necessary training	Personnel quality is not high and management is not reasonable	Backward management and no integrated management mechanism

Table 3.17 Evaluation criteria of wetland ecosystem health in mg/l

Index	The status of water environment quality			The characteristics of aquatic creatures	The quality of environmental habitat
	COD_{Mn}	TP	NH_3–N	Chl-a	DO
0	>30	>0.30	1.50 to ≤2.00	0.750 to ≤1.625	2.0 to <3.0
1	20 to ≤30	0.20 to ≤0.30	1.00 to ≤1.50	0.250 to ≤0.750	3.0 to <5.0
2	15 to ≤20	0.10 to ≤0.20	0.50 to ≤1.00	0.100 to ≤0.250	5.0 to <6.0
3	≤15	0.02 to ≤0.10	0.15 to ≤0.50	0.025 to ≤0.100	6.0 to <7.5
4	≤15	≤0.02	≤0.15	≤0.025	≥7.5

In the formula, *CI* stands for the health comprehensive evaluation index; n stands for the number of system evaluation index; I_i stands for the quantization value of the indexes; W_i stands for the weight.

(2) The evaluation criteria

According to the GB 3838–2002, *the quality standards for surface water environment*, we formulate the evaluation criteria of wetland ecosystem health in Liao River Conversation Area and it is seen in the Table 3.17.

(3) The evaluation level

According to the evaluation criteria, we grade the each index; we calculate the index value by the weighted average method; in order to distinguish the score differences between the sample, the score that we get by the weighted average method multiply by 20 to make sure the index score is between 0 and 20 and the full score of the wetland ecosystem health is 100 points. And the score is divided into 5 levels, including 0–20, >20–0, >40–0, >60–0 and >80–100, and they respectively stands for the health status of the river ecosystem, namely disease, general ill health, sub-health, health and good health (Table 3.18).

3.4.2.2 The Health Evaluation of the Wetland Ecosystem of the Tributary Entrance in the Liao River Conversation Area

The Layout of Sampling Point

The information of sampling points is shown in Table 3.19.

Table 3.18 Evaluation level of ecosystem health

Level	Score (%)	The level of ecosystem health
Good health	>80–100	The wetland landscape is in good condition; structure is reasonable; external pressure is low; ecosystem is extremely stable
Healthy	>60–80	The wetland landscape is in natural condition; structure is quite reasonable; external pressure is low; ecosystem is still stable
Less healthy	>40–60	The wetland landscape changes; structure is quite reasonable; external pressure is high; ecosystem is still stable
Morbidity	>20–40	The wetland landscape and structure are destroyed; external pressure is high; ecosystem begins to deteriorate
Illness	0–20	The wetland landscape and structure are completely destroyed; external pressure is high; ecosystem seriously deteriorate

Table 3.19 Latitude and longitude of sampling points

Number	Location	North latitude	East longitude	Administrative region
1	Zhaosutai River	42°38.023′	123°40.727′	Changtu
2	Liangzi estuary	42°27.689′	123°49.027′	Kaiyuan
3	Chaihe estuary	42°19.580′	123°51.600′	Tieling County
4	Fanhe estuary	42°15.544′	123°37.030′	Tieling County
5	Changgouzi River	42°18.993′	123°39.957′	Tieling County
6	Zuoxiaohe River Bridge	42°08.006′	123°23.174′	Shenbei
7	Yan Feili drained wetlands	41°57.475′	122°56.835′	Xinming
8	Xiaoliu estuary	41°11.589′	122°04.962′	Panjin
9	Yitong estuary	41°10.913′	122°00.485′	Panjin
10	Taiping estuary	41°08.614′	121°54.948′	Panjin
11	Raoyang estuary	41°07.035′	121°48.703′	Panjin
12	Pangxiegou	41°08.100′	121°56.350′	Panjin
13	Paihe River	41°03.396′	121°54.648′	Panjin

The Screening of Indexes

We analyze the data by the method of principal component analysis and on the principle of contribution rate of accumulated variance (>75%), we extract the principal components (Tables 3.20 and 3.21).

According to the principle that the factor loading value is over 0.8, the first principal component includes NH_3–N, DO, TP and NO_2–N; the second principal component includes pH, ORP and Chl-a; The third principal component includes NO_3–N. Because the variation range of pH is small, ORP was significantly correlated with

DO, NO_2–N, NO_3–N was significantly associated with NH_3–N. And according to the actual situation, we retain NH_3–N, TP, Chl-a and DO as the core indexes of wetland ecosystem health evaluation at the tributary estuary [24–27].

The Determination of Weight

According to the characteristic value, the initial factor loading value, the variance contribution rate and the cumulative variance contribution rate of the principal

Table 3.20 Explanation of total variance

Component	Initial eigenvalues			Extraction of squares and loading		
	1	2	3	1	2	3
1	4.393	39.935	39.935	4.120	37.454	37.454
2	2.842	25.834	65.769	2.925	26.595	64.049
3	1.586	14.422	80.192	1.776	16.143	80.192
4	0.981	8.915	89.106			
5	0.498	4.531	93.638			
6	0.251	2.281	95.919			
7	0.199	1.811	97.730			
8	0.152	1.385	99.115			
9	0.078	0.711	99.827			
10	0.019	0.173	100.00			
11	1.283E-5	0.000	100.00			

1. sum   2. variance (%)   3. accumulated variance (%)

Table 3.21 Initial factor loading

Index	First principal component	Second principal component	Third principal component
COD (mg/L)	0.534	0.667	−0.442
NH_3–N (mg/L)	0.915	0.005	0.179
DO (mg/L)	−0.873	0.132	0.082
pH	−0.411	0.867	0.159
EC (µs/cm)	0.483	0.273	−0.538
ORP (mV)	0.403	−0.870	−0.160
Watertemp. (°C)	0.711	0.261	0.203
TP (mg/L)	0.813	0.227	0.050
Chl-a (mg/L)	−0.194	0.811	0.061
NO_3–N (mg/L)	0.135	0.002	0.943
NO_2–N (mg/L)	0.850	0.136	0.275

Table 3.22 Weight

	NH$_3$–N	TP	Chl-a	DO
Estuarine wetland	0.2439	0.2437	0.2444	0.2098

Table 3.23 The results of the evaluation of the wetland ecosystem health of the tributary estuary

Sampling point	Comprehensive score	Scoring grade
1	24	Morbidity
2	11	Illness
3	54	Sub-health
4	48	Sub-health
5	23	Morbidity
6	48	Sub-health
7	70	Health
8	37	Morbidity
9	17	Illness
10	17	Illness
11	37	Morbidity
12	6	Illness
13	12	Illness

components that the four selected indexes correspond to, we calculate the weight of every index (Table 3.22).

The Results of Evaluation

The results of the evaluation are shown in Table 3.23.

Summary

(1) From the evaluation results, we can see that among the 13 sampling points, (7) is in the healthy state; (3), (4), (6) is in the sub-healthy state; (1), (5), (8) is in the general pathological state; (2), (9), (10) is in the disease state.
(2) The reason why (2), (9), (10), (12), (13) is in the disease state is that the tributaries of the river bring a large number of industrial and domestic sewage. And the content of COD$_{Mn}$, NH$_3$–N and TP is far beyond the standards of class V surface water, which makes the wetland water environment quality unhealthy and leads to the decreasing of the health level.
(3) The reason why (1), (5), (8), (11) is in the general pathological state is that the content of N, P and Chl-a in the tributaries of the river is too high, which will

promote the growth of algae and make the water in the eutrophication trend and results in the decline of biological diversity and water environmental quality and therefore it leads to a decreasing health level.

On the contrary, if the pollutants are purified inside the wetland, its contents significantly decrease in the outlet of wetland. So the wetland ecosystem health of the (7) sampling point is in good condition.

3.4.2.3 The Health Evaluation of the Wetland Ecosystem in the Liao River Conversation Area

The Layout of Sampling Point

The detailed information of sampling points is shown in Table 3.24.

The Screening of Indexes

We analyze the data by the method of principal component analysis and on the principle of the contribution rate of accumulated variance ($>75\%$), we extract the principal components (Tables 3.25 and 3.26).

According to the principle that the factor loading value is over 0.7, the first principal component includes NH_3–N, DO, TP, NO_3–N, EC, COD and Chl-a; the second principal component includes pH, ORP and TP. Because the variation range of pH is small, ORP was significantly correlated with DO; NO_3–N was significantly associated with NH_3–N; EC was significantly associated with TP. And according to the actual situation, we retain NH_3–N, TP, Chl-a, DO5 as the core indexes of wetland ecosystem health evaluation [28, 29].

The Determination of Weight

According to the characteristic value, the initial factor loading value, the variance contribution rate and the cumulative variance contribution rate of the principal

Table 3.24 Latitude and longitude of sampling points

Number	Location	N	E	Administrative division
1	Ford Dian	42°59.031′	123°33.481′	Changtu
2	Yubao platform	41°55.251′	122°53.309′	Xinmin
3	Liao wetland	41°31.528′	122°38.225′	Liaozhong

Table 3.25 Explanation of total variance

Component			Initial eigenvalues		
1	2	3	1	2	3
5.854	58.542	58.542	5.850	58.503	58.503
4.146	41.458	100.000	4.150	41.497	100.000
4.601E-16	4.601E-15	100.000			
2.729E-16	2.729E-15	100.000			
1.405E-16	1.405E-15	100.000			
8.577E-17	8.577E-16	100.000			
−3.095E-18	−3.095E-17	100.000			
−2.166E-16	−2.166E-15	100.000			
−2.933E-16	−2.933E-15	100.000			
−4.863E-16	−4.863E-15	100.000			

1. sum　2. variance (%)　3. accumulated variance (%)

Table 3.26 Initial factor loading

Index	First principal component	Second principal component
COD (mg/L)	0.781	0.625
NH_3–N (mg/L)	0.908	0.419
DO (mg/L)	−0.908	0.419
pH	0.124	0.992
EC (μ s/cm)	−0.999	−0.042
ORP (mV)	−0.123	−0.992
TP (mg/L)	0.975	−0.222
Chl-a (mg/L)	0.773	−0.634
NO_3–N (mg/L)	0.995	0.099
T (°C)	−0.168	0.986

Table 3.27 Weights

Area	Component				
Pond wetland	COD	NH_3–N	TP	Chl-a	DO
	0.20	0.30	0.28	0.20	0.30

components that the five selected indexes correspond to, we calculate the weight of every index (Table 3.27).

The Results of Evaluation

The results of the evaluation are shown in Table 3.28.

Summary

From the evaluation results, we can see that among the 3 sampling points, (1) is in the good health; (2) is in the healthy state; (3) is in the sub-healthy state.

The content of COD_{Mn} in the sampling point (3) meet the standards of class IV water quality, which makes the wetland water environment quality decline. So the wetland ecosystem is in the sub-healthy state.

(1) and (2) sampling points are in the good condition.

3.4.2.4 The Health Evaluation of the Oxbow Lake Wetland Ecosystem in the Liao River Conversation Area

The Layout of Sampling Point

The detailed information of sampling points is shown in Table 3.29.

The Screening of Indexes

We analyze the data by the method of principal component analysis and on the principle of the contribution rate of accumulated variance (>75%), we extract the principal components (Tables 3.30 and 3.31).

Table 3.28 Evaluation of wetland ecosystem health

Sampling point	Comprehensive score	Scoring grade
1	82.8	Good health
2	63.2	Health
3	43.6	Sub-health

Table 3.29 Evaluation of wetland ecosystem health

Number	Location	N	E	Administrative region
1	Juliu River rubber dam	42°00.754′	122°56.846′	Xinmin

According to the principle that the factor loading value is over 0.8, the first principal component includes NH_3–N, DO, pH, NO_2–N, EC, and Chl-a; the second principal component includes COD and TP. Because the variation range of pH is small, ORP was significantly correlated with DO; NO_2–N was significantly associated with NH_3–N; EC was significantly associated with TP. And according to the actual situation, we retain NH_3–N, TP, COD, Chl-a, DO5 as the core indexes of the Oxbow lake wetland ecosystem health evaluation.

The Determination of Weight

According to the characteristic value, the initial factor loading value, the variance contribution rate and the cumulative variance contribution rate of the principal components that the five selected indexes correspond to, we calculate the weight of every index (Table 3.32).

Table 3.30 Explanation of total variance

Component	Initial eigenvalues			The extraction of squares and loading		
	1	2	3	1	2	3
1	6.643	73.807	73.807	6.515	72.386	72.386
2	2.357	26.193	100.000	2.485	27.614	100.000
3	4.215E-16	4.683E-15	100.000			
4	2.505E-16	2.783E-15	100.000			
5	4.133E-17	4.592E-16	100.000			
6	−5.524E-17	−6.137E-16	100.000			
7	−1.478E-16	−1.642E-15	100.000			
8	−2.133E-16	−2.370E-15	100.000			
9	−2.824E-16	−3.137E-15	100.000			

1. sum 2. variance (%) 3. accumulated variance (%)

Table 3.31 Initial factor loading

Index	First principal component	Second principal component
COD (mg/L)	−0.217	0.976
NH_3–N (mg/L)	−0.999	0.046
DO (mg/L)	0.828	0.560
pH	0.989	−0.147
EC (μs/cm)	−0.995	−0.097
ORP (mV)	0.999	0.051
TP (mg/L)	−0.461	−0.888
Chl-a (mg/L)	−0.979	0.202
NO_2–N (mg/L)	0.880	−0.476

Table 3.32 Weights

Area	Component				
Pond wetland	COD	NH$_3$–N	TP	Chl-a	DO
	0.20	0.26	0.25	0.28	0.30

Table 3.33 Evaluation of wetland ecosystem health

Sampling point	Comprehensive score	Scoring grade
1	38.2	Morbidity

The Results of Evaluation

The results of the evaluation are shown in Table 3.33.

Summary

The reason why the wetland ecosystem of the sampling points is not in the good condition is that the content of N and P is too high, which will promote the growth of algae and make the water in the eutrophication trend. And this will also result in the decline of biological diversity and water environmental quality and therefore it leads to the decreasing of the health level.

3.4.2.5 The Health Evaluation of Qixing Wetland Ecosystem in the Liao River Conversation Area

The Layout of Sampling Points

In the study area, we set 13 sampling points (Fig. 3.54) and begin to investigate and monitor Qixing wetland ecosystem between August and October in 2012. And both of the situ collection and laboratory analysis refer to *the water and wastewater monitoring analysis method*. The name of sampling points: (1) the junction entrance of Xixiao River; (2) the junction entrance of Wanquan River; (3) the interchange of Xixiaohe River and Wanquan River; (4) the interchange of Xixiao River, Wanquan River and Yangchang River; (5) the junction entrance of Yangchang River; (6) the No. 1 steel dam gate (front); (7) the junction entrance of Chang River; (8) the location between the No. 1 steel dam gate and the No. 2 steel dam gate; (9) the No. 2 steel dam gate (front); (10) the water storage project the in the downstream of No. 2 steel dam gate (11) the submerged dam; (12) the front and back of Liao River mainstreams in Qixing wetland.

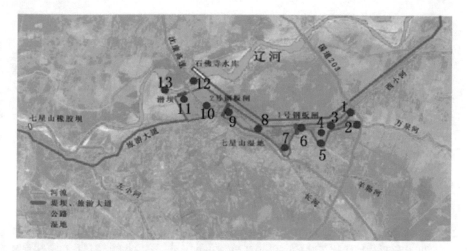

Fig. 3.54 Seven wetland distribution of sampling points

The Screening of Indexes

We analyze the data by the method of principal component analysis and on the principle of the contribution rate of accumulated variance (>75%), we extract the three principal components. The first principal component includes pH, DO and ORP; the second principal component includes EC and NH_3–N; the third principal component includes COD_{Mn} and TP. Because the variation range of pH in the sampling points is small, ORP was significantly correlated with DO and EC was significantly associated with TP. And according to the actual situation, we retain DO, NH_3–N, TP, Chl-a, COD_{Mn} as the core indexes of the Oxbow lake wetland ecosystem health evaluation. The quality of water environment is composed of NH_3–N concentration, COD_{Mn} concentration and TP concentration; the aquatic feature is composed of Chl-a concentration; the quality of habitat environment is composed of DO concentration (Table 3.34).

The Determination of Weight

The corresponding weights of the indexes that we calculate are seen in Table 3.35.

The Results of Evaluation

The results of the health evaluation of Qixing wetland ecosystem are shown in Table 3.36.

Summary

(1) At present, Qixing wetland ecosystem is in sub-healthy state. And the sampling points of 1, 4, 5, 6, 12 and 13 is in the sub-healthy state; the sampling points of 2, 3, 8, 9, 10 and 11 is in the general pathological state; the sampling point of 7 is in the disease state.

(2) The reason why the sampling point of 2 and 7 is in the good condition is that Wanquan River and Chang River bring a large number of industrial and domestic sewage. And the content of COD_{Mn}, NH_3–N and TP is far beyond the standards of class V surface water (GB 3838-2002), which makes the wetland water environment quality unhealthy and leads to the decreasing of the health level.

(3) Because of the influence of Chang River, the concentration of TP and Chl-a in the sampling point of 9 is too high, which will promote the growth of algae and make the water in the eutrophication trend. As a result, it results in the decline of biological diversity and water environmental quality and leads to the decreasing of the health level.

Table 3.34 Principal component analysis of candidate indicators

Index	First principal component	Second principal component	Third principal component
Watertemp (°C)	−0.268	−0.158	0.627
pH	0.027	0.505	−0.294
ORP (mV)	0.073	−0.525	−0.002
EC (S/m)	0.402	0.106	0.306
DO (mg/L)	−0.229	0.224	0.551
COD_{Mn} (mg/L)	0.359	0.033	0.237
NH_3–N (mg/L)	0.422	−0.131	0.118
NO_3–N (mg/L)	0.355	−0.209	−0.072
NO_2–N (mg/L)	0.208	−0.214	−0.108
TP (mg/L)	0.451	0.154	0.180
Chl-a (mg/L)	0.167	0.502	0.065
Variance contribution rate (%)	27.954	27.659	18.418
Accumulated Variance Contribution Rate (%)	27.954	55.613	74.030

Table 3.35 The weight of each index

Index	Component				
Weight function	COD	NH_3–N	TP	Chl-a	DO
	0.196	0.266	0.235	0.259	0.271

Table 3.36 Evaluation of Qixing wetland ecosystem health

Sampling point	Comprehensive score	Scoring grade
1	53	Sub-health
2	26	Morbidity
3	37	Morbidity
4	51	Sub-health
5	58	Sub-health
6	49	Sub-health
7	12	Illness
8	32	Morbidity
9	23	Morbidity
10	32	Morbidity
11	36	Morbidity
12	49	Sub-health
13	45	Sub-health
Qixing wetland (overall)	39	Sub-health

On the contrary, if the pollutants are purified inside the wetland, the concentrations of them significantly decrease in the outlet of wetland. So the wetland ecosystem health in sampling points of 12 and 13 is in good condition.

(4) The tributaries in the Qixing wetland bring a large number of industrial and domestic sewage. So the concentration of COD_{Mn}, NH_{s3}–N and TP is far beyond the standards of class V surface water (GB 3838-2002) [4], which significantly disturb the water environment quality and habitats. Therefore, we should strengthen the construction of vegetation buffer zone along the banks of the river.

References

1. DB34/T 732-2007 Technical guidelines for water function zoning.
2. DB43/T 432-2009 Technical guidelines for investigation of freshwater organisms.
3. GB 12997-91 Design technique standard for water quality sampling plan.
4. GB 3838-2002 Environmental quality standards for surface water.
5. GB/T 12997-91 The design rules of water quality sampling plan.
6. GB/T 12998-91 Water quality-guidance on sampling techniques.
7. GB/T 12999-91 The preservation and management techniques of water quality sampling.
8. GB/T 14529-93 Principle for categories and grades of nature reserves.
9. GB/T 14529-93 The division principle for the type and level of natural conservation area.
10. GB/T 14848-1993 Quality standard for ground water.
11. HJ 623-2011 Standard for the assessment of regional biodiversity.
12. HJ/T 192-2006 Technical criterion for eco-environmental status evaluation.
13. HJ/T 192-2006 The evaluation specification for ecological environment.
14. HJ/T 91-2002 Technical specifications requirements for monitoring of surface water and waste water.

15. HJ/T 91-2002 The environmental monitoring of surface water.
16. HJ/T Technical regulation for inland ecosystem function management zoning (pending).
17. HJ/T Technical regulation for river ecological investigation (pending).
18. HJ/T338-2007 Technical guideline for delineating source water protection areas.
19. SL 219-98 Regulation for water environmental monitoring.
20. The technical specification of basin water ecological function zoning (draft).
21. The technical guide of lake ecological security and evaluation, 2012.
22. Cai, Qinghua, Ming Cao, and Xiangfei Huang. 2007. *The observation method of water ecosystem*. China: Environmental Science Press.
23. Chen, Weimin, Xiangfei Huang, and Wanping Zhou. 2005. *The observation method of lake ecosystem*. China: Environmental Science Publishing House.
24. IUCN. 2000. *The endangered levels and standards of species red list IUCN 3.1 Edition*. World Nature Conservation Union.
25. Liu, Dachang, and Ai Che. 2014. *Guide for the preparation of biological diversity conservation planning*. China Environment Publishing House.
26. Meng, Wei. 2011. *The survey technology of river ecosystem*. Science Press.
27. Silk, Nicole, and Kristine Ciruna. 2013. *A practitioner's guide to freshwater biodiversity conservation*. Island Press.
28. Wan, Bentai. 2013. *The technical guide of the ecosystem health assessment*. China: Environmental Science Publishing House.
29. Wang, Sung, and Yan Xie. 2004. *The red list of Chinese species*. Higher Education Press.
30. Xie, Yuhao, and Xiaoping Pu. 1984. The biological aspects of pond smelt (Hypomesus olidus (Pallas)) in the Shui Feng Reservoir. *Acta Hydrobiologica Sinica* 84: 457–468.

Chapter 4
Management Technology and Strategy for Environmental Risk Sources and Persistent Organic Pollutants (POPs) in Liaohe River Basin

Lu Han, Bin Li, Ruixia Liu, Jianfeng Peng, Yonghui Song, Siyu Wang, Peng Yuan, Ping Zeng and Moli Zhang

4.1 Risk Source Identification Tool Method

by Lu Han, Yonghui Song, Peng Yuan, Moli Zhang

4.1.1 Introduction

Following the rapid development of science and technology in China, serious environment pollution accidents, especially the environment pollution accidents in chemical industry, frequently occur. All these serious pollution accidents turn out to be severe threats to the harmonious development of Chinese socio-economy, as well as to be devastating damages to the ecological environment. Therefore, concerns on environment risk surveillance management become more noticeable than ever. Therefore, in this project we aim to establish the risk source identification tool methods which will be applied to reduce the environmental risk level.

In this report we first reviewed the literature on risk sources and strategies for POPs management. Secondly we investigated the technical status and management of laws, regulations, methods and standards for the environmental risk. A statistical analysis of the historical emergency environmental incidents was conducted in order to identify the key factors that influence the environmental hazards. The enterprises environmental risk assessment process include 5 steps: data preparation and environmental risk identification, the possibility of abrupting environment affairs and its consequences, comparative analysis of existing environmental risk prevention and emergency management, improve the implementation of environmental risk prevention and emergency measures, classify the environmental risk.

L. Han (✉) · B. Li · R. Liu · J. Peng · Y. Song · S. Wang · P. Yuan
P. Zeng · M. Zhang
Chinese Research Academy of Environmental Sciences, Beijing, China
e-mail: hanlu@craes.org.cn

© Springer International Publishing AG, part of Springer Nature 2018
Y. Song et al. (eds.), *Chinese Water Systems*, Terrestrial Environmental Sciences,
https://doi.org/10.1007/978-3-319-76469-6_4

Delimitation of enterprise environmental risk level includes: (1) Calculating the ratio of quantity and its critical mass which is involved in environment risk (Q); (2) Calculating the enterprise process and environment risk control level (M), determine the process and environment risk control level; (3) Judging whether enterprise environment risk receptor is in line with the EIA and approval documents of health or atmosphere protection distance requirements, and determine the environment risk receptor type (E); (4) Determining the enterprise environment risk level and characterize the classification.

The environment risk assessment method will improve the environmental risk control level and improve the environmental contingency management ability. Meanwhile, it will provide technical support for the management of environmental protection agencies.

4.1.2   Range of Application for Risk Source Identification

Method

This method includes contents, procedures and methods of the identification and classification for environmental risk sources (hereinafter referred to as the environmental risk).

This method is used to carry out the assessment of environmental risk for enterprises in which emergency environmental accidents may happen. The enterprises either have been completed and put into operation or can be still in the stage of production. Assessment object are enterprises which impact production, usage, storage or release of abrupt environmental incidents risk substances or chemical substances (including raw materials, fuel, products, intermediate products, by-products, etc.) which are listed in the Appendix 2 or in the list of threshold quantities (Hereinafter referred to as environmental risk substance).

This method is not applicable to the following situations of environmental risk assessment: (1) Enterprises having nuclear facilities and/or process radioactive substances; (2) Enterprises engaged in hazardous collection, storage, utilization, disposal of business activities; (3) Enterprises or vehicles engaged in hazardous chemicals transportation; (4) Tailing ponds; (5) Oil and gas extraction facilities; (6) Military installations; (7) Oil and gas long-distance pipelines, municipal gas pipelines; (8) Gas stations and refueling stations; (9) harbors and docks.

4.1.3   General Requirements of the Environmental Risk Assessment

1. In the case of any of the following circumstances, enterprise shall timely classify or reclassify its environmental risk level and prepare or revise its environmental risk assessment:

- Enterprises have not classified its environmental risk level or have classified over 3 years;
- Enterprises have species and quantity of substance impacting environmental risk, production and processing of chemicals, precautionary measures or changes of surrounding receptors that may be influenced by environmental risk accidents.
- Enterprises have been involved in emergency environmental incidents and have polluted the environment.
- Changes in enterprises environmental risk assessment standards or normative documents to which the company has to adapt.

2. Enterprises can compile environmental risk assessment by itself or delegate relevant professional and technical services.
3. The content of environmental risk assessment in the environmental impact assessment report for new, change or expanded projects can be used as an important part of environmental risk assessment for the enterprise which the projects belong to.

4.1.4   Process of Environmental Risk Assessment

The enterprises environmental risk assessment is implemented according to 5 steps: data preparation and environmental risk identification, the possibility of abrupt environment affairs and its consequences, gap analysis of existing environmental risk prevention and emergency management, optimizing the implementation of environmental risk prevention and emergency measures, classify the environmental risk.

4.1.5   Content of Environmental Risk Assessment

4.1.5.1   Data Preparation and Environmental Risk Identification

On the basis of the collected relevant data we carry out environmental risk identification. The object of environmental risk identification includes: (1) enterprise basic information; (2) surrounding environment risk receptor; (3) environmental risk substance and its quantity; (4) production process; (5) management of safety production; (6) environment risk units and the existing environmental risk prevention and emergency measures; (7) the existing emergency resources.

The aforementioned steps (2)–(6) of environmental risk identification should be considered environmental risk enterprises, route of transmission of environmental risk and environmental risk receptors. This assessment enforces the enterprises to provide maps or diagrams of geographical location and plant floor, surrounding environmental risk receptor distribution, the rain and clean water collection and drainage pipe network, the sewage collection and drainage pipe network diagram and all the drainage final destinations. All these maps and diagrams will be accessory of the assessment.

Enterprise Basic Information

List the following elements:

1. Company name, organization institution bar code, legal representative, unit location, latitude, longitude, industry class, construction time, the latest date of reconstruction measures, the main contact person, enterprise scale, factory area, number of staff and workers etc. and if needed the names of its mother company or membership groups.
2. Terrain, landform (when it's flood area, river nearby or sloping fields), climate type, wind rose, recorded extremely weather conditions and natural disaster occurred in the specific area such as such as earthquake, typhoon, debris flow, floods.
3. Environmental function zones and surface water, ground water, atmosphere and soil environmental quality in a recent year.

The Existing Emergency Resources

The existing emergency resources are the enterprise internal and external emergency supplies including the rescue deal signed with other companies and the used urgent equipment and emergency rescue team system The emergency supplies includes various of flocculants, adsorbents, neutralizers, antidotes, redundant-oxidants, etc. The emergency equipment includes personal protective equipment, emergency monitory ability, emergency communication system, power supply and emergency power supply, lighting devices, etc.

List the following elements according to the emergency goods, equipments and rescue teams:

- Name and type of emergency resources (including goods, equipments or teams),
- number of people,
- period validity of goods,
- the name and contact and phone number of the external supply companies.

4.1.5.2 Possible Emergency Environmental Incidents and Related Impact Scenario Analysis

Data Base of Comparable Domestic and International Emergency Environmental Incidents

List and explain the following elements:

Year and date, address, equipment scale, root cause, material leakage, sphere of influence, emergency measures have taken, event loss, effects of the environments and human.

Analysing Worst Case Scenarios of Emergency Environmental Incidents

List and analyse the worst scenarios which could trigger an emergency environmental incident according to the following aspects:

1. Production safety accident like a fire, an explosion or a leakage could cause emergency environmental incident leading to outside environment pollution and casualties. Typical secondary incidents could be the poisonous and harmful gases diffusion, the fire fighting water, material leakage and resultant of reaction effuse the plant area from the rain, clean water, sewage discharge openings and from the plant door or walls to pollute the environment;
2. The environmental risk prevention and control facilities failed to work or have been abnormal used. For instance the rain valve cannot be normally closed, or the chemical industry torch is accidental quenched;
3. Abnormal operation modes (driving or parking);
4. Pollution control facilities abnormal operated;
5. Illegal release;
6. Power cut, water-break, gas cut;
7. Communication or transportation system failure;
8. Various natural disasters and extreme or adverse weather condition;
9. Other possible scenarios.

The Intensity of Pollution Analysis of Each Scenario

Analyze the intensity of pollution according to the aforementioned scenarios, including the types of environmental risk materials, physicochemical properties, minimum and maximum release quantity, diffusion range, concentration distribution, duration and hazard.

The calculation of the intensity of pollution can be reference to <Technical Guidance of Environmental Risk Assessment for Construction Projections> [25].

Environmental Risk Materials Release Route, Environmental Risk Prevention and Emergency Control Measures, and Emergency Resources Analyze at Each Scenario

The distribution of environmental risk materials may cause ground water and surface water and soil pollution. Analyze the possibility of influence on the environmental risk receptor, the release condition and the emission pathways. Furthermore the emergency goods and emergency equipments and emergency teams are analysed.

In a scenario causing air pollution, the possible sphere of influence of leaking risk material has to be analysed at day and night conditions considering wind speed and direction. The sphere of influence includes emergency isolation distance and the leeward personal protective distance to the accident location.

Analysis of Direct, Secondary and Derive Consequence at Each Scenario

According to the analysis of sections "The Intensity of Pollution Analysis of Each Scenario" and "Environmental Risk Materials Release Route, Environmental Risk Prevention and Emergency Control Measures, and Emergency Resources Analyze at Each Scenario", considering and giving the influence degree and scope of emergency environmental accidents to environmental risk receptors, including the evacuation number, influence of drinking water sources, the transboundary influence, the ecological function of ecological sensitive area, and estimating the abrupt environment affairs level regarding those aspects as surface water, ground water, solid, atmosphere, population, property and society.

4.1.6 Comparative Analysis of Existing Environment Risk Prevention and Control and Emergency Measures

According to the analysis of Sects. 4.1.5.1 and 4.1.5.2, the completeness, reliability and effectiveness of the existing environment risk prevention and control and emergency measures from the following five aspects are analyzed. To find and subsequently solve gaps and problems, it is proposed to demand rectification on a short, intermediate and long-term scale

4.1.6.1 Environmental Risk Management System

Check if the enterprise has done the work as follows:
1. Whether environment risk prevention and control system and emergency measures are established or not, whether the responsible person or agency of environment risk prevention and control position is clear or not, whether the institution of regular visit and responsibility of maintenance is implemented or not;
2. Whether the environment risk prevention and control and emergency measures of EIA and approval documents are implemented or not;
3. Whether to carry out the publicity and training to employees about environment risk and environment emergency management regularly or not;
4. Whether the abrupt environment affair information reporting system is established or not, whether the system is effective executed or not.

4.1.6.2 Environmental Risk Prevention and Control and Emergency Measures

Check if the enterprise has done the work as follows:
1. Whether has set monitoring and controlling equipments according to the characteristics and harm of the environment risk substances which may be expelled from

waste gas discharge outlet, waste water and rain water and clean sedge discharge outlets or not. Analyzing each measure above, and analyzing the implementation of responsibility and the effectiveness of those measures;

2. Whether has taken measures including river closure, accident drainage collecting, the clean water system prevent and control, rainwater system control, production wastewater treatment system control, and so on, for preventing accident drainage and pollutant diffusion. Analyzing each measure above, and analyzing the implementation of responsibility and the effectiveness of those measures;
3. Whether has set toxic gas leak emergency disposal, whether has set toxic gas leak monitor-control and early warning system in production area, whether has measures and means to remind surrounding public to emergency evacuation. Analyzing each measure above, and analyzing the implementation of responsibility and the effectiveness of those measures.

4.1.6.3 Environment Emergency Resources

Check if the enterprise has done the work as follows:

1. Whether has equipped with the necessary emergency supplies and equipment (including emergency monitor);
2. Whether has set up emergency rescue teams consisted of professionals and part-timers;
3. Whether has signed the agreement of emergency rescue and communal agreement with other organization (including emergency supplies, emergency equipment and rescue teams).

4.1.6.4 Historical Experience and Lessons

Analyzing and summing up the experience of the same type enterprise which has had abrupt environment accident in history, and checking whether the enterprise under assessment has set up measures to prevent the same affairs.

4.1.6.5 Short-Term, Medium-Term and Long-Term Rectification

Put forward the rectification period as short-term 3 month, medium-term 3 to 6 month and long-term above 6 month according to the harmfulness, urgency and time allowed to the gaps and hidden trouble of the screening of each item aforesaid. Listing the content need to be corrected including environment risk unit, environment risk substance, current problems (including environment risk management system, environment risk prevention and control emergency measures, emergency resources), and environment risk receptor.

4.1.7   Improve the Implementation Plan of Environment Risk Prevention and Control and Emergency Measures

A plan to improve environment risk prevention, control and the emergency measures is set out, respectively, to the short-term, medium-term and long-term item. The plan should clearly define the environment risk management system, the environment risk prevention and control measures and the environment emergency ability, and so on. The plan should strengthen the destination of environment risk prevention and control measures and emergency management, and the dead line to complete.

The schedule attainment should be archivist files to future reference after each complement of the plan.

The scenarios that cannot be improved because of the external factors such as the problem of distance and protection of environment risk receptor should be reported to the authorities at or above the county level and the related departments in the region. The enterprise should cooperate to take measures to eliminate hidden dangers.

4.1.8   Delimit Enterprise Environmental Risk Level

The enterprise should reformulate the emergency environment incident response according to the short, medium and long-term plan. Enterprise should reclassify its environmental risk level according to the Appendix 1 and record levels defined process, including:

1. Calculate the ratio of quantity and its critical mass which involved in environment risk (Q);
2. Calculate the enterprise process and environment risk control level (M), determine the process and environment risk control level;
3. Judge whether enterprise environment risk receptor is in line with the EIA and approval documents of health or atmosphere protection distance requirements, and determine the environment risk receptor type (E);
4. Determine the enterprise environment risk level and characteristic the classification as request.

4.2   Source Inventory of Water Environmental Risk

by Lu Han, Jianfeng Peng, Yonghui Song, Siyu Wang, Peng Yuan

4.2.1   Introduction

With rapid economy growth and development, sudden environmental pollution accidents have become a major cause to environmental health and risk in China. According to the environment statistics data from 2002 to 2011, over 9,000 environmental pollution accidents occurred during ten years, which lead to huge economic losses and adverse social impacts (Ministry of Environmental Protection of China, 2003–2012). Water and air pollution accidents are the main types of all accidents, accounting for 55% and 34% respectively. Pollution accident occurred mainly due to production safety accidents and chemical transportation accidents.

The pollution accident prevention and emergency response were attracted a great attention by the Chinese government, especially after the Songhua River major chemical pollution incident in 2005. Environmental risk prevention was proposed as a strategic mission in China's national environmental protection plan of the 12th Five-year plan (2011–2015).

The management of the environmental risk sources has been the most effective approach of environmental accidents prevention. Liao River basin is China's old industrial base with a large number of petrochemical, chemical, pharmaceutical and metallurgy industries. How to identify and to classify the high-risk and heavily polluting industries is the key issue for environmental risk prevention and management in Liao River basin.

The purpose of the report was to investigate data of key petrochemical, chemical, pharmaceutical industries and enterprises, to identify and to classify the water environmental risk sources, and to put forward a key risk source inventory in the Liao River Basin.

4.2.2   Study Area and Data Investigation

4.2.2.1   Data Investigation

Major chemical enterprises were investigated. Investigation data included among others used chemicals, production process, environmental risk control measures, waste water discharge management and environment risk acceptors.

The data of 1008 enterprises were collected supported by the Liaoning province inspection on major enterprises environmental risk and chemicals in the key industries.

4.2.3   Water Risk Sources Identification and Classification

4.2.3.1   Result of Water Risk Sources Classification

Based on the risk sources identification and level classification method, the environmental risk sources were classified into three categories, i.e. high risk, medium risk and low risk.

By applying the method, 1008 enterprises in Liao River basin were analyzed and evaluated. The quantity of environmental risk substances (Q), the technical level of environmental risk control in the enterprises (M) and the vulnerability of the environmental receptors (E) were calculated. According to the grade evaluation matrix, 1008 enterprises were classified according to the three risk levels.

The key risk source inventory identified 117 enterprises as high risk sources, which were 11.6% of all evaluated enterprises. Furthermore 183 medium risk sources could be identified, which account for 18.2% of all evaluated enterprises. The majority of enterprises, 53.8% or 542 in total, could be classified as sources providing a low risk. The remaining 166 enterprises were not evaluated because of a lack in data assessment.

Figure 4.1 showed the percentage of different risk levels for the water environmental risk sources.

The source inventory for risk enterprises in LRB will be shown in the Sect. 4.4 named "List of Environmental Risk Sources". The results provide necessary information for risk management and will support the strategy development for Liao River Basin

The numbers of high risk and medium risk sources are shown in Table 4.1.

Fig. 4.1 Statistics for the different risk levels for the water environmental risk sources

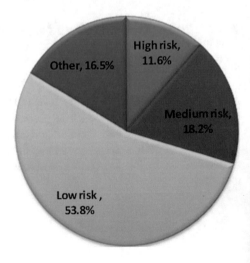

Table 4.1   Result of risk sources classification of the cities in Liao river

City	High risk sources		Medium risk sources	
	Number	Percentage (%)	Number	Percentage (%)
Shenyang	28	23.9	56	30.6
Anshan	5	4.3	21	11.5
Fushun	23	19.7	17	9.3
Benxi	1	0.9	17	9.3
Jinzhou	1	0.9	3	1.6
Yingkou	8	6.8	20	10.9
Fuxin	4	3.4	2	1.1
Liaoyang	16	13.7	25	13.7
Panjin	23	19.7	12	6.6
Tieling	8	6.8	9	4.9
Chaoyang	0	0.0	1	0.5
Total	117	100.0	183	100.0

4.2.3.2   Spatial Distribution of High Risk Sources

Spatial distribution of high risk sources and medium risk sources were shown in Fig. 4.2 and Fig. 4.3, respectively.

4.2.4   Suggestions for Risk Sources Management

To achieve the gains at low cost, different risk management measures should be taken for the facilities at different risk levels. For the low risk sources, manufactures should pay attention to internal risk management, such as risk information reporting, major hazards registration, monitoring and detection system establishment and vocational training. For the high risk sources, environmental risk receptor investigation, public participation and preparation of emergency response plans should also be strength-ened besides the daily risk management, which were helpful for the prevention of water environmental pollution accidents and the reduction of environmental hazards.

Develop the management strategy for environmental risk sources will be the next task in WP3.

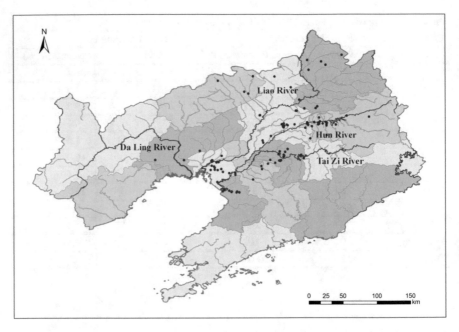

Fig. 4.2 Spatial distribution of high risk sources in Liao River Basin

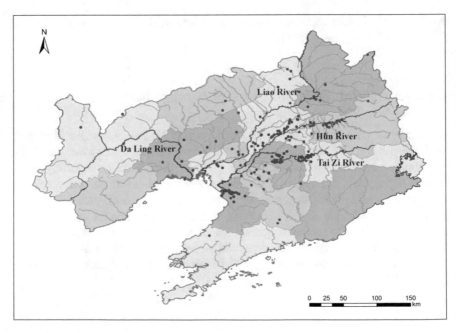

Fig. 4.3 Spatial distribution of medium risk sources in Liao River Basin

4.3   Priority Persistent Organic Pollutants

by Bin Li, Ruixia Liu, Lu Han, Yonghui Song

4.3.1   Introduction

The water quality, described by the chemical oxygen demand (COD) as an index, has been significantly improved in the Liao River Basin (LRB) in recent years. Toxic organic pollutants and heavy metals, however, are still present. Due to their bio-accumulative, persistent and poorly biodegradable properties as well as carcinogenic, teratogenic and mutagenic effects, these pollutants have caused severe water environment pollution in the LRB and threatened to human health. This reports provided the pollution levels of polycyclic aromatic hydrocarbons (PAHs), organic chlorinated pesticides (OCPs), polychorinated biphenyls (PCBs), emerging persistent organic pollutants (POPs) and heavy metals in the LRB. The quantitative source apportionment methods and their application in LRB were summarized in detail. It was indicated that the organic pollutants and heavy metals in the LRB mainly originated from industrial sources, municipal wastewaters and agricultural nonpoint sources. The main problems and some suggestions for future study were proposed in order to acquire systematic data sets and comprehensive information on water pollution characteristics, which would provide valuable basis for government to establish an efficient strategy on prevention and control of highly toxic organic pollutants and trace heavy metals in the LRB.

There are various types of toxic and hazardous substances (THSs) present in the diverse environment. In terms of their properties, these THSs can be classified as chemical contaminants including natural and synthetic compounds, physical pollutants derived from excessive sediment load and rubbish from human manufacturing activity e.g. plastic bags, bottles, as well as biological contaminants covering bacteria, virus and parasite. Of the THSs the most concern is chemical contaminant. It was reported that more than 70,000 chemicals were globally in common use and about 1,000 new chemicals were produced and registered per year [6, 16]. During their manufacture, storage, selling, transportation, service and waste disposal, the chemicals may release to the environmental systems due to their abuse, misuse and leakage. These chemicals, to some extent are toxic and hazardous to the public health and eco-systems [27, 50]. The typical THSs are composed of inorganic heavy metals and organic pollutants.

As reported by previous studies, the surface water of rivers and lakes in China was subjected to severe heavy metal pollution. Owing to the accumulation and enrichment of heavy metals in suspended particulate matter (SPM) or sediments, about 80.1% of sediments were polluted by heavy metals [83]. The survey on Yellow River Basin demonstrated that 16.7% of the state-controlled sampling sites showed higher concentrations of Cd than Chinese surface water standard level [84]. Lower concentrations of Cu, Pb and Cd were found in the sediments of Taihu Lake [14]. As one type of widespread pollutants, heavy metals were also observed in the global water bodies from other countries with serious pollution level, such as Hg was leaked to water system from the industrial wastewater occurred in Japan, causing Minamata disease [44]; about 50% of surface water in Poland was polluted by heavy metals released from mining and metallurgical industries, leading to water quality worse than the national water quality standard class-III [13].

Organic pollutants include traditional persistent organic pollutants (POPs), such as polychlorinated biphenyls (PCBs) (Jaward et al. 2005), polycyclic aromatic hydrocarbons (PAHs) [4, 28], organic chlorine pesticides (OCPs) [4], emerging POPs and other substituted benzene [42]. Although these pollutants are at a trace level or less in water environment, they can be accumulated in aquatic biota, migrate and transform, and further result in massive threat to ecological safety and human health because of their high toxicity and bioaccumulation, and poor degradability. Additionally, most of these organic pollutants have "carcinogenic, teratogenic and mutagenic" effects and some act as endocrine disrupting chemicals. According to the 10-year-term monitoring of Hudson River by US Environmental Protection Agency (EPA), the concentration of PCBs in the surface water, sediments and fish were found to be over the standard set by US Food and Drug Administration (FDA) [56, 57]. The level of PCBs in Seine River (France) was 50–150 ng/L in the dry season and 500 ng/l in the wet season [5]. The PCBs were also detectable in most of seven major water basins in China with higher levels in the Huai River, Yangtze River, Pearl River and Songhua River [15, 40, 59, 62].

The occurrence of PAHs in water systems has attracted attention globally. As reported by previous researches, the most heavily polluted water bodies were River Thames and River Trent (UK), as well as Ache River (Germany) with levels of PAHs exceeding the magnitude of $\mu g/L$ [45, 52]. The detection of PAHs in the major rivers of China has been conducted from 1990s [10, 33, 39, 60]. It was found that the concentrations of PAHs were mainly in the range of 10–1000 ng/L in water environment. OCPs have been monitored continuously all over the world because of their proven toxicity to human beings, animal, plant life and food chain. The pollution of OCPs in water originated from point and nonpoint sources, such as the wastewaters discharged directly to the river during the manufacture and use of pesticides. A survey on the level of OCPs showed that the concentration in global marine environment ranged from 0.1 to 4.4 ng/g in sediments and the heavily polluted level could achieve the magnitude of $\mu g/g$ in some water bodies [11]. The concentrations of OCPs in water systems in China varied highly depending on geo-locations with the concentration less than 200 ng/L in most sample sites [34, 36, 63].

Therefore, the pollution level and data sets on spatial distribution of POPs will gain systematic information in natural aquatic systems and the list of priority persistent organic pollutants will improve the environmental risk management ability.

4.3.2   Occurrence of POPs and Heavy Metals in LRB

The LRB is one of the important heavy industry base in China. The traditional industries have been developed for 60 years. Industrial clusters including chemical, petrochemical, pharmaceutical, metallurgical, dyeing and other sectors have been formed in the LRB, particularly in the middle and lower reaches of water systems near teh cities Fushun and Shenyang.

Since 2008, some efforts have been made for the pollution control of point sources in the LRB and a series of wastewater treatment techniques have been developed for the removal of Chemical Oxygen Demand (COD), Biochemical Oxygen Demand (BOD) and NH3-N [8, 49, 55, 65, 68]. A large-scale constructed wetland in Liao River Conservation Area (LRCA) has been also built for water purification [81].

The monitoring results for water bodies indicated that the water quality, in terms of indexes such as COD, BOD and NH3-N, has been improved in the LRB year by year [12, 43, 48]. However, there have been more concerns regarding to the level and ecological risk of trace pollutants, in particular POPs and heavy metals in water bodies. Although some efforts have been made to determine the level and identify pollution sources of the trace pollutants in the LRB [75], there are few comprehensive overviews regarding the pollution level and source apportionment of trace THSs in the LRB [38]. The objective of the present study is to summarize the pollution level and discuss the potential pollution sources of POPs and heavy metals in the LRB, so as to improve the understanding of environmental status at the regional level. The challenges and suggestions are proposed for further developing an efficient strategy in the control and reduction of trace pollutants in the LRB.

4.3.2.1   PAHs

Wastewater in the industrial areas of the Liao River basin often contains PAHs discharged to the basin from these industries due to the less advanced industrial technology and lack of efficient wastewater treatment and management approaches. In particular, waste water from from coking and gasworks, organic chemical, petrochemical, steel and power industries, causing the higher concentration of PAHs in the river basin. Thus, PAHs have attracted considerable attention in the LRB.

The concentration levels and pollution characteristic of PAHs in water body and sediments of the LRB have been investigated since 2000. It was shown that the concentration of total PAHs in sediments collected from Xinming section of Liao River mainstream ranged from 27 to 198 ng/g [70], and the highest level (μg/g) was found in Tieling section of the mainstream [21, 72]. The PAHs with high-molecular-weight were dominant in the sediments from rivers Hun-Taizi-Daliao mainstreams and the levels of total PAHs were in the range of 62–841 ng/g, which were obviously higher than those of Xinming section in Liao River mainstream [17]. Within the three riverine mainstreams, the Daliao River showed the highest level of the PAHs in sediments, followed by the Taizi River and Hun River.

One-year-term monitoring in Hun River and Daliao River indicated that the mean concentration of the PAHs compounds in the dry season was higher than that in the wet season, and the order of the PAHs concentration in surface sediments from different districts was Shenyang > Fushun > Yingkou [37, 82]. Xi River is the major sewage canal in Shenyang City. All kinds of sewage and pollutants flow into Hun River through Xi River. The concentrations of PAHs in the sediments detected in 2005 was 340–5070 ng/g [19], which was comparable to other investigated tributaries even though a large amount of domestic sewage and industrial effluents were discharged directly to Xi River.

PAHs can be highly absorbed by suspended particulate matter (SPM) and surface sediments because of their high octanol-water partition coefficients. Water samples collected from rivers Hun-Taizi-Daliao mainstreams showed that the level of PAHs in SPM (ranged from 318 to 238519 ng/g) was much higher than that in the sediments (62–841 ng/g). The concentration of PAHs in water ranged from 950 to 13,456 ng/L with level order of Taizi River > Daliao River > Hun River [18, 82]. Surprisingly, it was found by Wu and others that the higher level of PAHs at the magnitude of mg/L occurred in the waters collected from Shenyang and Fushun sections in Hun River mainstream, which was significantly higher than that (magnitude of μ close to the industrial area are more easily and severely PAHs-polluted than the residential and non-industrial regions) [82].

4.3.2.2   OCPs and PCBs

In 1998, thirteen OCPs and four PCBs were primarily identified in surface water and sediments of Liao River mainstream, of which dichlorodiphenyltrichloroethane (DDTs), hexachlorcyclohexane (HCHs) and PCBs were dominant pollutants. The results showed that the concentration of each OCPs or PCBs was below 91.3 ng/L in water and 29 ng/g in sediments respectively, but slightly higher than the levels in water bodies and comparable to those in sediments from global water systems monitored in the 1990s [76]. In 2002, the monitoring results for OCPs in Liao River mainstream showed that the concentration of total OCPs ranged from 7.59 to 34.98 ng/L in water and from 0.45 to 7.26 ng/g in sediments, respectively. The concentration of OCPs in water from Liao River mainstream were close to the level in Xiamen port, Nanjing section of Yangtze River and Haihe estuary in China [69].

The monitoring results for OCPs and PCBs in 2011 indicated that the concentration of total OCPs in Liao River mainstream was from 3.06 to 23.24 ng/L and the level of PCBs was lower than 115.3 ng/L [79]. The pollution levels of OCPs in the surface sediments from Hun-Taizi-Daliao mainstream were in the range of 3.06–23.24 ng/g. The HCHs compounds were the main pollutants of OCPs and the level of γ-HCH surpassed the threshold effects level (TEL), which could cause the toxic effects on aquatic organism and ecosystem [61].

Dahuofang Reservoir is the drinking water source for nine cities in the central of Liaoning Province. It is situated in the upstream of Hun River in Fushun. The OCPs and the PCBs compounds were detected in surface water and sediment from the reservoir. The result showed that the levels of OCPs and PCBs in water bodies were in the medium or slightly lower than those in other national water systems, the concentrations of HCHs, DDTs, PCBs in sediments, however, were in the range from 0.70 to 3.48 ng/g, 0.85 to 4.94 ng/g and 1.46 to 3.52 ng/g, respectively, of which DDTs and γ-HCH in sediments were at middle/high ecological risk level [35].

4.3.2.3 Other Organic Pollutants

Table 4.3 presented the concentration of other organic pollutants in LRB. Phthalic acid esters (PAEs), environmental endocrine disrupting chemicals, have been widely used as plastic addictives for 80 years. The continued long-term application of these chemicals has been accompanied by their emission and pollution to aquatic environment. The surface waters in Liao River and some tributaries were collected in 2011 for the determination of PAEs and the results showed that the average concentration of the PAEs was 61.6 μg/L, implying the water body has been severely contaminated by industrial effluents [79]. Our previous study indicated the concentration of PAEs was in the range of 6.3–33.9 ng/L in the Xi River and Pu River of Shenyang (data has not been published).

The dioxins and perfluorinated compounds (PFCs) were also detectable in the waters and sediments in the LRB. The concentrations of seventeen polychlorinated dibenzo-p-dioxins and furans (PCDD/Fs) in Daliao-Hun-Taizi mainstream ranged from 0.108 to 87.6 pg/L in surface water and 29.1–3039 ng/kg in the sediment (dry weight), respectively [53, 78]. The total concentrations of PCDD/Fs in the surface sediments of the middle and lower reaches in the Liao River mainstream were in the range of 13.7–458.5 ng/kg (dry weight), suggesting that PCDD/Fs pollution in most sediment samples might originated from the local combustion processes, including coal burning, agricultural straw open burning, iron orc sintering, cement production and secondary Al and Cu metallurgy [75].

Perfluorooctanoic acid (PFOA) and perfluorodecanoic acid (PFHxA) were identified to be the dominant PFCs in Hun River mainstream with levels of 1.83–10.7 ng/L and 1.26–37.6 ng/L, respectively, while perfluorooctane sulfonic acid (PFOS) was measured at low levels. PFHxA, PFOA, PFOS, perfluorohexane sulfonic acid (PFHxS) and perfluoroheptanoic acid (PFHpA) were found to be major contributors to the pollution of PFCs in the Xi River of Shenyang with concentrations of 19.2, 14.5, 10.7, 34.1 and 38.1 ng/L, respectively [54]. The concentration of PFCs varied in the range of 1.4–131 ng/L in the surface water and 0.26–1.11 ng/g in the sediment collected from Hun River and Liao River mainstreams [73]. North, South and Weigong canals in Shenyang are typical channels conveying industrial and domestic effluents. It was showed that the concentration of PFOA, PFHxA, 8:2 fluorotelomer unsaturated carboxylic acids (FTUCA) in North canal were 18, 8.8 and 4.9 ng/l, respectively. The pollution level of PFOA in South canal was 9.4 ng/L and the main contaminants, PFHxA and PFOA in Weigong channel were at the level of 28.4 and 13.7 ng/L, respectively [54]. Our monitoring results showed that the level of PFCs was between 1.8–13.0 ng/L in the waters and between 0.13–0.5 ng/g in the sediments in Hun River and Daliao River (data has not been published).

A survey report on water quality declared that there were 14 categories of 403 pollutants, including PAHs, OCPs, PAEs and phenols detected in the surface waters and sediments from Hun River in Shenyang and the priority pollutants list was proposed using potential hazardous index method [58]. A scientific investigation on the occurrence of THSs in the LRB, supported by the Major Science and Technology Program for Water Pollution Control and Treatment, was carried out from 2008 to 2012. It revealed that approximately 33–67 organic pollutants including 14 Chinese Priority Pollutants, such as alkanes, amines, phenols, carboxylic acids, N-heterocyclics and PAEs, were identified in 13 sampling sites of Hun River and Xi River, and the wastewater discharge from pharmaceutical, chemical, petrochemical and other industries in the upstream were estimated to be the prominent sources of these organic contaminants.

4.3.2.4 Heavy Metals

The pollution level and ecological risk assessment of heavy metals have been the considerable issues in the LRB. The spatial distribution and pollution characteristics of heavy metals in surface sediments of the LRB were investigated from 2008 to 2011. The results showed that due to the accumulation and deposition of heavy metals onto the sediments, the mean concentrations of As, Cu, Ni, Cd, Cr and Pb were obviously higher than those determined in 1998. The pollution of the heavy metals in Daliao River was worse than in Liao River, which supposed that tons of wastewaters containing heavy metals from chemical, metallurgical and other industries contributed to the more serious pollution in Daliao River [7, 9, 77]. Among the studied heavy metals, the order of pollution extent for these heavy metals in Daliao River was Pb > Zn > Co > Cu [66].

The quantitative assessment of heavy metal contamination in the sediments of LRB was carried out for the investigation of geochemical background of heavy metals and sediment heterogeneity [26]. The average background levels of Cr, Cu, Ni, Pb and Zn were 32.6, 11.1, 13.1, 16.3 and 37.8 mg/kg, respectively; while the upper limit of background level were 60, 21, 27, 23 and 96 mg/kg. The river sediments adjacent to big cities and mining areas were contaminated by these heavy metals. The study on the metal bioavailability (Ni, Cu, Zn, and Pb) showed that the highest bioavailability values for metals could be found in Daliao River mainstream, followed by Taizi River [22].

4.3.3 Potential Source of THSs

The source identification of THSs plays significant role in proposing efficient pollution control strategies for these pollutants. The source apportionment of atmospheric aerosol constituents has been extensively studied in the past decades [23, 29]. Several approaches including Chemical Mass Balance model (CMB) and multivariate statistical methods, such as UNMIX and positive matrix factorization (PMF), Factor Analysis with Multiple Linear Regression (FA-MLR) have been developed and widely employed to calculate the source contribution or chemical composition of pollutants [30, 46, 74].

The CMB model combines the chemical and physical characteristics of chemical species measured at sources and receptors to quantify the source contributions to the receptor. It requires detailed knowledge of source types and uses measured fingerprints of source profiles to reconstruct concentrations of chemical species [64, 74]. UNMIX, PMF and FA-MLR are commonly used multivariate statistical models. They are capable of simultaneously analyzing a series of observations and quantitatively assessing the contributions of various sources to each observation without prior knowledge about number and nature of sources. The Unmix model estimates the number of sources using a singular value decomposition method to reduce the dimensionality of data [24, 30]. The PMF model is a bilinear model with non-negativity constraints, by oblique solutions in a reduced dimensional space to quantitatively apportion pollution source [46, 51].

The FA-MLR can be used to describe variability among observed variables in terms of fewer unobserved variables called factors [1, 47]. It offers valuable qualitative information about potential pollution sources instead of adequate quantitative information concerning the contribution of each pollution source. The stepwise combination with MLR can be used to determine the mass apportionment of each source to total concentration.

The accurate source apportionment of THSs in aquatic system is rather scarce because of the complexity of water matrix, the incomplete pollution source profiles, lack of historical data and the limited analytical techniques. Given advances in source sampling, analytical techniques and predetermined source profiles, the most concern is the detailed source identification of PAHs in atmospheric and aquatic environment and soil [2, 3, 31].

Since 2006, the pollution source of heavy metals and organic pollutants in the LRB, based on the primary monitoring data, has been identified using several methods, such as UNMIX, PMF and FA-MLR.

4.3.3.1 Pollution Sources of Heavy Metals

It has been ascertained that the water pollution by heavy metals, e.g. Pb, As, Cd and Hg, is mainly caused by the emission of wastewater from pesticide production, apparatus manufacture, chemical and metallurgical industries. Heavy metals can be dispersed to water systems and adsorbed onto SPM in various binding forms, under hydrodynamic force, migrated in water across a long distance and eventually aggregated to the sediment [47, 72]. During transportation, heavy metals may undergo numerous changes such as dissolution, precipitation, sorption and complexation phenomena.

According to our monitoring data during 2012–2014, the important regional sources of heavy metal pollution in the South Sha River, one tributary of Taizi River, were apportioned to the metallurgical, constructional and chemical industries by the FA coupled with multivariable regression. The contributions of the three types of industries to heavy metal pollution accounted for approximately 54.42%, 25.32% and 20.26% in water, and 52.22%, 34.19% and 13.59% in sediments, respectively. The pollution by target Hg and Pb was more severely in surface water samples in comparison to sediments. The source identification of heavy metal pollution in Hun River by FA method indicated that the outflows from mining and electroplating.

Industries were the important sources of Cu and Cd in the river, particularly in the reaches of Qingyuan County, close to Fushun City. A large amount of As and Hg in the riverine water could be debouched during the manufacture and usage of pesticides and fertilizers. The wastewater discharge from fertilizer production and agricultural runoff was the predominant pollution source of Pb and Cr in the Hun River. Owing to the confluence of urban sewage, industrial wastewater from Fushun City and the upstream of Hun River, the sampling site at the downstream of Fushun City was identified as the most damaged areas by heavy metals. The extremely high concentration of Pb in Daliao River mainstream was associated with the wastewater discharged from printing and dyeing industries nearby and the Pb confluence from Taizi River and Hun River [9, 66].

The headwaters of Liao River is one of the important agricultural bases in China, where there are also many mechanical manufactural, chemical and metallurgical industries. The release of significant amount of Hg-containing wastewater from these industrial activities caused the higher Hg concentration in upper reaches of the river. In the meantime, Cu-containing effluents from electroplating, metallurgical and machine factories led to the highest level of Cu in Tieling and Shenyang sections of Liao River mainstream. In addition, severe Zn pollution was also presented in Tieling section of the mainstream, where mining industry, machine and paper plants were located.

4.3.3.2 Sources of Organic Pollutants

Due to the diversity of organic pollutants in category, property and source, the pollution source apportionment of organic pollutants is critical but arduous to be solved. Currently, much more attention has been given to the source identification of PAHs in the environment [32, 71]. It is well known that PAHs mainly originate from incomplete combustion of petrochemical fuels, coke ovens, domestic heating, transportation vehicles, the leakage of petroleum- and coal-related products, incineration of municipal and industrial wastes, and biomass burning.

Using UNMIX and PMF models, four source categories, including petrogenic source, biomass burning, diesel emission and coal combustion, were identified to be major contributors to PAHs sources in Liao River reed wetland soils [31]. The gasoline and diesel engine emissions contributed the most to the toxic equivalent quantity (TEQBaP) and mutagenic equivalent quantity (MEQBaP), while petrogenic source, the largest contributor to PAHs, made lower contribution to TEQBaP and MEQBaP. An extended fit measurement mode for CMB model was applied to estimate source contributions for PAHs in sediment of the Daliao-Hun-Taizi Rivers. Apportionment results showed that for high molecular weight PAHs, power plant (45.75%), biomass burning (29.34%) and traffic tunnel (10.59%) were identified as the major sources of sediment PAHs from these rivers [3].

In our study, the MLR and eigenvalue methods were combined to identify the sources of PAHs in Hun River mainstream, indicating that the emission of petrochemicals was the major contributor to the PAHs in the dry season, while petrochemicals and fuel combustion or fuel pyrolysis were identified as the mixing source of PAHs in the wet and normal seasons.

COD is an indicator of organic pollution in Taizi River, including PAHs, volatile phenols, nitrobenzene and others. Approximately 60% of the total wastewaters and more than 70% of the total COD in Taizi River were emitted through metallurgical industries, chemical fiber mill, printing and dyeing factories. By comparison with other rivers, the Taizi River showed severe PAHs pollution with seasonal variations. The result revealed that the emission of petrochemicals was the dominant source of PAHs in the dry season, while fuel pyrolysis other than petrochemical was the contributor in the wet and normal seasons.

The monitoring survey on West Liao River indicated that the major organic pollutants, petroleum hydrocarbon with straight-chain alkanes, originated from the petroleum production nearby and application of petroleum-related products. OCPs and PCBs were mainly from the usage of pesticides and the PAEs pollution was derived from domestic wastewaters and industrial effluents [79].

Overall, the source of organic pollutants and heavy metals in the LRB was apportioned to industrial pollution sources, municipal wastewaters and agricultural nonpoint sources. The annual report on environmental data statistics in 2012 demonstrated that industrial, domestic and agricultural activities were responsible to the COD discharge in the LRB with 87,900, 262,300 and 902,700 t/a, respectively, and to the total NH3-N emission with 6,000, 57,600 and 33,600 t/a, respectively [43]. The discharge of wastewaters with high level of COD from petrochemicals, pharmaceutical, textile dyeing, and metallurgical industries was accompanied by the emission of high toxic organic pollutants. Due to the less advanced treatment approaches, partial combination with industrial wastewater and ineffective management, the municipal wastewaters could not be effectively treated, which would be one of THSs sources in the LRB. Moreover, as a key agricultural region, the LRB was subjected to severe THSs pollution from agricultural activities which required more attention.

4.3.4   List of Organic Pollutants

4.3.4.1   Levels of PAHs in Liao River Basin

The concentration levels and pollution characteristic of PAHs in water body and sediments of the LRB have been investigated since 2000 (Table 4.2). It was shown that the concentration of total PAHs in sediments collected from Xinming section of Liao River mainstream ranged from 27 to 198 ng/g. The PAHs with high-molecular-weight were dominant in the sediments from the mainstreams of Hun, Taizi and Daliao river with total PAH concentrations of 62–841 ng/g. Within the three riverine mainstreams, the Daliao River showed the highest level of the PAHs in sediments, followed by the Taizi River and Hun River. The concentrations of PAHs in the sediments detected in 2005 was 340–5070 ng/g.

4.3.4.2   List of OCPs and PCBs in Liao River Basin

Based on the reference, in 2002, the monitoring results for OCPs in Liao River mainstream showed that the concentration of total OCPs ranged from 7.59 to 34.98 ng/L in water and from 0.45 to 7.26 ng/g in sediments, respectively (Table 4.3). The concentration of OCPs in water from Liao River mainstream were close to the level in Xiamen port, Nanjing section of Yangtze River and Haihe estuary in China. The pollution levels of OCPs in the surface sediments from Hun-Taizi-Daliao mainstream were in the range of 3.06–23.24 ng/g. The HCHs compounds were the main pollutants of OCPs and the level of γ-HCH surpassed the threshold effects level (TEL), which could cause the toxic effects on aquatic organism and ecosystem.

4.3.5   List of PAEs, PFCs and PCDD/Fs in Liao River Basin

Table 4.4 presented the concentration of other organic pollutants in LRB. Our previous study indicated the concentration of PAEs was in the range of 6.3–33.9 ng/L in the Xi River and Pu River of Shenyang (data has not been published).

The dioxins and perfluorinated compounds (PFCs) were also detectable in the waters and sediments in the LRB. The concentrations of seventeen polychlorinated dibenzo-p-dioxins and furans (PCDD/Fs) in Daliao-Hun-Taizi mainstream ranged from 0.108 to 87.6 pg/L in surface water and 29.1–3039 ng/kg in the sediment (dry weight), respectively. The total concentrations of PCDD/Fs in the surface sediments of the middle and lower reaches in the Liao River mainstream were in the range of 13.7–458.5 ng/kg, while perfluorooctane sulfonic acid (PFOS) was measured at low levels. PFHxA, PFOA, PFOS, perfluorohexane sulfonic acid (PFHxS) and perfluoroheptanoic acid (PFHpA) were found to be major contributor to the pollution of PFCs in the Xi River of Shenyang. The concentration of PFCs varied in the range

Table 4.2 List of PAHs in Liao River Basin

PAHs

Rivers	Sampling time (site No)	Water (μg/L)	SPM (μg/g)	Sediments (μg/g)	Reference
Liao river mainstream (Xinming section)	2000 (4)	–	–	0.027–0.20	[70]
Hun-Taizi-Daliao river mainstream	Wet season 2005 (12)	–	–	0.062–0.84	[17]
Hun-Taizi-Daliao river mainstream	2005 (16)	0.95–13. 5	0.32–238.5	–	[18]
Liao river mainstream (Tieling, Shenyang, Anshan, Panjin)	Dry and wet season 2007 (12)	0.22–1.36	–	–	[20]
Liao river mainstream	2007 (15)	–	–	0.025–1.48	[72]
Liao river mainstream	2007 (12)	–	–	0.18–2.26	[21]
Hun river mainstream (including four samples in Liao river)	Dry season 2009 (19)	–	–	0.021–22.1	[37]
Xi river (Shenyang section)	Wet season 2011	0.039–0.25		0.34–5.07	[19]
Liao-Hun-Daliao mainstream	Dry and wet season 2010 (19)	–	–	0.038–21.2	[82]
Liao-Hun-Daliao mainstream	Dry and wet season, 2010 (19)	34–20.2 × 103	–	–	[67]
West and east Liao-Daliao rivers	Dry and wet season, 2010	0.033–0.108	0.41–76.5	0.033–0.108	[80]
Liao river mainstream	2011 (14)	3.32–5.28	–	–	[79]

Table 4.3 Levels of OCPs and PCBs in Liao River Basin

Pollutants	River	Sampling time (site No)	Water (ng/L)	Sediments (ng/g)	Reference
OCPs	Liao river mainstream (middle and lower reaches)	1988 (12–24)	7.59–34.98	0.45–7.26	[69, 76]
	Liao Rive mainstream	2011 (14)	9.48–87.78	–	[79]
	Hun-Taizi-Daliao river mainstream	Wet season 2005 (22)	–	3.06–23.24	[61]
	Dahuofang reservoir	Frozen season 2010 (14)	–	1.6–8.4	[35]
PCBs	Liao river mainstream	2011 (14)	≤115.3	–	[79]
	Dahuofang reservoir	Frozen season 2010 (14)	–	1.46–3.52	[35]

Table 4.4 List of PAEs, PFCs and PCDD/Fs in Liao River Basin

Pollutants	River	Sampling time (site No)	Water (ng/L)	Sediments (ng/g)	Reference
PAEs	Liao river mainstream and some tributaries	2011	6.16×10^4	–	[79]
	Xi River and Pu river	2013 (19)	$6.3–33.9 \times 10^3$	–	–
PFCs	Hun river mainstream and some tributaries	2009 (12)	4.1–55.9	–	[54]
	Hun River and Liao river mainstream	2009 (20)	1.4–131	0.26–1.11	[73]

of 1.4–131 ng/L in the surface water and 0.26–1.11 ng/g in the sediment collected from Hun River and Liao River mainstreams. There were 14 categories of 403 pollutants, including PAHs, OCPs, PAEs and phenols detected in the surface waters and sediments from Hun River in Shenyang and the priority pollutants list was proposed using potential hazardous index method.

4.3.6 Challenges, Suggestions and Summary

The water quality indicated by COD in the LRB has been significantly improved in recent years, the pollution of trace organic pollutants and heavy metals, however, is still severe because of bio-accumulative, persistent and poorly biodegradable properties of these pollutants as well as their carcinogenic, teratogenic and mutagenic effects to human health. Although some efforts have been made for the determination and monitoring of specific pollutants in the LRB so far [41], there is a lack of the knowledge on sources and pathways of pollutants including the quantitative source apportionment, transportation and fate of diverse pollutants, and systematic information on pollution characteristics of pollutants in the LRB is still unclear, particularly for highly toxic pollutants such as the typical α-/β-HCHs, PFCs and polybrominated diphenyl ethers (PBDEs). Thus, further work is needed in terms of pollution characteristics within the river basin in order to propose regional effective strategy for prevention and control of these pollutants in the LRB.

First of all, the spatial and temporal distribution of typical pollutants and detailed monitoring survey on these pollutants should be further investigated. On the basis of pollutant levels, distribution profiles and toxicity of pollutant in water and sediment, the time-dependent priority pollutant list will be proposed regionally. Secondly, the composition of effluents from industrial, agricultural and domestic activities should be qualitatively and quantitatively determined so that the complete components of each potential pollution source will be provided. Thirdly, the quantitative method for source identification of specific organic pollutants in SPM, surface water and sediments should be developed and modified, in order to apportion the primary contributions to the pollution and further to provide a technical and theoretical support for water quality management in the LRB. Finally, the macro- and micro- source-sink relationship in water pollution should be explored, which is crucial to predict accurate response of water quality to pollutant loads. The suggested studies will provide data support for decision-making of the pollutant prevention and control in the LRB.

In summary, this deliverable has provided more detailed information on the pollution level of POPs and heavy metals in the LRB. In particular, the quantitative methods of pollution source apportionment and their application for these pollutants in the LRB were reviewed. These would improve the comprehensive understanding of environmental pollution status to further develop an efficient strategy in the control and reduction of trace pollutants in the LRB. Although the water quality (referring to COD) in the LRB markedly improved in recent years [43], the occurrence of trace organic pollutants and heavy metals in the water bodies has hindered the local economic development. Thus, the source control and management, through reducing the industrial discharge and increasing the treatment capacity of domestic sewage, combined with remediation technologies for polluted river water will be essential for the effective and thoroughly pollution control in the LRB.

4.4   Control Strategy for Priority Persistent Organic Pollutants and Water Environment Risk Sources

by Lu Han, Yonghui Song, Peng Yuan, Ping Zeng

4.4.1   Priority Persistent Organic Pollutants Management

Persistent organic pollutants (POPs) including many kinds of chemicals, such as poly-chlorinated biphenyls (PCBs), polycyclic aromatic hydrocarbons (PAHs), organic chlorine pesticides (OCPs), emerging POPs and other substituted benzene. Although these pollutants are at a trace level or less in water environment, they can be accumulated in aquatic biota, migrate and transform, and further result in massive threat to ecological safety and human health because of their high toxicity and bioaccumulation, and poor degradability. Additionally, most of these organic pollutants have "carcinogenic, teratogenic and mutagenic" effects and some act as endocrine disrupting chemicals. According to the 10-year-term monitoring of Hudson River by US Environmental Protection Agency (EPA), the concentration of PCBs in the surface water, sediments and fish were found to be over the standard set by US Food and Drug Administration (FDA). The level of PCBs in Seine River (France) was 50–150 ng/L in the dry season and 500 ng/l in the wet season. The PCBs were also detectable in most of seven major water basins in China with higher levels in the Huai River, Yangtze River, Pearl River and Songhua River.

The occurrence of PAHs in water systems has attracted attention globally. As reported by previous researches, the most heavily polluted water bodies were River Thames and River Trent (UK), as well as Ache River (Germany) with the level of PAHs exceeding the magnitude of μg/L. The detection of PAHs in the major rivers of China has been conducted from 1990s. It was found that the concentrations of PAHs were mainly in the range of 10–1000 ng/L in water environment. OCPs have been monitored continuously all over the world because of their proven toxicity to human beings, animal, plant life and food chain. The pollution of OCPs in water originated from point and nonpoint sources, such as the wastewaters discharged directly to the river during the manufacture and use of pesticides. A survey on the level of OCPs showed that the concentration in global marine environment ranged from 0.1 to 4.4 ng/g in sediments and the heavily polluted level could achieve the magnitude of μg/g in some water bodies. The concentrations of OCPs in water systems in China varied highly depending on geo-locations with the concentration less than 200 ng/L in most sample sites. Therefore, the pollution level and data sets on spatial distribution of POPs should be investigated in detail to gain the systematic information in natural aquatic systems.

Liao River Basin (LRB) is one of seven major river basins in China, located in northeast of China. It has been the greatest cluster of heavy industry in China, including chemical, petrochemical, pharmaceutical, metallurgical, dyeing and other sectors. The discharge of wastewater from these industrial sectors has damaged the natural water bodies within the basin. Since 2008, some efforts have been made for the pollution control of point sources in the LRB and a series of wastewater treatment techniques have been developed for the removal of Chemical Oxygen Demand (COD), Biochemical Oxygen Demand (BOD) and NH_3–N. A large-scale constructed wetland in Liao River Conservation Area (LRCA) has been also built for the water purification. The monitoring results for water bodies indicated that the water quality, in terms of indexes such as COD, BOD and NH_3–N, in the LRB has been improved year by year. However, there have been more concerns regarding to the level and ecological risk of trace pollutants, in particular POPs in water bodies. Although some efforts have been made to determine the level and identify pollution sources of the trace pollutants in the LRB, there are few comprehensive overviews regarding the pollution level and source apportionment of trace THSs in the LRB. Work package 3 summarized the pollution level and discussed the potential pollution sources of POPs in the LRB, so as to improve the understanding of environmental status at the regional level.

4.4.2   Water Environment Risk Sources Management

The project group investigated literature about environmental risk prevention, risk management of the US, EU and the situation in China. Research content includes "The Emergency Planning and Community Right-To-Know Act (EPCRA)", "chemical accident prevention regulations" and the risk management plan (RMP). The Work group mastered the method and basis of environmental risk source identification and classification management of US. Furthermore, the Seveso "III" was investigated which is the basis for management the major environmental disasters, accident, as well as enterprise in the EU. The list of hazardous substances according to the "environmental emergency management regulations" of Canada is also taken into account.

1. Carry out the research of environmental risk source identification and classification method.
2. Compare with the technique of environmental risk source identification, grading and index.
3. Analyse the advantages and disadvantages of each management method.

Based on the framework of environment risk source identification by the "source-pathway-risk acceptor" concept, the project group have established environmental risk source identification and classification in accordance with the work plan. It includes the range of application, normative documents, terms and definitions, the general requirements of the environmental risk assessment, process of environmental risk assessment and content of environmental risk assessment.

We investigated the EU risk management. We overviewed the framework of water law which included European level (Federal Ministry for the Environment, Nature Conservation, Building and nuclear safety (BMUB)), Federal level (Federal Environmental Agency (UBA)), State level (State Ministry for Environment-Water Department Federal State owned company for Environment and Agriculture, Federal State Reservoir Management Authority) and Municipal level (Municipal Water Authority, Municipal Water Association). The European water law will be helpful for environment risk management.

To establish and improve the environmental risk management laws and regulations system is the basis and guarantee for the enterprise environment risk supervision, and there are many successful foreign experiences in this respect for reference.

4.4.2.1 Accident Risk Prevention and Management of United States

The relevant research on environmental risk prevention and emergency management of United States from the beginning of 1970s, which has formed a more perfect environmental risk management laws, regulations and standards system.

Environmental emergency management: "Emergency Planning and Community Right-to-know Act" (EPCRA) corresponding with the two laws.

Chemical accidents risk prevention and risk management: "Chemical Accident Prevention Regulations" and "Risk Management Plan" (RMP).

In the 1980s, in response to major hazardous chemicals emergency accidents, the security issues on storage and use of toxic chemicals, then the US Congress passed the "emergency plan and the public's right to know Act" (EPCRA) in 1986.

The bill identified the requirements for toxic and hazardous chemicals emergency plan and community right-to-know act reports of federal, state, local governments and enterprises. It increases the related knowledge of chemical facilities, the use and the emissions to the public, as well as the chemical safety, public health and environmental protection.

The bill mainly provides emergency plans, emergency notification, community right-to-know act requirements, toxic chemical release list and other terms.

The local government shall formulate a plan for the emergency response of chemicals, and at least once a year, the state government shall be responsible for the supervision and coordination of local plans.

The facility which storage of hazardous substances (EHS) exceed the critical amount (TPQ) must cooperate with the preparation of emergency plans.

The United States established emergency response requirements of State Council, and asked the local community to set up local emergency plan committee.

In 1990, the "US Clean Air Act Amendments (CAAA)" requires to implement risk management plan for enterprises who use or storage of toxic and harmful substances, and establish emergency response to the emissions of toxic substances. The United States Environmental Protection Agency promulgated the "Regulations on the prevention of chemical accidents" and "risk management plan (RMP)" in 1999.

The legislation is the first federal regulation in the United States that to prevent the possible hazards of chemical accidents for the public and environment.

The regulations have list control inventory and critical weight standards about 77 kinds of toxic substances and 63 kinds of flammable substances, and set up the specific requirements of development, submission, modifition and updation the "risk management plan (RMP)" for risk enterprises in detail.

RMP provides that the owner and operator shall submit relevant information about the use of hazardous substance enterprise equipment and facilities surrounding functional areas and sensitive areas.

The system provides the content of submit information for enterprises, including: the prevention and emergency policy of fixed pollution source emission accident, the use information for fixed source and controlled substances, the alternative plan for the maximum of non normal discharge pollutants leak and pollutants release, the reduction methods and technical measures of normal discharge of pollutants, pollution accident records in past five years, emergency system, increase measures related to the safe operation of the business.

In addition, the environmental protection agency issued a series of technical documents related to environmental risk management, including "chemical accident prevention comprehensive guide to risk management plan", "chemicals warehouse risk management planning guide", "off-site consequence analysis risk management plan guide", and constructing the integrated risk information system (IRIS), in order to achieve the environmental risk of comprehensive and effective management.

According to the consequences of accident, the RMP divided business risk into three levels. It provides the specific requirements of development, submission, modifition and updation for the risk management plan at different risk levels. The EPA Office of emergency management through on-site inspection and supervision to perform enterprise RMP, according to the United States Code, Environmental Protection Agency (EPA) can be sentenced to the high of $32500/day of civil penalties for illegal enterprises; people who convicted by intentionally violation or submitted false information can be sentenced at the top for two years in prison.

4.4.2.2 Accident risk management of the European Union and its member states

"Seveso Directive III":
It defined the critical quantity of 30 kinds of chemicals.

Enterprises involved in hazardous substances are required to comply with the directive, including industrial production activities, but also the storage of dangerous chemicals.

The critical quantity standard is divided into two level threshold, so the directive provides the enterprise with 3 levels of control, enterprises shall be made more strict management which corresponding to high critical quantity.

Italy formulated the domestic laws of Law 238/05 on the basis of EU "Seveso Directive III" for enterprise risk classification, the method can be points into two steps: one is to determine the amount of the hazardous substances that enterprises involved; the second is to assess the level of industrial enterprises based on environmental receptors fragile characteristic of sensitive areas.

The EU set up a major accident Seveso Disaster Management Authority responsible for the execution of instruction.

4.4.2.3 Environmental Emergency Management of Canada

In order to implement the "Environmental Protection Act" (1999), Canada promulgated the "Environmental emergency regulations" in 2003, which has proposed a list of 174 chemicals. In 2011, Canada issued the "Environmental Emergencies Management Ordinance". The list increased to 215 kinds of chemical substances, and divided into three parts:

The first part provides 80 kinds of chemical substances and mainly are potentially explosive material.

The second part provides 101 kinds of chemical substances, most of them are substances which harmful to people after inhalation.

The third part provides 34 other harmful chemical substances, mainly about the heavy metal compounds and nonylphenol other endocrine disruptors.

4.4.2.4 Related Research Work in China

In the related fields of the supervision and administration of production safety, the State Administration of work safety developed the "production safety accident investigation and risk management Interim Provisions", "Implementation Guidelines of Dangerous Chemicals Business Potential Accidents Investigation and Management" and "Hazardous Chemicals of Major Hazard Installations Supervision and Management Interim Provisions".

"Hazardous Chemicals of Major Hazard Installations Supervision and Management Interim Provisions":

1. It put forwards the identification, classification, evaluation, filing and verification of major dangerous chemical hazard source, monitoring, inspection and testing, emergency system, emergency drills and so on a series of contents.
2. It identified the responsibilities of safety supervision departments for supervision and inspection.

"Production Safety Accident Investigation and Risk Management Interim Provisions" has put forward the general requirements of strengthening the supervision and management of accident hidden danger:

1. Responsibility for production and business units include operate in accordance with law, the establishment of a legal system, regular investigation and management, periodic reports and other content.
2. Responsibility for safety supervision department include the establishment of rules and regulations, regular supervision and inspection, supervision and rectification of hidden danger, information reports and other contents.

"Implementation Guidelines of Dangerous Chemicals Business Potential Accidents Investigation and Management": For hazardous chemicals enterprises, the assessment methods on the investigation ways and frequency of hidden danger, the investigation contents of hidden danger and so on were refined.

In the field of environmental protection and management, our country currently has involved in different aspects on environmental impact assessment system, environmental pollution liability insurance and environmental management of dangerous chemicals, and provides technical basis for the work carried out at present.

In the environmental risk assessment system, the Ministry of environmental protection promulgated the "Construction Project Environmental Risk Assessment Technology and Guidelines" in 2004:

It proposes the evaluation work level of the first and the two levels.

It provides the basic content of the environmental risk assessment: risk identification, source analysis, consequences calculation, risk calculation and evaluation, risk management, put forward whether construction project environmental risk lower than the industry risk level.

In terms of environmental pollution liability insurance, the Ministry of environmental protection in conjunction with the China Insurance Regulatory Commission released the "Environmental Risk Assessment Technical Guidelines-Chlor Alkali Enterprises Environmental Risk Classification Method" in 2010:

1. The environmental risk assessment index system consists of two parts: the reference value and the correction value. Reference values reflect the environmental risk factors of production, site environmental sensitivity and other universality, general index may lead to the environmental risk, which are intrinsic factors of enterprise environmental risk; Correction value reflect the level of environmental risk control and emergency rescue ability of specific targets, which are exogenous factors.
2. According to the assessment scores, the level of chlor-alkali enterprises environmental risk was divided into five.

In the aspects of environmental management of hazardous chemicals, the fouling prevention department formulated and promulgated the "Key Environmental Management of Hazardous Chemicals Environmental Risk Assessment Report Preparation Guide (Trial)" in 2013:

1. Mainly focus on the production/use process of key environmental management of hazardous chemicals, which may have long-term, potential adverse effects on external environment and human health.
2. It determines the corporate environmental risk level is significant, large, medium and general.

4.4.3   Control Strategy for Priority Persistent Organic Pollutants

The water quality indicated by COD in the LRB has been significantly improved in recent years, the pollution of trace organic pollutants, however, is still severe because of bio-accumulative, persistent and poorly biodegradable properties of these pollutants, as well as their carcinogenic, teratogenic and mutagenic effects to human health.

Although some efforts have been made for the determination and monitoring of specific pollutants in the LRB so far, there is a lack of the knowledge on sources and pathways of pollutants including the quantitative source apportionment, transportation and fate of diverse pollutants, and systematic information on pollution characteristics of pollutants in the LRB is still unclear, particularly for highly toxic pollutants such as the typical α-/β-HCHs, PFCs and polybrominated diphenyl ethers (PBDEs). Thus, further work is needed in terms of pollution characteristics within the river basin in order to propose regional effective strategy for prevention and control of these pollutants in the LRB.

First of all, the spatial and temporal distribution of typical pollutants and detailed monitoring survey on these pollutants should be further investigated. On the basis of pollutant levels, distribution profiles and toxicity of pollutant in water and sediment, the time-dependent priority pollutant list will be proposed regionally.

Secondly, the composition of effluents from industrial, agricultural and domestic activities should be qualitatively and quantitatively determined so that the complete components of each potential pollution source will be provided.

Thirdly, the quantitative method for source identification of specific organic pollutants in SPM, surface water and sediments should be developed and modified, in order to apportion the primary contributions to the pollution and further to provide a technical and theoretical support for water quality management in the LRB.

Finally, the macro and micro-source-sink relationship in water pollution should be explored, which is crucial to predict accurate response of water quality to pollutant loads. The suggested studies will provide data support for decision-making of the pollutant prevention and control in the LRB.

In summary, the ESP work group has provided the more detailed information on the pollution level of POPs in the LRB. In particular, the quantitative methods of pollution source apportionment and their application for these pollutants in the LRB have been reviewed. These would improve the comprehensive understanding of environmental pollution status so as to further develop an efficient strategy in the control and reduction of trace pollutants in the LRB. Although the water quality (referring to COD) in the LRB has been markedly improved in recent years, the occurrence of trace organic pollutants in the water bodies has hindered the local economic development. Thus, the source control and management, through reducing the industrial discharge and increasing the treatment capacity of domestic sewage, combined with remediation technologies for polluted riverine water will be essential for the effective and thoroughly pollution control in the LRB.

4.4.4 Management Strategy for Water Environment Risk Sources

4.4.4.1 Building Enterprise Environmental Risk Classification and in Dynamic Management System in Liaohe River

Based on the requirement of the "Measures for the Administration of the Environmental Emergency Accidents" (Ministry of Environmental Protection Order No. 34), according to the environmental risk level of the enterprise, to achieve the different management mode and requirements, and to further study and establish the "Measures for the management of enterprise environmental risk classification".

Realize the management cost and benefit optimization on environmental risk declaration, environment risk assessment, plan formulation, emergency drills, reserves of emergency supplies, pollution liability insurance, listing environmental inspection, clean production, green credit verification and so on.

Regularly carry out statistical analysis, and grasp our country enterprise environmental risk level more comprehensive and dynamic, gradually promote the fine management.

4.4.4.2 To Strengthen the Chemical Environment Risk Assessment and Prevention Research and Control Technology

Strengthening research on chemical accident risk assessment methods and technology, the difficulty of risk assessment is uncertainty analysis, setting accident source scientifically and the extreme weather conditions such as natural disasters, water power, communications and transportation system fault chain, the domino effect of situational factors such as comprehensive consideration have direct impact on the environment risk assessment results, prevention and control measures.

1. Making further research on technical standards of environmental risk prevention and control measures, departments related to the construction develop chemical petrochemical industry emergency pool (pollutants, fire water collection system and other measures to prevent and design construction standards as soon as possible).
2. To avoid and reduce the impact of fire, explosion, leakage and other production safety accidents caused by the secondary environmental events.

4.4.4.3 Strengthening Contingency Plans for Environmental Emergency Risk Sources

For local governments:

1. Speed up the editing work of contingency plans for environment emergency risk sources;

2. Expedite the completion of contingency plans for local environmental emergency risk sources;
3. Develop and improve special plans for drinking water and key watersheds environmental emergency risk sources, and submit for the record;

For enterprises and institutions:

1. Carry out the editing work of contingency plans for environmental emergency risk sources;
2. Strengthen the management of emergency plan for the record, and supervise enterprises to do environmental risk assessment;
3. Carry out the case analysis, and put up with the paradigm of environmental contingency plans.

4.4.4.4 Control Environmental Risk Management Strictly

Strengthen the environmental risk management and supervision

1. Improve governance and establish investigation files for environmental risk, approved basic information of important risk source;
2. Timely organize the inspection work, enterprises which exist significant environmental risk to be included in the social credit system;
3. Periodically assess environmental and health risks are focus on rivers, reservoirs industrial enterprises and industrial zone, promote the establishment of water environmental risk prevention system.

Establish and improve the environmental emergency warning system

1. Explore the grading standards of early warning, achieve early detection, early warning, early preparation, early response and early cancelation of environmental emergencies;
2. Carry out informational warning for region where environmental emergencies happened frequently;
3. Timely access to the information which was related to drinking water safety and the release of toxic gases, to achieve early warning of major environmental emergencies.

Control the emergency environmental risks strictly, strengthen environmental emergency management of key areas:

1. Conduct emergency drills,to effectively prevent and respond to environmental emergencies properly;
2. Carry out poisonous gas environmental risk early warning system for key industrial areas, to promote biological toxicity early warning of drinking water sources.

4.4.4.5   Continue to Strengthen Risk Prevention Mechanism for Public Participation

1. To promote public participation in risk prevention work.
2. Make full use of environmental reporting data networking.
3. To do networking well between environmental reporting data integration and environmental reporting reception system.
4. Ministry of Environmental Protection will be released early warning notification for different environmental problems, to detect environmental hot issues through statistical analysis, to provide basis for decision making in environmental management.
5. Strengthen the investigative work of inspection and supervision, to solve the problem of environmental pollution from the source.
6. Give full play to the role of environmental protection WeChat report.

4.4.4.6   Make a Deep Summary of Environmental Emergency Cases Experience

Make a good job of case assessment and warning education of environmental emergencies:

1. Carry out case studies on typical environmental emergencies;
2. Analyses the problems and lessons in-depth;
3. Propose corrective measures and policy recommendations for environmental management.

Establish evaluation mechanism to promote the implementation of corrective measures:

Carry out warning education through the issuance of notification or case playback, to prevent similar incidents from recurring.

The whole process of environmental emergency information should be opened:

1. Improve the mechanisms of environmental emergency information disclosure and public opinion response;
2. Improve the spokesman system, strengthen expert interpretation, make use of television, radio, wechat and other news carriers rationally;
3. Establishment linkage mechanism between environmental emergencies investigation and public interest.

4.5 Dissemination and Training of the Environmental Risk Management Strategy

by Lu Han

4.5.1 Key Technical Results

4.5.1.1 Classification of Enterprise Abrupt Environment Incident Risk

Based on source-pathway-risk method framework, provided the method framework of environmental risk source identification and classification. According to the environmental risk acceptor sensitivity, atmospheric, water and soil environmental risk acceptor have been classified and quantified.

Environmental risk evaluation indicators were established and quantified. Assessment indexes include production process, environmental risk control measures, wastewater discharge etc.

Environmental risk substances inventory has been developed. Based on the case study of environmental pollution accidents and pollutant analysis, combining the environmental risk substances list of America, Canada and EU, the team put forward 246 kinds of environmental risk substances inventory and threshold quantities.

According to the matrix method to divide the enterprise abrupt environmental incident risk (hereinafter referred to as environmental risk) level by means of analyzing all the environmental risk substance that involve enterprise production, process, utilize and storage quantities and its critical ratio (Q), assessing process and environment risk control level (M) and sensitiveness of environment risk receptor (E). Environmental risk level is divided into general, higher and major environmental risk level, those three levels are marked using blue, yellow and red color. The assessment process is showed in Fig. 4.4.

4.5.2 Put Forward the List of Toxic and Hazardous Substances in Liao River Basin

The identification of POPs and screening of industrial pollution sources in Liao River Basin have been carried out. The potential risk sources in demonstration area were investigated. The pollution characteristics and the inventory of emissions of persistent organic pollutants were determined. The spatial distribution characteristics of potential hazardous substances in demonstration area has obtained, and the risk assessment techniques and methods for toxic and hazardous pollutants were studied.

Priority control pollutant inventory of key pollution river in typical industry of Liao River basin was put forward. According to the improved potential hazard index, on the basis of the chemicals toxic, comprehensive consideration of chemical detection

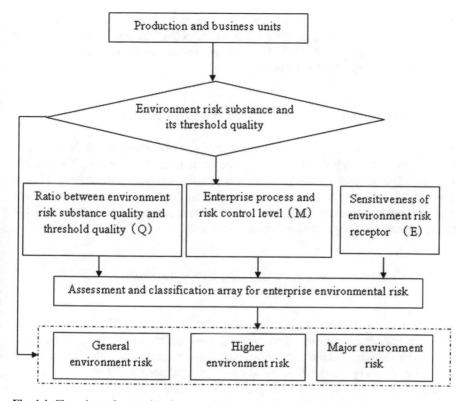

Fig. 4.4 Flow chart of enterprise abrupt environment incident risk classification

concentration, detection rate, the weighted average value, to determine whether the chemical is watershed pollutant by comparing with score. Priority control pollutants inventory for key polluting river in each typical industries were obtained.

4.5.3   Political Contributions

The management strategies of persistent organic pollutants and environmental risk sources were put forward. This work would provide support for the improvement of water environment quality in Liao River Basin.

4.5.3.1   Strength the Watershed Environmental Quality Target Management

Setup priority control unit according to the characteristic of the control unit, which including the water quality, ecology, water source, accident risk. According to the requirement of Ministry of Environmental Protection, select the control unit which cannot meet the water discharge standard, with important rolls in ecology and water supply, with the risk of depraved water quality as priority control unit based on the three stage system of Liao river basin, water ecological control area and water environmental control unit. Setup the toxic substance control target, carry out the overall control of the toxic substance.

4.5.3.2 Strengthen Macro-Plan, Adjust Industrial Pattern

The plan made based on the considering of water source protection, water pollution control, domestic, industry and ecology. In the area of water resource protection, key ecological protection, new projects or companies are prohibited to setup. For the company already stay in the area, the adjust plan should be made, to require them make treatment, produce product with less pollution or move out of the area. Prohibit the construction of high environmental risk project.

4.5.3.3 Control the Environmental Permission, Eliminate the Low Level Productivity Legally

According to the Action Plan for Prevention and Treatment of Water Pollution, and the requirement of Liao river Basin water quality target and main functional block plan, make clear the regional environmental permission, distinguish the dysfunctional block deeply, carry out the differential environmental permission strategy. For area within the environmental carrying capacity, the carrying capacity pre-warning system are setup; For area over the environmental carrying capacity, the water pollutants reducing plan will be carry out. Control the POPs discharge from priority control unit. Encourage the clean production to discharge fewer pollutants.

4.5.3.4 Develop Recycling Economy, Encourage Clean Production

Based on the rule of "minimizing, resource, innoxiously", implement the comprehensive POPs and risk source control scheme including clean production and reduce the production of POPs and risk pollutants. Require the company to make clean production review.

Encourage the company using industrial chain design, comprehensive recycling, production relative extension to produce less pollutants and improve the resource efficiency.

4.5.3.5 Well Establish the Watershed Ecological Environmental Function Division Management System

Based on the watershed ecological environmental function division, combine the evaluation of typical company distribution, POPs discharge and key polluted reach, make sure the control unit risk level and the performance. Make more concern of the priority control unit.

Make sure the risk level and risk grade of control area and control unit to make clear the target, task, project and performance of the toxic pollutants control.

4.5.3.6 Strengthen the Monitoring on the Key Pollution Source and Risk Pollutants, Deepen the Overall Pollution Control

Based on the list of key pollution discharge of Liaohe Basin 2016, set up the discharge list of typical technics source and discharge list of toxic pollutants. Estimate the pollution load of main toxic pollutants to control the discharge of pollutants.

Make the dynamic management of the key pollutants source, mark the key pollution sources, key company, main pollutants and project of wastewater discharge on the electrical map. Require the company to report the change of sort, concentration, quantity of pollutants.

4.5.3.7 Strengthen the Pollution Control of Wastewater Containing POPs

Carry out the toxic pollutants discharge standard, strengthen the pollution control of wastewater containing POPs. Encourage the study of POPs control technology with Chinese characteristics and low cost, especially the maintaining and management of wastewater disposal equipment. Besides, the facilities and equipment should be developed and be maintained and managed during the operation.

4.5.3.8 To Build Enterprise Environmental Risk Classification and in Dynamic Management System in Liao River

Based on the requirement of the "Measures for the Administration of the Environmental Emergency Accidents" (Ministry of Environmental Protection Order No. 34), according to the environmental risk level of the enterprise, to achieve the different management mode and requirements, and to further study and establish the "Measures for the management of enterprise environmental risk classification".

Realize the management cost and benefit optimization on environmental risk declaration, environment risk assessment, plan formulation, emergency drills, reserves of emergency supplies, pollution liability insurance, listing environmental inspection, clean production, green credit verification and so on.

Regularly carry out statistical analysis, and grasp our country enterprise environmental risk level more comprehensive and dynamic, gradually promote the fine management.

4.5.3.9 To Strengthen the Chemical Environment Risk Assessment and Prevention Research and Control Technology

Strengthening research on chemical accident risk assessment methods and technology, the difficulty of risk assessment is uncertainty analysis, setting accident source scientifically and the extreme weather conditions such as natural disasters, water power, communications and transportation system fault chain, the domino effect of situational factors such as comprehensive consideration have direct impact on the environment risk assessment results, prevention and control measures.

4.5.3.10 Strengthen Public Participate and Social Supervision

Encourage public list of the internet environmental protection management. Give encourage for the company with good effort on environmental protection; For the company with illegal discharge, a "yellow" card will be sent to make warning; For the company can not improve within limited time, a "black" card will be sent with the disclose on internet.

Try to resolve environmental pollution by economic lever. Improve the cost of pollutants discharge. Improve the material maybe cause heavy pollution to push them using materials without pollution.

Encourage the disclose the enterprise information; Set up the reward mechanism for environmental pollution to promote public participation.

4.5.3.11 Sustainable Contributions to the Local Water Management

The environment risk assessment method will improve the environmental risk control level and improve the environmental contingency management ability. Meanwhile, it will provide technical support for the management of environmental protection agencies.

The actions have effective impact on local and national environmental risk management policy. The developed environmental risk source methodologies and action plan will support local risk management and environmental protection.

Appendix 1: Classification of Enterprise Abrupt Environment Incident Risk

Ratio Between Environment Risk Substance Quality and Threshold Quality (Q)

The first step of risk classification is to list and illustrate the following information to the used raw materials, fuels, end-products, intermediate products, by-products, catalysts, auxiliary raw materials and "three waste" pollutants (waste water, waste gas, and waste solid):

1. Name of substance
2. CAS Number
3. Quantity at present and the largest existing quantity possible
4. Physical, chemical and toxically character of each substance in the state of the normal use and at the accident situation
5. The Acute and the chronic toxicity to humans and the environment of the environment risk substance and its secondary materials.
6. The basic emergency disposal methods to every hazard caused by the environment risk substance

The indication whether the involved substances are environment risk substances can be found in Appendix 2.

Calculating the total maximum in factory of each environment risk substance involved, if the total dynamic changes, calculating the total maximum in the day. Calculating the ratio Q between the existing quantities and the critical mass given by Appendix 2:

1. When the enterprise only have one kind of environment risk substance, calculating the ratio between its total qualities and its critical quality, Q;
2. When the enterprise have many kinds of environment risk substance, calculating the ratio between substance quality and the threshold quality, Q, according to the following formula:

$$Q = q_1/Q_1 + q_2/Q_2 + \cdots + q_n/Q_n \qquad (4.1)$$

In the formula:
(1) $q1, q2, ..., qn$—the largest amount of each existing environment risk substance in tons; $Q1, Q2, ..., Qn$—the critical mass of each environment risk substance in tons. When $Q < 1$, enterprise is evaluated as general environment risk level directly, expressed in Q. When $Q \geq 1$, divide it into: (1) $1 \leq Q < 10$; (2) $10 \leq Q < 100$; (3) $Q \geq 100$, respectively represented it by Q1, Q2 and Q3.

Production Process and Environment Risk Control Level (M)

The enterprise production process, safety production control, environmental risk prevention and control measure, EIA and approval, wastewater discharge is assessed and subsequently summarized by the scoring method. Determine the enterprise production process level and environment risk control level, and it will be described as M. With the M, the higher the better (Tables 4.5 and 4.6).

Table 4.5 Enterprise production process and environment risk control level assessment index

Assessment index		Score
Production process		20
Production safety control (8)	Fire acceptance	2
	Dangerous chemicals safety assessment	2
	Safe production permit	2
	Dangerous chemicals major hazards for the record	2
Water environment risk prevention and control measures (40)	River closure measures	8
	Accident drainage collected measures	8
	Clean water system prevention and control measures	8
	Rainwater system prevention and control measures	8
	Production wastewater system prevention and control measures	8
Atmospheric Environment Risk Prevention and Control Measures (12)	Toxic gas leakage emergency handling equipment	8
	Toxic gas leakage monitoring and early warning system in production area or factory bound	8
EIA and approval of the other environment risk prevention and control measures to carry out the situation		10
Wastewater discharge whereabouts		10

Table 4.6 Enterprise production process and environmental risk prevention and control level

The score of process and environment risk control level (M)	Process and environment risk control level
$M \leq 25$	M1
$25 < M \leq 45$	M2
$45 < M \leq 60$	M3
$M \geq 60$	M4

Production Process

List the enterprise production process and its characters: name of production process, reaction condition (including high temperature, high pressure, and inflammable, explosive), whether to belong to the "Key Supervision and Dangerous Chemical Process Directory" or to the state regulations that have eliminate term.

Assess the enterprise production process according to Table 4.3. To the enterprise which has multiple sets of process units, assess each process unit respectively and do

Table 4.7 Enterprise production process

Assessment basis	Score
Involving the phosgene and optical gasification process, electrolysis process (Chlor-alkali), chlorination process, fluoride process, hydrogenation process, diazotization process, oxidation process, peroxide process, amino process, sulfonation process, polymerization process, alkylation process, the new coal chemical industry process, calcium carbide production process, diazotization process	10/Set
Other process of high temperature or high pressure and involving substance that be inflammable and explosive1	5/Set
Process and equipment eliminated within a time limit prescribed by law	5/Set
Process not involve dangerous process or process/equipment forbidden by law	0

the summation. The highest score of enterprise production process is 20, if someone's is higher than 20 will be summed as its highest score. List of chemical process will be adjustment by the status of abrupt environment incident and the relevant provision (Table 4.7).

Notice 1: High temperature means that the temperature of process is or higher than 300 °C. High pressure means that the design pressure of pressure vessel $(p) \geq$ 10.0 MPa. Flammable and explosive substance is the chemical belongs to the regulation GB20576 to GB 20602 "Classification, Warming-Label and Warming Instructions of Chemicals";

Notice 2: The backward production technology equipments have eliminated the deadline according to the "Industrial Structure Adjustment Guidance Catalogue" (Latest version) by NDRC.

Security Production Management

Assesses existing safety production management of enterprise according to the Table 4.8, and attach relevant certificate documents to the enterprise's assessment.

Environment Risk Prevention and Control and Emergency Measures

List each environment risk unit that deals with environment risk substance, and investigate the implementation and management of their environment risk prevention and control measures according to the following aspects: production equipment, storage system, utility system, support facilities and environment protect facilities.

List water and atmosphere environment risk prevention and control measures applied by each unit according to Table 4.9, including river closure measures, accident drainage collection measures, clean water drainage system prevention and control measures, rainwater system prevention and control measures, production wastewater

Table 4.8 Enterprise safety production control

Assessment index	Assessment basis	Score
Fire acceptance	Fire acceptance opinion is qualified, and the latest opinion is qualified	0
	Fire acceptance opinion is disqualification, or the latest opinion is disqualification	2
Safe production permit	Not is dangerous chemicals production enterprise, or dangerous chemicals production enterprise that doesn't have safety production permit	0
	Dangerous production enterprise have not safety production permit	2
Production evaluation	Carry out the assessment for dangerous chemicals safeness; through the completion inspection and dangerous chemicals acceptance of safety facilities, or no requirement	0
	Not to carry out the dangerous chemicals safety evaluation, or not through the completion inspection and acceptance of safety facilities	2
Dangerous chemicals major hazards for record	No major hazards, or all the major hazards had been recorded	0
	Major hazards haven't been recorded	2

treatment system prevention and control measures; toxic gas leakage emergency disposal equipment and toxic gas leakage monitoring and early-warning measures; EIA and approval involving other risk prevention and control measures.

The enterprise environment risk prevention and control and emergency measures implementation is assessed. An Enterprise should be scored according to Table 4.9 as relevant function respectively when it has a set of collection measures, has the function of collecting half and all the leakage, collecting the clean water pollutant, collecting rainwater and fire water.

Rainwater, Clean Sewage, and Wastewater Discharge Whereabouts

The enterprise rainwater and treated wastewater discharge system, the location and average and maximum velocity of related receiving waters are assessed and scored according to Table 4.10.

Table 4.9 Enterprise environment risk prevention and control and emergency measures

Assessment index	Assessment basis	Score
River closure measures	Various environment risk unit set up measures of anti-seep, anti-corrosion and anti-leaching, collecting measures of early rainwater, leakage, pollutant fire water that drain into the rain and lean water drainage system. All the measures complied with the design specification; and Set up cofferdams and drain valve switch outside the tank section fire dike. The valve to the rainwater system is closed and the valve to the accident pool, emergency accident pool and clean water drainage pool or wastewater treatment system is opened under normal circumstances; and The daily management and maintenance is good of above-mentioned measures. Have someone who's in charge of the valve switch to ensure the early rainwater, leakage and pollutant fire water discharged into the sewage system	0
	Not meeting any of above requirement	8
Accident drainage collected measures	Set up emergency accident pool, accident pool or clean water drainage pool according to the regulations of the relevant design. Set up accident drainage collected facilities according to downstream environmental risk receptor sensitivity and extreme weather conditions; and All sorts of accident drainage collected facility such as accident storage pool, emergency accident pool and clean water drainage buffer pool location is reasonable, to be able to scoop or ensure the collection of leakage and fire water, and ensure enough accident drainage buffering capacity; and (3) Set water intake facility that combined to sewage pipeline	0
	Not meeting any of above requirement	8
Clean water drainage system prevention and control measures	Do not involve clean water; or Clean water in the factory all drain into waste water treatment system; Set the clean-up distributaries device. Clean water drainage system has all measures as follows: Set clean water drainage buffering pool (or rainwater collected pool) that has functions that collecting pollutant water, early rain and fire water. The pool keeps sufficient accident drainage buffering capacity daily and sets upgrading facilities that be able to send all the collecting to the factory sewage treatment facilities; and Set clean water system (or rainwater system) totally discharge outlet monitor and closing device, special-assign person response the closing totally discharge outlet of clean water system under emergency situation to prevent the pollutant rainwater, clean water, fire water and leakage diffuse into out-environment	0
	Involved clean water drainage, but dose not meet the requirement of above (2)	8

(continued)

Table 4.9 (continued)

Assessment index	Assessment basis	Score
Rainwater drainage system prevention and control measures	Rainwater in the factory all goes into wastewater treatment system or into clean-up distributaries device. The rainwater drainage system carries all the measures as follows: Set rain collecting pool or rain monitoring pool which has function of collecting early rainwater. The pool discharging tube has stop valve that is closed under the normal situation to prevent pollutant water diffusion. The pool has set grade-up facility to send all the collecting to the factory wastewater treatment facilities; Set rainwater system (including flood control and drainage canal) totally discharge outlet monitor and closing device, special-assign person response the closing discharge outlet of the rainwater drainage system under emergency situation (including when using the same drainage system with clean water) to prevent the rainwater, fire water and leakage diffuse into the out-environment; If has flood discharge trend and it does not through the production area and terminal. There are facilities prevent leakage and pollutant fire water into flood discharge trend	0
	Do not conform to the requirement above any one	8
Production Wastewater Treatment System Prevention and Control Measures	(1) No production wastewater or drainage; or (2) Production wastewater or drainage: Such as pollutant circulating cooling water, rainwater, fire water discharge into production wastewater treatment system or independent treatment system; and Set monitoring pool before discharging production wastewater to sent the unqualified wastewater to wastewater treatment facility to retreat; and Set accident water buffering pool in the wastewater treatment system if enterprise pollutant clean water or rainwater discharged into the wastewater treatment system; Set production wastewater totally discharge outlet monitor and closing device, special-assign person response opening and closing to prevent leakage, pollutant fire water and unqualified wastewater diffuse to the out of factory	0
	Involving production wastewater or drainage but not conform to requirement of anyone of above (2)	8
Toxic Gas Leakage Emergency Disposal Equipment	No toxic harmful gas; or Set up leak emergency disposal measures against to poisonous harmful gas (such as hydrogen disulfide, hydrogen cyanide, hydrogen chloride, phosgene, chlorine, ammonia and benzene) according to the actual conditions	0
	No toxic gas leakage emergency disposal equipment	8

(continued)

Table 4.9 (continued)

Assessment index	Assessment basis	Score
Toxic gas leakage monitoring and early-warning measures	(1) No toxic harmful gas; or (2) Set up production area or factory bound leak early warning measures against to poisonous harmful gas (such as hydrogen disulfide, hydrogen cyanide, hydrogen chloride, phosgene, chlorine, ammonia and benzene) according to the factual conditions	0
	No production area or factory bound leak monitoring early warning measures against to poisonous harmful gas	4
EIA and Approval Involving Other Risk Prevention and Control Measures	EIA and approval of other construction environmental risk prevention and control measures to carry out	0
	Not to carry out the requirements of EIA and approval of other environmental risk prevention and control measures	10

Table 4.10 Enterprise rainwater, clean water, and production wastewater discharge whereabouts

Assessment basis	Score
No wastewater of full recycling of wastewater after treatment	0
Into the urban sewage treatment plant or concentrated industrial wastewater treatment plant (like industrial park wastewater treatment plant)	7
Enters the other units	
Others (including back-eject, recharge, recycling, etc.)	
Directly into the sea or river lake water environment such as library	10
In the urban sewage into the rivers and lakes library again or in the urban sewage into the sea area again	
Direct access into the sewage irrigation of farmland or enter to seepage or evaporation	

Sensitiveness of Environment Risk Receptor

List all the environment risk receptors around enterprise:

Firstly protection goods in a range less than 5 Km from the production site have to be assessed according to their potential to be a risk receptor.

The following information have to be enlisted for each potential risk receptor: name, scale (population, level or area), latitude, longitude, distance to enterprise, orientation to enterprise, service scope, contact and contact phone number.

Enterprise rainwater outlet (including flood way), clean water outlet, wastewater total outlet, the downstream of which 10 Km scope with water environmental risk receptor (including drinking water source protection area, the water inlet of the water plant, nature conservation area, important wetland, special eco-system, aquaculture, fish spawning grounds, etc.) should be concerned. In terms of maximum flow velocity, within the scope of water flowing through the 24 h involved in border, boundary, city boundary, and so on and so forth. Then listing the contents as follows: name, scale (level or area), and center latitude, longitude, distance to enterprise (meter), and orientation to enterprise, service scope, contact and contact phone number.

The environment risk receptor is arranged style1, style2 and style3 from high to low according to the importance and sensitiveness of environmental risk receptor which are expressed as E1, E2 and E3 Listing the more important and sensitiveness one if there are various styles of environmental risk receptors (Table 4.11).

Classification of Enterprise Environment Risk Level

Finally, the enterprise environment risk level can determined using Tables 4.12, 4.13 and 4.14 in accordance to the ratio between environmental risk substance quality and its threshold (Q), the production process and environment risk control level (M) matrix and the three classes of enterprise surrounding environmental risk receptors.

Ratio Between Environmental Risk Substance Quality and Its Threshold (Q)

Production Process and Environment Risk Control Level (M)
(Table 4.12).

Ratio Between Environmental Risk Substance Quality and Its Threshold (Q)

Production Process and Environment Risk Control Level (M)
(Table 4.13).

Ratio Between Environmental Risk Substance Quality and Its Threshold (Q)

Production Process and Environment Risk Control Level (M)
(Table 4.14).

Table 4.11   Classification of Enterprise Environmental Risk Receptors

Scene	Environmental risk receptor
Scene1 (E1)	Enterprise rainwater outlet, clean water outlet, sewage outlet, the downstream of which 10 Km scope with one kind or kinds of environmental risk receptor as follows: drinking water (surface water and ground water) sources at urban or above urban level protection area; the water inlet of water plant; water conservation area; nature conservation area; important wetland; rare and endangered species of natural concentrated distribution area; the natural spawning and feeding grounds of important aquatic organisms and their wintering grounds and migration routes; landscape and famous scenery; special eco-system; the world cultural and natural heritage sites; mangroves, coral reefs and coastal wetland ecosystem; natural concentrated distribution area of rare and endangered marine life; special marine protection area; salt works reserve; bathing beach; marine natural monuments; or Considered the enterprise rainwater outlet (including flood way), clean water outlet, wastewater total outlet, in terms of maximum flow velocity, within the scope of water flowing through the 24 h involved in border or boundary; or Enterprise surroundings do not conform to the requirement of health prevention distance or atmosphere environmental prevention distance according to EIA and approval; or Within the scope of enterprise around 5 Km of which total population including residential area, institution of medical health, culture education, science research and administration is higher than 50 thousands, or within the scope of enterprise around 500 m population is higher than 1 thousand, or enterprise around 5 Km involved in military reservation, military administrative zones, and relevant national security area
Scene2 (E2)	Enterprise rainwater outlet, clean water outlet, sewage outlet, the downstream of which 10 Km scope with one kind or kinds of environmental risk receptor as follows: aquaculture, fishing grounds, farmlands, basic farmland protection area, eutrophication of waters area, basic prairie, forest park, geography park, natural forest, coastal scenery area, marine living area with important economic value; or Within the scope of enterprise around 5 Km of which total population including residential area, institution of medical health, culture education, science research and administration is higher than 10 thousands and less than 50 thousands, or within the scope of enterprise around 500 m population is higher than 500 hundreds and less than 1 thousand; or Enterprise located in the lava landscape, flood, and debris flow happens areas
Scene3 (E3)	Enterprise downstream with scope of 10 Km do not involving neither of style1 nor style2 above; or Within the scope of enterprise around 5 Km of which total population including residential area, institution of medical health, culture education, science research and administration is less than 10 thousands, or within the scope of enterprise around 500 m population is less than 500 hundreds

Table 4.12 Style1 (E1)-Enterprise environment risk level matrix

	M1 level	M2 level	M3 level	M4 level
$1 \le Q < 10$	Higher environment risk	Higher environment risk	Major environment risk	Major environment risk
$10 \le Q < 100$	Higher environment risk	Major environment risk	Major environment risk	Major environment risk
$100 \le Q$	Major environment risk	Major environment risk	Major environment risk	Major environment risk

Table 4.13 Style2 (E2)—Enterprise environment risk level matrix

	M1 level	M2 level	M3 level	M4 level
$1 \le Q \le 10$	General environment risk	Higher environment risk	Higher environment risk	Major environment risk
$10 \le Q < 100$	Higher environment risk	Higher environment risk	Major environment risk	Major environment risk
$100 \le Q$	Higher environment risk	Major environment risk	Major environment risk	Major environment risk

Table 4.14 Style3 (E3)—Enterprise environment risk level matrix

	M1 level	M2 level	M3 level	M4 level
$1 \le Q < 10$	General environment risk		Higher environment risk	
$10 \le Q < 100$	General environment risk	Higher environment risk nt risk		Major environment risk
$100 \le Q$	Higher environment risk		Major environment risk	

Level Characterization

Enterprise environmental risk level can be expressed as "level (Q code + process and environmental risk control level code + environmental risk receptor style code)", for instance: if Q is in the scope $1 \le Q < 10$, environmental risk receptor is style1, and the process and environmental risk control level is M3, so the enterprise environmental risk level of abrupt environment incident can be described as "major (Q1M3E1)".

Appendix 2: Abrupt Environment Incidents Risk Substance and Critical Mass List

See Tables 4.15 and 4.16.

Table 4.15 Risk substances and their critical mass

Serial number	Substance name	CAS number	Critical mass (t)	Remarks
Part one environment risk substance				
1	Formaldehyde	50-00-0	0.5	
2	Carbon tetrachloride	56-23-5	7.5	
3	1,1-Dimethylhydrazine	57-14-7	7.5	
4	Ethyl ether	60-29-7	10	
5	Methyl hydrazine	60-34-4	7.5	The key environmental management of dangerous chemicals
6	Aniline	62-53-3	5	
7	Dichlorvos	62-73-7	2.5	
8	Methanol	67-56-1	500*	
9	Isopropyl alcohol	67-63-0	5	
10	Acetone	67-64-1	10	
11	Chloroform	67-66-3	10	
12	Butylalcohol	71-36-3	5	
13	Benzene	71-43-2	10	The key environmental management of dangerous chemicals
14	Methane	74-82-8	5	
15	Methyl bromide	74-83-9	7.5	
16	Ethylene	74-85-1	5	
17	Acetylene	74-86-2	5	
18	Chloromethane	74-87-3	10	
19	Methyl iodide	74-88-4	10	
20	Monomethylamine	74-89-5	5	
21	Hydrocyanic acid	74-90-8	2.5	
22	Methanethiol	74-93-1	5	
23	Propane	74-98-6	5	
24	Vinyl chloride	75-01-4	5	The key environmental management of dangerous chemicals
25	Ethylamine	75-04-7	10	
26	Acetonitrile	75-05-8	10	
27	Acetaldehyde	75-07-0	5	
28	Ethyl mercaptan	75-08-1	10	
29	Dichloromethane	75-09-2	10	
30	Carbon disulfide	75-15-0	10	
31	Dimethyl sulphide	75-18-3	10	

(continued)

Table 4.15 (continued)

32	Cyclopropane	75-19-4	5	
33	Ethylene oxide	75-21-8	7.5	The key environmental management of dangerous chemicals
34	Isobutane	75-28-5	5	
35	2-Chloropropane (isopropyl chloride)	75-29-6	5	
36	Isopropylamine	75-31-0	5	
37	Vinylidene chloride	75-35-4	5	
38	Phosgene	75-44-5	0.25	
39	Trimethylamine	75-50-3	2.5	
40	Propyleneimine	75-55-8	10	
41	Propylene oxide	75-56-9	10	
42	Tert-butylamine (2-amino-2-methylpro pane)	75-64-9	5	
43	Trimethylchlorosilane (chlorotrimethylsilane)	75-77-4	7.5	
44	Dimethyldichlorosilane (dichlorodimethylsilane)	75-78-5	2.5	
45	Methyltrichlorosilane	75-79-6	2.5	
46	Acetone cyanohydrin	75-86-5	2.5	The key environmental management of dangerous chemicals
47	Chloropicrin (trichloronitromethane)	76-06-2	0.25	
48	Dimethyl sulfate	77-78-1	0.25	
49	Tetraethyl lead	78-00-2	2.5	The key environmental management of dangerous chemicals
50	Isobutyronitrile	78-82-0	10	
51	1,2-Dichloropropane	78-87-5	7.5	
52	Trichloroethylene	79-01-6	10	
53	Methyl acetate	79-20-9	5	
54	Peracetic acid	79-21-0	5	
55	Methyl chloroformate	79-22-1	2.5	
56	Trifluorochloroethylene (chlorotrifluoroethylene)	79-38-9	5	
57	Methyl methacrylate	80-62-6	5	
58	Dibutyl phthalate	84-74-2	10	
59	Toluene-2,6-diisocyanate	91-08-7	5	

(continued)

Table 4.15 (continued)

60	Naphthalene	91-20-3	5	The key environmental management of dangerous chemicals
61	Benzidine	92-87-5	0.5	
62	1,2-Dichlorobenzene	95-50-1	10	
63	3,4-Dichlorotoluene	95-75-0	10	
64	Methyl chloroacetate	96-34-4	7.5	
65	Nitrobenzene	98-95-3	10	
66	2,6-Dichloro-4-nitroaniline	99-30-9	5	
67	P-nitroaniline	100-01-6	5	
68	Ethylbenzene	100-41-4	10	
69	Phenylethylene	100-42-5	10	
70	1,4-Dichlorobenzene	106-46-7	10	
71	Monomethylaniline	100-61-8	5	
72	1,4-Benzoquinone	106-51-4	1	
73	Epichlorohydrin	106-89-8	10	
74	Butane	106-97-8	5	
75	1-Butene (alpha-butylene)	106-98-9	5	
76	1,3-Butadiene	106-99-0	5	
77	2-Butene	107-01-7	5	
78	Acrolein	107-02-8	2.5	
79	Allyl chloride	107-05-1	5	
80	1,2-Dichloroethane	107-06-2	7.5	
81	Ethylene chlorohydrin (2-chloroethanol)	107-07-3	5	
82	Allylamine	107-11-9	5	
83	Propionitrile	107-12-0	5	
84	Acrylonitrile	107-13-1	10	
85	Ethylenediamine	107-15-3	10	
86	Allyl alcohol	107-18-6	7.5	
87	Vinyl methyl ether	107-25-5	5	
88	Chloromethyl methyl ether	107-30-2	2.5	
89	Methyl Chloroformate	107-31-3	5	
90	Vinyl acetate	108-05-4	7.5	
91	Isopropyl chloroformate	108-23-6	7.5	
92	Cyanuric chloride	108-77-0	10	
93	Toluene	108-88-3	10	
94	Chlorobenzene	108-90-7	5	
95	Cyclohexylamine	108-91-8	10	

(continued)

Table 4.15 (continued)

96	Cyclohexanone	108-94-1	5	
97	Phenol	108-95-2	5	
98	1-Propyl acetate	109-60-4	5	
99	n-Propyl chloroformate (propyl chloroformate)	109-61-5	5	
100	Ethyl nitrite	109-95-5	5	
101	Furan	110-00-9	2.5	
102	n-hexane	110-54-3	500*	
103	Cyclohexane	110-82-7	10	
104	Piperidine	110-89-4	7.5	
105	Hexanedinitrile	111-69-3	2.5	
106	Octanol	111-87-5	7.5	
107	Propylene	115-07-1	5	
108	Methyl ether	115-10-6	5	
109	Isobutylene (2-methylpropene)	115-11-7	5	
110	Tetrafluoroethylene	116-14-3	5	
111	Hexachlorobenzene	118-74-1	1	
112	Trinitrotoluene	118-96-7	5	
113	2,4-Dichlorophenol	120-83-2	5	
114	2,4-Dinitrotoluene	121-14-2	5	
115	Trans-crotonaldehyde	123-73-9	10	
116	Dimethylamine	124-40-3	5	
117	Methylacrylonitrile	126-98-7	2.5	
118	2-Chloro-1,3-butadiene	126-99-8	5	
119	Tetrachloroethylene	127-18-4	10	
120	Phenylacetonitrile	140-29-4	1	
121	n-Butyl acrylate	141-32-2	5	
122	Butyryl chloride	141-75-3	5	
123	Ethyl acetate	141-78-6	500*	
124	2,4,6-Tribromoaniline	147-82-0	5	
125	Ethyleneimine	151-56-4	5	
126	Disyston/Dsulfoton	298-04-4	0.5	
127	Hydrazine	302-01-2	7.5	
128	Methyl fluoroacetate	453-18-9	0.25	
129	Propadiene	463-49-0	5	
130	Ketene	463-51-4	2	
131	Carbonyl sulphide (carbon oxysulfide)	463-58-1	2.5	
132	1,3-Pentadiene	504-60-9	5	
133	Cyanogen bromide	506-68-3	2.5	

(continued)

Table 4.15 (continued)

134	Cyanogen chloride	506-77-4	7.5	The key environment al management of dangerous chemicals
135	Tetranitromethane	509-14-8	5	
136	Methyl thiocyanate	556-64-9	10	
137	2-Chloropropene (2-chloropropylene)	557-98-2	5	
138	Toluene 2,4-Diisocyanate	584-84-9	5	
139	1-Chloropropene (1-chloropropylene)	590-21-6	5	
140	Potassium cyanate	590-28-3	2.5	
141	Perchloromethyl mercaptan	594-42-3	5	
142	Bromotrifluoroethylene	598-73-2	5	
143	Trans-2-butene (2-butene-trans)	624-64-6	5	
144	Methyl isocyanate	624-83-9	5	
145	Carbon monoxide	630-08-0	7.5	
146	1-Buten-3-yne (vinyl acetylene)	689-97-4	5	
147	Acryloyl chloride (acrylyl chloride)	814-68-6	1	
148	Cadmium oxide	1306-19-0	0.25	
149	Phosphorus pentoxide	1314-56-3	10	
150	Methylnaphthalene	1321-94-4	5	
151	Arsenic trioxide	1327-53-3	0.25	The key environmental management of dangerous chemicals
152	Xylene	1330-20-7	10	
153	Methyl tert-butyl ether	1634-04-4	5	
154	Sodium dichloroisocyanurate	2893-78-9	2.5	
155	Epibromohydrin	3132-64-7	2.5	
156	Nickel carbonate	3333-67-3	0.25	
157	Potassium chlorate	3811-04-9	100*	
158	Dichlorosilane	4109-96-0	5	
159	Crotonaldehyde	4170-30-3	10	
160	Ammonium nitrate	6484-52-2	50	
161	Mercury	7439-97-6	0.5	The key environmental management of dangerous chemicals

(continued)

Table 4.15 (continued)

162	Arsenic	7440-38-2	0.25	The key environmental management of dangerous chemicals
163	Sulfur dioxide	7446-09-5	2.5	
164	Sulfur trioxide	7446-11-9	2.5	
165	Titanium tetrachloride	7550-45-0	1	
166	Perchloryl fluoride (trioxychlorofluoride)	7616-94-6	2.5	
167	Boron trifluoride	7637-7-2	2.5	
168	Hydrogen chloride (gas only)	7647-01-0	2.5	
169	Phosphoric acid	7664-38-2	2.5	
170	Hydrogen fluoride	7664-39-3	5	
171	Ammonia	7664-41-7	7.5	
172	Nitric acid	7697-37-2	7.5	
173	Nickel chloride	7718-54-9	0.25	
174	Thionyl chloride	7719-09-7	5	
175	Phosphorus trichloride	7719-12-2	7.5	
176	Bromine	7726-95-6	2.5	
177	Chromic acid	7738-94-5	0.25	
178	Sodium chlorate	7775-09-9	100*	
179	Disodium chromate	7775-11-3	0.25	
180	Sodium arsenate	7778-43-0	0.25	
181	Fluorine	7782-41-4	0.5	
182	Chlorine	7782-50-5	1	
183	Hydrogen sulfide	7783-06-4	2.5	
184	Hydrogen selenide	7783-07-5	0.25	
185	Ammonium sulfate	7783-20-2	10	
186	Difluorine monoxide	7783-41-7	0.25	
187	Sulphur tetrafluoride	7783-60-0	1	
188	Arsenic trichloride (arsenous trichloride)	7784-34-1	7.5	
189	Arsine	7784-42-1	0.5	The key environment al management of dangerous chemicals
190	Nickelous sulfate	7786-81-4	0.25	
191	Potassium chromate	7789-00-6	0.25	
192	Chlorosulphonic acid	7790-94-5	0.5	
193	Ammonium perchlorate	7790-98-9	5	
194	Chlorine monoxide (dichlorine oxide)	7791-21-1	5	

(continued)

Table 4.15   (continued)

195	Phosphine	7803-51-2	2.5	
196	Antimonous hydride	7803-52-3	2.5	
197	Silane	7803-62-5	2.5	
198	Sulfuric acid	8014-95-7	2.5	
199	Disulfur dichloride	10025-67-9	2.5	
200	Trichlorosilane	10025-78-2	5	
201	Phosphorus oxychloride	10025-87-3	2.5	
202	Silicon tetrachloride	10026-04-7	5	
203	Hydrogen bromide (hydrobromic acid)	10035-10-6	2.5	
204	Sodium hydrogen arsenate heptahydrate	10048-95-0	0.22	
205	Chlorine dioxide	10049-04-4	0.5	
206	Nitric oxide (nitrogen monoxide)	10102-43-9	0.5	
207	Nitrogen dioxide	10102-44-0	1	
208	Cadmium chloride	10108-64-2	0.25	
209	Cadmium sulfate	10124-36-4	0.25	
210	Boron trichloride	10294-34-5	2.5	
211	Phosphorus, white	12185-10-3	5	
212	Nickel carbonyl	13463-39-3	0.5	
213	Iron pentacarbonyl	13463-40-6	1	
214	Ammonium nickel sulfate	15699-18-0	0.25	
215	Sodium hydrosulfide	16721-80-5	2.5	
216	Fluosilicic acid	16961-83-4	5	
217	Diborane	19287-45-7	1	
218	Pentaborane	19624-22-7	0.25	
219	Osmium tetroxide	20816-12-0	0.25	
220	Butylene (butene)	25167-67-3	5	
221	Chloronitrobenzene	25167-93-5	5	
222	Diphenylmethane diisocyanate	26447-40-5	0.5	
223	Toluene diisocyanate	26471-62-5	2.5	
224	Acephate	30560-19-1	0.25	
225	Sulfur	63705-05-5	10	
226	Liquefied petroleum gas	68476-85-7	5	
227	Gas	/	7.5	
228	Copper and its compounds	/	0.25	
229	Antimony and its compounds	/	0.25	
230	Thallium and its compounds	/	0.25	
231	Molybdenum and its compounds	/	0.25	
232	Vanadium and its compounds	/	0.25	

(continued)

Table 4.15 (continued)

233	Manganese and its compounds	/	0.25	
234	Oil substance	/	2500**	
235	The rank poison chemical substances	/	5	
236	The toxic chemical substances	/	50	
237	Organic waste liquid which CODCr concentration ≥10000 mg/L	/	10	
238	Waste liquid which NH3-N concentration ≥2000 mg/L	/	1	

Part two major environment management dangerous chemicals

239	1,2,3-Trichlorobenzene	87-61-6	5	
240	1,2,4-Trichlorobenzene	120-82-1	2.5	
241	1,2,4,5-Tetrachlorobenzene	95-94-3	5	
242	1,2-Dinitrobenzene	528-29-0	0.5	
243	1,3-Dinitrobenzene	99-65-0	0.5	
244	1-Chloro-2,4-dinitrobenz ene	97-00-7	5	
245	5-Tart-butyl-2,4,6-trinitro-m-xylene	81-15-2	5	
246	Pentachloronitrobenzene	82-68-8	0.5	
247	2-Toluidine	95-53-4	7.5	
248	o-Chloroaniline	95-51-2	5	
249	Nonyl phenol	25154-52-3	1	
250	4-n-Nonylphenol	84852-15-3	1	
251	Hexachloro-3-butadiene	87-68-3	2.5	
252	Fluoranthene	206-44-0	0.5	
253	Anthracene	120-12-7	5	
254	Anthraxcene		5	
255	Chloroacetone	78-95-5	2.5	
256	1-Octanesulfonic acid	1763-23-1	5	
257	Ammonium perfluorooctanesulfonate	29081-56-9	5	
258	Didecyldimethylammoni um perfluorooctane sulfonate	251099-16-8	5	
259	1-Octanesulfonic acid	70225-14-8	5	
260	Perfluorooctanesulfonic acid potassium salt	2795-39-3	5	
261	1-Octanesulfonic acid	29457-72-5	5	
262	Heptadecafluorooctanes-ulfonic acid tetraethy-lammonium salt	56773-42-3	5	
263	Perfluoro-1-octanesulfonyl fluoride	307-35-7	5	
264	Hexabromocyclododecane	25637-99-4	5	
266	Sodium cyanide	143-33-9	0.25	
267	Potassiumtetracyanonick elate	14220-17-8	0.25	
268	Potassium dicyanoargentate	506-61-6	0.25	
269	Copper cyanide	544-92-3	0.25	

(continued)

Table 4.15 (continued)

270	Orthoarsenicacid	7778-39-4	0.25	
265	Potassium cyanide	151-50-8	0.25	
271	Arsenic(V) oxide	1303-28-2	0.25	
272	Arsenenous acid	7784-46-5	0.25	
273	Cobalt nitrate	10141-05-6	0.25	
274	Nitric acid	13138-45-9;14216-75-2	0.25	
275	Mercury chloride	7487-94-7	0.25	
276	Aminomercuric chloride	10124-48-8	0.25	
277	Mercury nitrate	10045-94-0	0.25	
278	Mercuric acetate	1600-27-7	0.25	
279	Mercuric oxide	21908-53-2	0.25	
280	Mercurousbromide	10031-18-2	0.25	
281	Phenyl mercuric acetate	62-38-4	0.25	
282	Mercury, (nitrato-kO)phenyl-	55-68-5	0.25	
283	Ammonium dichromate	7789-9-5	0.25	
284	Potassium dichromate	7778-50-9	0.25	
285	Sodium dichromate	10588-01-9	0.25	
286	Chromium trioxide	1333-82-0	0.25	
287	Tetramethyllead	75-74-1	2.5	
288	Acetic acid	301-04-2	0.25	
289	Lead silicate	10099-76-0;11120-22-2	0.25	
290	Lead fluoride	7783-46-2	0.25	
291	Lead oxide	1314-41-6	0.25	
292	Lead monoxide	1317-36-8	0.25	
293	Sulfuric acid	7446-14-2	0.25	
294	Lead nitrate	10099-74-8	0.25	
295	Dibutyltin dilaurate	77-58-7	0.5	
296	Dibutyltin oxide	818-08-6	0.25	
297	Selenium dioxide	7446-8-4	0.25	
298	Cadmium selenide	1306-24-7	0.25	
299	Leadmonoselenide	12069-00-0	0.25	
300	Cadmium fluoroborate	14486-19-2	0.25	
301	Cadmium telluride	1306-25-8	0.25	
302	Paraquat	4685-14-7	1	
303	Malathion	121-75-5	10	
304	Tetramethylthiuram disulfide	137-26-8	0.25	
305	Zinc dimethyldithiocarbamate	137-30-4	0.25	
306	Alachlor	15972-60-8	5	
307	Acetochlor	34256-82-1	5	
308	Thiosulfan	115-29-7	0.25	
309	Cypermethrin	52315-07-8	5	
310	Fentin hydroxide	76-87-9	0.25	

Table 4.16 Source inventory of high risk enterprises in LRB

City	District (Town)	Enterprise code	Industrial classification	River basin/Subbasin	Wastewater discharge pathway	E	Q	M
Shenyang	Tiexi	HR001	Inorganic salts manufacturing	Xi river	Enter the municipal wastewater treatment plant	E3	1,720.95	M3
Shenyang	Tiexi	HR002	Chemical pharmaceutical manufacturing	Xi river	Enter the municipal wastewater treatment plant	E2	983.76	M3
Shenyang	Sujiatun	HR003	Chemical reagent manufacturing	Sha river	Enter other units	E3	102.01	M3
Shenyang	Shenbei new district	HR004	Chemical reagent manufacturing	Liao river	Enter the municipal wastewater treatment plant	E1	1.5	M3
Shenyang	Shenbei new district	HR005	Chemical raw material and chemical manufacturing	Liao river	Enter the municipal wastewater treatment plant	E1	7.35	M3
Shenyang	Shenbei new district	HR006	Chemical raw material and chemical manufacturing	Liao river	Enter the municipal wastewater treatment plant	E1	1.08	M3
Shenyang	Shenbei new district	HR007	Other special chemical manufacturing	Liao river	Enter the municipal wastewater treatment plant	E1	11.5	M3
Shenyang	Yuhong	HR008	Paint manufacturing	Pu river	Enter the rivers, lakes and reservoirs	E1	14	M4
Shenyang	Yuhong	HR009	Paint manufacturing	Pu river	Enter the rivers, lakes and reservoirs	E1	37.2	M4
Shenyang	Yuhong	HR010	Paint manufacturing	Pu river	Enter the rivers, lakes and reservoirs	E1	1.7	M4
Shenyang	Yuhong	HR011	Sealing packing and similar products manufacturing	Pu river	Enter the rivers, lakes and reservoirs	E1	5.33	M4
Shenyang	Yuhong	HR012	Paint Manufacturing	Pu river	Enter the rivers, lakes and reservoirs	E1	19.6	M4

(continued)

Table 4.16 (continued)

City	District (Town)	Enterprise code	Industrial classification	River basin/Subbasin	Wastewater discharge pathway	E	Q	M
Shenyang	Yuhong	HR013	Paint manufacturing	Pu river	Enter the rivers, lakes and reservoirs	E1	3.76	M4
Shenyang	Yuhong	HR014	Special chemical manufacturing	Pu river	Enter the rivers, lakes and reservoirs	E1	4.14	M3
Shenyang	Yuhong	HR015	Paint manufacturing	Pu river	Enter the rivers, lakes and reservoirs	E1	6.4	M4
Shenyang	Yuhong	HR016	Paint manufacturing	Pu river	Enter the rivers, lakes and reservoirs	E1	13.86	M4
Shenyang	Liaozhong	HR017	Organic chemical raw material manufacturing	Liao river	Enter other units	E3	242.4	M3
Shenyang	Liaozhong	HR018	Chemical raw material and chemical manufacturing	Liao river	Enter the seepage basin or evaporation basin	E2	30.81	M3
Shenyang	Kangping	HR019	Organic chemical raw material manufacturing	Liao river	Enter other units	E2	29.36	M3
Shenyang	Faku	HR020	Organic chemical raw material manufacturing	Lama river	Enter other units	E1	80.91	M3
Shenyang	Xinmin	HR021	Basic chemical raw material manufacturing	Pu river	Enter the rivers, lakes and reservoirs	E3	90	M4
Shenyang	Xinmin	HR022	Crude oil processing and petrochemical manufacturing	Liao river	Enter the rivers, lakes and reservoirs	E3	250	M3
Shenyang	Xinmin	HR023	Other special chemical manufacturing	Liao river	Enter the seepage basin or evaporation basin	E3	10	M4
Shenyang	Economic and technological development zone	HR024	Other special chemical manufacturing	Xi river	Enter the seepage basin or evaporation basin	E3	550.17	M4

(continued)

Table 4.16 (continued)

City	District (Town)	Enterprise code	Industrial classification	River basin/Subbasin	Wastewater discharge pathway	E	Q	M
Shenyang	Economic and technological development zone	HR025	Special chemical manufacturing	Xi river	Enter the seepage basin or evaporation basin	E1	1.94	M4
Shenyang	Economic and technological development zone	HR026	Crude oil processing and petrochemical manufacturing	Xi river	Enter the municipal wastewater treatment plant	E2	7,368.58	M2
Shenyang	Hunnan new district	HR027	Chemical raw material and chemical manufacturing	Hun river	Enter the seepage basin or evaporation basin	E3	28.89	M4
Shenyang	Qipanshan	HR028	Organic chemical raw material manufacturing	Hun river	Enter the municipal wastewater treatment plant	E2	414.33	M2
Anshan	Tiexi	HR029	Coking	Yunliang river	Enter the municipal wastewater treatment plant	E3	7,500.00	M3
Anshan	Lishan	HR030	Chemical reagent manufacturing	Sha river	Enter industrial wastewater treatment plant	E1	370	M3
Anshan	Qianshan	HR031	Other basic chemical manufacturing	Nansha river	Enter the rivers, lakes and reservoirs	E2	5.4	M4
Anshan	Qianshan	HR032	Chemical reagent manufacturing	Yangliu river	Enter the rivers, lakes and reservoirs	E1	1.05	M3
Anshan	Hancheng	HR033	Pesticide manufacturing	Haicheng river	Enter the rivers, lakes and reservoirs	E3	12.1	M4
Fushun	Xinfu	HR034	Chemical raw material and chemical manufacturing	Hun river	Enter the urban sewage system firstly, then enter the rivers, lakes and reservoirs	E2	6	M4

(continued)

Table 4.16 (continued)

City	District (Town)	Enterprise code	Industrial classification	River basin/Subbasin	Wastewater discharge pathway	E	Q	M
Fushun	Xinfu	HR035	Crude oil processing and petrochemical manufacturing	Hun river	Enter the urban sewage system firstly, then enter the rivers, lakes and reservoirs	E3	320	M3
Fushun	Dongzhou	HR036	Crude oil processing and petrochemical manufacturing	Taizi river	Enter the urban sewage system firstly, then enter the rivers, lakes and reservoirs	E2	3,432.18	M3
Fushun	Dongzhou	HR037	Plastics and resin manufacturing	Sha river	Enter the urban sewage system firstly, then enter the rivers, lakes and reservoirs	E2	254.68	M3
Fushun	Dongzhou	HR038	Explosives and related product manufacturing	Hun river	Enter the urban sewage system firstly, then enter the rivers, lakes and reservoirs	E3	463.33	M4
Fushun	Dongzhou	HR039	Other synthetic material manufacturing	Taizi river	Enter the urban sewage system firstly, then enter the rivers, lakes and reservoirs	E1	3,973.98	M3
Fushun	Dongzhou	HR040	Organic chemical raw material manufacturing	Hun river	Enter the urban sewage system firstly, then enter the rivers, lakes and reservoirs	E1	94.29	M3
Fushun	Dongzhou	HR041	Chemical reagent manufacturing	Hun river	Enter the urban sewage system firstly, then enter the rivers, lakes and reservoirs	E3	252	M3
Fushun	Dongzhou	HR042	Chemical reagent manufacturing	Hun river	Enter the urban sewage system firstly, then enter the rivers, lakes and reservoirs	E2	24.33	M3
Fushun	Dongzhou	HR043	Crude oil processing and petrochemical manufacturing	Hun river	Enter the urban sewage system firstly, then enter the rivers, lakes and reservoirs	E2	14.29	M3

(continued)

Table 4.16 (continued)

City	District (Town)	Enterprise code	Industrial classification	River basin/Subbasin	Wastewater discharge pathway	E	Q	M
Fushun	Dongzhou	HR044	Organic chemical raw material manufacturing	Hun river	Enter the urban sewage system firstly, then enter the rivers, lakes and reservoirs	E1	6	M3
Fushun	Dongzhou	HR045	Organic chemical raw material manufacturing	Hun river	Enter the urban sewage system firstly, then enter the rivers, lakes and reservoirs	E1	5.1	M3
Fushun	Wanghua	HR046	Crude oil processing and petrochemical manufacturing	Taizi river	Enter other units	E3	7,971.00	M3
Fushun	Wanghua	HR047	Chemical raw material and chemical manufacturing	Hun river	Enter the municipal wastewater treatment plant	E3	1,203.09	M3
Fushun	Wanghua	HR048	Petroleum refineries	Taizi river	Enter the urban sewage system firstly, then enter the rivers, lakes and reservoirs	E3	586.97	M3
Fushun	Xinbin manchu autonomous county	HR049	Explosives and related product manufacturing	Hun river	Enter the rivers, lakes and reservoirs	E2	109.09	M3
Fushun	Xinbin manchu autonomous county	HR050	Special chemical manufacturing	Hun river	Enter the municipal wastewater treatment plant	E2	51.24	M3
Fushun	Economic and technological development zone	HR051	Other basic chemical manufacturing	Hun river	Enter the municipal wastewater treatment plant	E1	31.39	M3

(continued)

Table 4.10 (continued)

City	District (Town)	Enterprise code	Industrial classification	River basin/Subbasin	Wastewater discharge pathway	E	Q	M
Fushun	Economic and technological development zone	HR052	Petroleum refineries	Hun river	Enter the municipal wastewater treatment plant	E1	1,428.57	M3
Fushun	Economic and technological development zone	HR053	Petroleum refineries	Hun river	Enter the municipal wastewater treatment plant	E1	73.29	M3
Fushun	Economic and technological development zone	HR054	Chemical pharmaceutical manufacturing	Hun river	Enter the municipal wastewater treatment plant	E1	119.4	M3
Fushun	Economic and technological development zone	HR055	Synthetic material manufacturing	Hun river	Enter the municipal wastewater treatment plant	E2	19.31	M3
Fushun	Economic and technological development zone	HR056	Chemical reagent manufacturing	Hun river	Enter the municipal wastewater treatment plant	E1	45.54	M2
Benxi	Benxi	HR057	Nitrogenous fertilizer manufacturing	Xi river	Enter the rivers, lakes and reservoirs	E3	1,214.08	M3
Jinzhou	Beizhen	HR058	Pesticide manufacturing	Xisha river	Enter the municipal wastewater treatment plant	E1	3.57	M3

(continued)

Table 4.16 (continued)

City	District (Town)	Enterprise code	Industrial classification	River basin/Subbasin	Wastewater discharge pathway	E	Q	M
Yingkou	Xishi	HR059	Paint manufacturing	Hun river	Enter the municipal wastewater treatment plant	E1	24	M3
Yingkou	Laobian	HR060	Chemical reagent manufacturing	Hun river	Enter other units	E3	132.4	M3
Yingkou	Laobian	HR061	Chemical reagent manufacturing	Hun river	Enter other units	E3	100.12	M3
Yingkou	Laobian	HR062	Inorganic salts manufacturing	Hun river	Enter other units	E3	100.33	M3
Yingkou	Laobian	HR063	Chemical reagent manufacturing	Hun river	Enter other units	E3	177.6	M3
Yingkou	Laobian	HR064	Other special chemical manufacturing	Hun river	Enter the rivers, lakes and reservoirs	E3	100	M3
Yingkou	Dashiqiao	HR065	Other special chemical manufacturing	Hun river	Enter other units	E3	20	M4
Yingkou	Dashiqiao	HR066	Paint manufacturing	Hun river	Enter the urban sewage system firstly, then enter the rivers, lakes and reservoirs	E2	1	M4
Fuxin	Zhangwu	HR067	Organic chemical raw material manufacturing	Erdao river	Enter other units	E1	44.74	M3
Fuxin	Zhangwu	HR068	Organic chemical raw material manufacturing	Yangximu river	Enter other units	E1	57.65	M3
Fuxin	Zhangwu	HR069	Organic chemical raw material manufacturing	Erdao river	Enter other units	E1	114.69	M3
Fuxin	Zhangwu	HR070	Organic chemical raw material manufacturing	Hun river	Enter other units	E1	57.59	M3
Liaoyang	Baita	HR071	Other basic chemical manufacturing	Taizi river	Enter the municipal wastewater treatment plant	E2	277.62	M3

(continued)

Table 4.16 (continued)

City	District (Town)	Enterprise code	Industrial classification	River basin/Subbasin	Wastewater discharge pathway	E	Q	M
Liaoyang	Baita	HR072	Organic chemical raw material manufacturing	Taizi river	Enter the municipal wastewater treatment plant	E2	500.4	M3
Liaoyang	Wenshen	HR073	Chemical reagent manufacturing	Taizi river	Enter the urban sewage system firstly, then enter the rivers, lakes and reservoirs	E1	12.87	M3
Liaoyang	Wenshen	HR074	Chemical pharmaceutical manufacturing	Taizi river	Enter other units	E1	2.07	M3
Liaoyang	Wenshen	HR075	Chemical raw material and chemical manufacturing	Taizi river	Enter industrial wastewater treatment plant	E1	34.66	M3
Liaoyang	Wenshen	HR076	Organic chemical raw material manufacturing	Taizi river	Enter the rivers, lakes and reservoirs	E1	5.71	M3
Liaoyang	Wenshen	HR077	Explosives and related product manufacturing	Taizi river	Enter the rivers, lakes and reservoirs	E1	2,576.71	M3
Liaoyang	Wenshen	HR078	Special chemical manufacturing	Taizi river	Enter the municipal wastewater treatment plant	E1	33.33	M3
Liaoyang	Hongwei	HR079	Crude oil processing and petrochemical manufacturing	Taizi river	Enter the rivers, lakes and reservoirs	E1	200	M3
Liaoyang	Hongwei	HR080	Organic chemical raw material manufacturing	Taizi river	Enter the rivers, lakes and reservoirs	E1	3	M3
Liaoyang	Hongwei	HR081	Crude oil processing and petrochemical manufacturing	Taizi river	Enter industrial wastewater treatment plant	E1	44,366.60	M3
Liaoyang	Hongwei	HR082	Chemical reagent manufacturing	Taizi river	Enter the urban sewage system firstly, then enter the rivers, lakes and reservoirs	E3	2,000.00	M3
Liaoyang	Hongwei	HR083	Chemical raw material and chemical manufacturing	Taizi river	Enter other units	E3	209.64	M3

(continued)

Table 4.16 (continued)

City	District (Town)	Enterprise code	Industrial classification	River basin/Subbasin	Wastewater discharge pathway	E	Q	M
Liaoyang	Hongwei	HR084	Special chemical manufacturing	Taizi river	Enter other units	E1	10.81	M3
Liaoyang	Hongwei	HR085	Organic chemical raw material manufacturing	Taizi river	Enter the rivers, lakes and reservoirs	E1	5.9	M3
Liaoyang	Dengta	HR086	Chemical reagent manufacturing	Sha river	Enter other units	E3	128.57	M3
Panjin	Shuangtaizi	HR087	Chemical reagent manufacturing	Liao river	Enter the rivers, lakes and reservoirs	E2	7.75	M4
Panjin	Shuangtaizi	HR088	Plastics and resin manufacturing	Liao river	Enter the rivers, lakes and reservoirs	E2	1,613.70	M3
Panjin	Shuangtaizi	HR089	Plastics and resin manufacturing	Liao river	Enter the rivers, lakes and reservoirs	E2	160	M3
Panjin	Shuangtaizi	HR090	Nitrogenous fertilizer manufacturing	Liao river	Enter the rivers, lakes and reservoirs	E2	2,480.00	M3
Panjin	Shuangtaizi	HR091	Other basic chemical manufacturing	Liao river	Enter the rivers, lakes and reservoirs	E2	12.32	M3
Panjin	Xinglongtai	HR092	Special chemical manufacturing	Liao river	Enter the rivers, lakes and reservoirs	E2	21	M3
Panjin	Xinglongtai	HR093	Crude oil processing and petrochemical manufacturing	Liao river	Enter other units	E1	68.8	M3
Panjin	Xinglongtai	HR094	Crude oil processing and petrochemical manufacturing	Liao river	Enter the rivers, lakes and reservoirs	E3	250	M3
Panjin	Xinglongtai	HR095	Crude oil processing and petrochemical manufacturing	Liao river	Enter the rivers, lakes and reservoirs	E2	14,285.71	M4

(continued)

Table 4.16 (continued)

City	District (Town)	Enterprise code	Industrial classification	River basin/Subbasin	Wastewater discharge pathway	E	Q	M
Panjin	Xinglongtai	HR096	Nitrogenous fertilizer manufacturing	Liao river	Enter the municipal wastewater treatment plant	E1	155.7	M2
Panjin	Dawa	HR097	Other basic chemical manufacturing	Liao river	Enter the rivers, lakes and reservoirs	E1	15.95	M3
Panjin	Dawa	HR098	Other basic chemical manufacturing	Liao river	Enter the rivers, lakes and reservoirs	E1	161.4	M4
Panjin	Dawa	HR099	Chemical raw material and chemical manufacturing	Liao river	Enter the rivers, lakes and reservoirs	E1	11.41	M3
Panjin	Dawa	HR100	Petrochemical manufacturing and coking	Liao river	Enter the rivers, lakes and reservoirs	E1	120.4	M4
Panjin	Dawa	HR101	Chemical raw material and chemical manufacturing	Liao river	Enter the rivers, lakes and reservoirs	E3	50	M4
Panjin	Dawa	HR102	Chemical raw material and chemical manufacturing	Liao river	Enter the rivers, lakes and reservoirs	E1	60.6	M4
Panjin	Dawa	HR103	Chemical raw material and chemical manufacturing	Liao river	Enter the rivers, lakes and reservoirs	E1	1	M3
Panjin	Dawa	HR104	Chemical raw material and chemical manufacturing	Liao river	Enter the rivers, lakes and reservoirs	E1	1.67	M3
Panjin	Dawa	HR105	Pesticide manufacturing	Liao river	Enter the rivers, lakes and reservoirs	E1	10.86	M3
Panjin	Dawa	HR106	Organic chemical raw material manufacturing	Liao river	Enter the rivers, lakes and reservoirs	E1	60.4	M3

(continued)

Table 4.16 (continued)

City	District (Town)	Enterprise code	Industrial classification	River basin/Subbasin	Wastewater discharge pathway	E	Q	M
Panjin	Dawa	HR107	Other basic chemical manufacturing	Liao river	Enter the rivers, lakes and reservoirs	E1	202	M4
Panjin	Dawa	HR108	Crude oil processing and petrochemical manufacturing	Liao river	Enter the rivers, lakes and reservoirs	E1	25	M3
Panjin	Panshan	HR109	Potassium fertilizer manufacturing	Liao river	Enter the rivers, lakes and reservoirs	E3	520	M3
Tieling	Tieling	HR110	Synthetic material manufacturing	Wanquan river	Enter the rivers, lakes and reservoirs	E3	22.5	M4
Tieling	Tieling	HR111	Other synthetic material manufacturing	Liao river	Enter the rivers, lakes and reservoirs	E3	67.5	M4
Tieling	Tieling	HR112	Organic chemical raw material manufacturing	Liao river	Enter the rivers, lakes and reservoirs	E3	127.86	M3
Tieling	Changtu	HR113	Organic chemical raw material manufacturing	Zhaosutai river	Enter the rivers, lakes and reservoirs	E1	59.14	M3
Tieling	Changtu	HR114	Organic chemical raw material manufacturing	Zhaosutai river	Enter the rivers, lakes and reservoirs	E2	29.07	M3
Tieling	Changtu	HR115	Organic chemical raw material manufacturing	Erdao river	Enter the rivers, lakes and reservoirs	E1	29.57	M3
Tieling	Changtu	HR116	Other basic chemical manufacturing	Erdao river	Enter the rivers, lakes and reservoirs	E1	43.36	M4
Tieling	Changtu	HR117	Organic chemical raw material manufacturing	Mazhong river	Enter the urban sewage system firstly, then enter the rivers, lakes and reservoirs	E1	43.16	M3

References

1. Belis, C.A., F. Karagulian, B.R. Larsen, and P.K. Hopke. 2013. Critical review and meta-analysis of ambient particulate matter source apportionment using receptor models in Europe. *Atmospheric Environment* 69: 94–108.
2. Callén, María Soledad, Amaia Iturmendi, and José Manuel López. 2014. Source apportionment of atmospheric PM2.5-bound polycyclic aromatic hydrocarbons by a PMF receptor model. *Environmental Pollution* 195: 167–177.
3. Chen, Hai-Yang, Yan-Guo Teng, and J.S. Wang. 2012. Source apportionment for sediment PAHs from the Daliao River (China) using an extended fit measurement mode of chemical mass balance model. *Ecotoxicology and Environmental Safety* 88 (11): 148–154.
4. Chen, Xiaoqiu. 2006. Discussing on the filtering method of prior controlled organic pollutants in water (in Chinese). *Fujian Analysis & Testing* 15 (1): 15–17.
5. Chevreuil, M., A. Chesterikoff, and R. Letolle. 1987. PCB pollution behaviour in the river Seine. *Water Research* 21 (4): 427–434.
6. China Council for International Cooperation on Environment and Development. 2007. Reports on environmentally sound management and strategies of chemicals in China (in Chinese).
7. Deng, Baole, Lingyan Zhu, Man Liu, Nannan Liu, Liping Yang, and yang Du. 2011. Sediment quality criteria and ecological risk assessment for heavy metals in Taihu Lake and Liao River (in Chinese). *Research of Environmental Sciences*, 24 (1): 33–42.
8. Duan, Liang, Lu Shan Li, Yonghui Song Han, Beihai Zhou, and Jing Zhang. 2015. Comparison between moving bed-membrane bioreactor and conventional membrane bioreactor systems. Part I: membrane fouling. *Environmental Earth Sciences* 73 (9): 4881–4890.
9. Fan, Y.H., C.Y. Lin, M.C. He, and Z.F. Yang. 2008. Transport and bioavailability of Cu, Pb, Zn and Ni in surface sediments of Daliao River watersystem (in Chinese). *Journal of Environmental Sciences* 29 (12): 3469–3476.
10. Feng, Chenglian, Xinghui Xia, Zhenyao Shen, and Zhui Zhou. 2007. Distribution and sources of polycyclic aromatic hydrocarbons in Wuhan section of the Yangtze River China. *Environmental Monitoring and Assessment* 133 (1): 447–458.
11. Fowler, Scott W. 1990. Critical review of selected heavy metal and chlorinated hydrocarbon concentrations in the marine environment. *Marine Environmental Research* 29 (1): 1–64.
12. Gao, Hongjie, Chunjian Lv, Yu. Yonghui Song, Lijie Zheng Zhang, Yujie Wen, Jianfeng Peng, and Yu. Huibin. 2015. Chemometrics data of water quality and environmental heterogeneity analysis in Pu River China. *Environmental Earth Sciences* 73 (9): 5119–5129.
13. Glasby, G.P., P. Szefer, J. Geldon, and J. Warzocha. 2004. Heavy-metal pollution of sediments from Szczecin Lagoon and the Gdansk Basin, Poland. *Science of the Total Environment* 330 (1): 249–269.
14. Gu, Z.F., and Y. Wu. 2005. The occurrence and assessment of heavy metal pollution in Taihu Lake (in Chinese). *Gansu Science and Technology* 21 (12): 21–23.
15. Guan, Y.F., Q. Yue, X.Y. Tu, and H.H. Wu. 2011. Distribution and sources of polychlorinated bisphenyls (PCBs) in riverine water of the outlets of the Pearl River Delta (in Chinese). *Research of Environmental Sciences* 24 (8): 865–872.
16. Guerrero, P.F. 1994. *Toxic Substances Control Act: Preliminary Observations on Legislative Changes to Make TSCA More Effective*. General Accounting Office, United States. Congress. Senate. Committee on Environment and Public Works. Subcommittee on Toxic Substances, Research, and Development.
17. Guo, Wei, M.C. He, Z.F. Yang, C.Y. Lin, X.C. Quan, and H.Z. Wang. 2007. Contamination characters of polycyclic aromatic hydrocarbons in Daliao River system of China (in Chinese). *The Journal of Applied Ecology* 18 (7): 1534–1538.
18. Guo, Wei, H.E. Mengchang, Zhifeng Yang, and Chunye Lin. 2007. Distribution and sources of petroleum hydrocarbons and polycyclic aromatic hydrocarbons in sediments from Daliao River watershed, China (in Chinese). *Acta Scientiae Circumstantiae* 27 (5): 824–831.

19. Guo, Wei, Mengchang He, Zhifeng Yang, Chunye Lin, and Xiangchun Quan. 2011. Aliphatic and polycyclic aromatic hydrocarbons in the Xihe River, an urban river in China's Shenyang city: Distribution and risk assessment. *Journal of Hazardous Materials* 186 (2): 1193–1199.

20. Han, Fei, Xiping Ma, and Yingying Liu. 2009. Pollution level and source apportionment of polycyclic aromatic hydrocarbons (PAHs) in Liaohe River (in Chinese). *Journal of Meteorology and Environment* 25 (6): 68–71.

21. Han, F., B.D. Guo, and Y.Y. Wang. 2011. Distribution, sources and risk assessment of polycyclic aromatic hydrocarbons (PAHs) in sediments of Liaohe River (in Chinese). *Environmental Protection & Re-cycling Economy* 30 (12): 62–66.

22. Han, Shuping, Ying Zhang, Shigeki Masunaga, Siyun Zhou, and Wataru Naito. 2014. Relating metal bioavailability to risk assessment for aquatic species: Daliao River watershed China. *Environmental pollution* 189: 215–222.

23. Henry, Ronald C., Charles W. Lewis, Philip K. Hopke, and Hugh J. Williamson. 1984. Review of receptor model fundamentals. *Atmospheric Environment (1967)*, 18 (8):1507–1515.

24. Henry, Ronald C. 2003. Multivariate receptor modeling by N-dimensional edge detection. *Chemometrics and Intelligent Laboratory Systems* 65 (2): 179–189.

25. HJ/T169-2004. Technical guidance of environmental risk assessment for construction projections.

26. Jiang, Jianbin, Jing Wang, Shaoqing Liu, Chunye Lin, Mengchang He, and Xitao Liu. 2013. Background, baseline, normalization, and contamination of heavy metals in the Liao River Watershed sediments of China. *Journal of Asian Earth Sciences* 73: 87–94.

27. Kalkbrenner, Amy E., Rebecca J. Schmidt, and Annie C. Penlesky. 2014. Environmental chemical exposures and autism spectrum disorders: a review of the epidemiological evidence. *Current Problems in Pediatric and Adolescent Health Care* 44 (10): 277–318.

28. Kanaki, M., A. Nikolaou, C.A. Makri, and D.F. Lekkas. 2007. The occurrence of priority PAHs, nonylphenol and octylphenol in inland and coastal waters of Central Greece and the Island of Lesvos. *Desalination* 210 (1–3): 16–23.

29. Keats, Andrew, Eugene Yee, and Fue-Sang Lien. 2007. Bayesian inference for source determination with applications to a complex urban environment. *Atmospheric Environment* 41 (3): 465–479.

30. Lang, Yin-hai, and Wei Yang. 2014. Source apportionment of PAHs using Unmix model for Yantai costal surface sediments, China. *Bulletin of Environmental Contamination and Toxicology* 92 (1): 30–35.

31. Lang, Yin-Hai, Guo-Liang Li, Xiao-Mei Wang, Peng Peng, and Jie Bai. 2015. Combination of Unmix and positive matrix factorization model identifying contributions to carcinogenicity and mutagenicity for polycyclic aromatic hydrocarbons sources in Liao River delta reed wetland soils China. *Chemosphere* 120: 431–437.

32. Li, An, Jae-Kil Jang, and Peter A. Scheff. 2003. Application of EPA CMB8. 2 model for source apportionment of sediment PAHs in Lake Calumet, Chicago. *Environmental Science & Technology* 37 (13): 2958–2965.

33. Li, G.C., X.H. Xia, R. Wang, M.C. He, and X.Q. Xiao. 2006. Pollution of polycyclic aromatic hydrocarbons (PAHs) in middle and lower reaches of the Yellow River. *Environmental Science* 27 (9): 1738–1743.

34. Lin, Hai Tao, Gan Zhang, Zhang Dong Jin, Xiang Dong Li, and Hui Bin Sun. 2007. Sedimentary records of organochlorine pesticides in the Taihu Lake (in Chinese). *China Environmental Science* 27 (4): 441–444.

35. Lin, Tian, Yanwen Qin, Lei Zhang, Binghui Zheng, Yuanyuan Li, and Zhigang Guo. 2011. Distribution, sources and ecological risk assessment of organochlorine pesticides and polychlorinated biphenyl residues in surface sediment from Dahuofang reservoir Liaoning. *Environmental Science* 32 (11): 3294–3299.

36. Liu, Jian Guo, Jian-Xin Hu, and X.Y. Tang. 2006. Global governance on environmentally sound management of chemicals and improvement requirements for China's system (in Chinese). *Research of Environmental Sciences* 19 (6): 121–126.

37. Liu, N.N., P. Chen, S.Z. Zhu, and L.Y. Zhu. 2011. Distribution characteristics of PAHs and OCPs in sediments of Liao River and Taihu Lake and their risk evaluation based on sediment quality criteria. *Zhongguo Huanjing Kexue/china Environmental Science* 31 (2): 293–300.
38. Liu, Rui Xia, L.I. Bin, Yong Hui Song, and Ping Zeng. 2014. Occurrence and pollution sources of toxic and hazardous substances in Liao River Basin (in Chinese). *Journal of Environmental Engineering Technology* 4 (4): 299–305.
39. Lu, J.J., C. Yang, R.Q. Lu, and S.R. Hong. 2009. The pollution of polycyclic aromatic hydrocarbon in the Pearl River around Guangzhou higher education mega center (in Chinese). *Environmental Monitoring in China* 25 (5): 86–89.
40. Luo, H., X.H. Wang, L. Tang, L.Y. Hong, S.P. Wu, and W. Xie. 2010. Distributions of dissolved organochlorine pesticides and polychlorinated biphenyls in China coastal waters (in Chinese). *Marine Environmental Science* 29 (1): 115–120.
41. Lv, Jiapei, Xu Jian, Changsheng Guo, Yuan Zhang, Yangwei Bai, and Wei Meng. 2014. Spatial and temporal distribution of polycyclic aromatic hydrocarbons (pahs) in surface water from Liaohe River Basin, Northeast China. *Environmental Science and Pollution Research* 21 (11): 7088–7096.
42. Mahmoud, Manal A.M., Anna Kärrman, Sayoko Oono, Kouji H. Harada, and Akio Koizumi. 2009. Polyfluorinated telomers in precipitation and surface water in an urban area of Japan. *Chemosphere* 74 (3): 467–472.
43. Ministry of Environmental Protection of the People's Republic of China. 2014. Weekly monitoring report on water quality in main water basins (in Chinese).
44. Minstry of the Environment of Japan. 2013. Lessons from minamata disease and mercury management in Japan.
45. Moore, James W., and Subramaniam Ramamoorthy. 2012. *Organic chemicals in natural waters: Applied monitoring and impact assessment*. Springer Science & Business Media.
46. Paatero, Pentti, and Unto Tapper. 1994. Positive matrix factorization: A non-negative factor model with optimal utilization of error estimates of data values. *Environmetrics* 5 (2): 111–126.
47. Pekey, Hakan, Duran Karakaş, and Mithat Bakoglu. 2004. Source apportionment of trace metals in surface waters of a polluted stream using multivariate statistical analyses. *Marine Pollution Bulletin* 49 (9): 809–818.
48. Peng, Jianfeng, Zhaoyong Ren, Yonghui Song, Yu. Huibin, Xiaoyu Tang, and Hongjie Gao. 2015. Impact of spring flooding on DOM characterization in a small watershed of the Hun River China. *Environmental Earth Sciences* 73 (9): 5131–5140.
49. Qin, Weiwei, Yonghui Song, Yunrong Dai, Guanglei Qiu, Meijie Ren, and Ping Zeng. 2015. Treatment of berberine hydrochloride pharmaceutical wastewater by O3/UV/H2O2 advanced oxidation process. *Environmental Earth Sciences* 73 (9): 4939–4946.
50. Schultz, M.M. 2014. Chemicals of environmental concern. In *Encyclopedia of Toxicology (Third Edition)*, 3rd ed, ed. Philip Wexler, 805–809. Academic Press.
51. Sofowote, Uwayemi M., Brian E. McCarry, and H. Marvin Christopher. 2008. Source apportionment of PAH in Hamilton Harbour suspended sediments: Comparison of two factor analysis methods. *Environmental Science & Technology* 42 (16): 6007–6014.
52. Sorrell, R. Kent, Herbert J. Brass, and Richard Reding. 1980. A review of occurrences and treatment of polynuclear aromatic hydrocarbons in water. *Environment International* 4 (3): 245–254.
53. Su, F., H.J. Zhan, H.P. Yuan, C.J. Liu, P.P. Qi, and J.P. Chen. 2011. A linkage sampling equipment for collecting trace POPs in high-volume water and its performance evaluation (in Chinese). *Environmental Monitoring in China* 27 (10): 15–18.
54. Sun, Hongwen, Fasong Li, Tao Zhang, Xianzhong Zhang, Na He, Qi Song, Lijie Zhao, Lina Sun, and Tieheng Sun. 2011. Perfluorinated compounds in surface waters and WWTPs in Shenyang, China: Mass flows and source analysis. *Water Research* 45 (15): 4483–4490.
55. Tian, Zhiyong, Wang Xin, Yonghui Song, and Fayun Li. 2015. Simultaneous organic carbon and nitrogen removal from refractory petrochemical dry-spun acrylic fiber wastewater by hybrid A/O-MBR process. *Environmental Earth Sciences* 73 (9): 4903–4910.
56. US EPA. 2008. Hudson river PCBs.

57. US EPA. 2014. Hudson river PCBs: Background and site information.
58. Wang, Shaoyu, and Baixia Feng. 2005. *Emergency response and management of city disaster*. Chongqing Press.
59. Wang, Y., Z.J. Wang, Ja Liu, M. Ma, and C.X. Wang. 1999. Monitoring toxic and organic pollutants in the Huai River using Triolein-SPMD (in Chinese). *Environmental Monitoring in China*, 15 (4): 8–11.
60. Wang, Dong Hui. 2006. Study on the pollution of PAHs in Songhua River (in Chinese). *Environmental Science & Management* 31 (9): 69–70.
61. Wang, H.Z., M.C. He, C.Y. Lin, X.C. Quan, and Wei Guo. 2007. Distribution characteristics of organochlorine pesticides in river surface sediments in Song-Liao Watershed (in Chinese). *The Journal of Applied Ecology* 18 (7): 1523–1527.
62. Wang, Jingxian, Yonghong Bi, Gerd Pfister, Bernhard Henkelmann, Kongxian Zhu, and Karl-Werner Schramm. 2009. Determination of PAH, PCB, and OCP in water from the three gorges reservoir accumulated by semipermeable membrane devices (SPMD). *Chemosphere* 75 (8): 1119–1127.
63. Wang, Bin, M.I. Juan, Xue Jun Pan, Jin Yang Zhang, and Y.U. Fang. 2010. Contamination of organochlorine pesticides in waters and sediments in China (in Chinese). *Journal of Kunming University of Science & Technology* 35: 93–99.
64. Watson, John G. 1984. Overview of receptor model principles. *Journal of the Air Pollution Control Association* 34 (6): 619–623.
65. Wei, Jian, Yonghui Song, Xiaoguang Meng, and Jean-Stéphane Pic. 2015. Combination of fenton oxidation and sequencing batch membrane bioreactor for treatment of dry-spun acrylic fiber wastewater. *Environmental Earth Sciences* 73 (9): 4911–4921.
66. Wu, Jiang-Yue, Zheng-Tao Liu, Jun-Li Zhou, and Fu Gao. 2012. Spatial distribution and risk assessment of polycyclic aromatic hydrocarbons in partial surface sediments of Liao River (in Chinese). *Environmental Sciences* 33 (12): 4244–4250.
67. Wu, J.Y., Z.T. Liu, L. Feng, J.L. Zhou, and F. Gao. 2012. Distribution characteristics and risk assessment of PAHs in surface water of Liao River (in Chinese). *Environmental Chemistry* 31 (7): 1116–1117.
68. Xiao, Shuhu, Yonghui Song, Zhiyong Tian, Tu Xiang, Hu Xinqi, and Ruixia Liu. 2015. Enhanced mineralization of antibiotic berberine by the photoelectrochemical process in presence of chlorides and its optimization by response surface methodology. *Environmental Earth Sciences* 73 (9): 4947–4955.
69. Xiu, Fang Zhang, and Li Dong Xiao. 2002. Organic chlorinated pesticides in middle and lower reaches of Liao River (in Chinese). *Journal of Dalian Institute of Light Industry* 21 (2): 102–104.
70. Xu, Shi Fen, Xin Jiang, Lian Sheng Wang, Xie Quan, D. Martens, et al. 2000. Polycyclic aromatic hydrocarbons (PAHs) pollutants in sediments of the Yangtse River and the Liao River. *China Environmental Science* 20 (2): 128–131 (in Chinese).
71. Xue, Lidong, Yinhai Lang, Aixia Liu, and Jie Liu. 2010. Application of CMB model for source apportionment of polycyclic aromatic hydrocarbons (PAHs) in coastal surface sediments from Rizhao offshore area China. *Environmental Monitoring and Assessment* 163 (1): 57–65.
72. Yang, M., Y.W. Ni, F. Su, Q. Zhang, and J.P. Chen. 2007. Distribution and sources of polycyclic aromatic hydrocarbons(PAHs) in sediments of Liao River, China (in Chinese). *Environmental Chemistry* 26 (2): 217–220.
73. Yang, Liping, Lingyan Zhu, and Zhengtao Liu. 2011. Occurrence and partition of perfluorinated compounds in water and sediment from Liao River and Taihu Lake China. *Chemosphere* 83 (6): 806–814.
74. Yin, Jianxin, Roy M. Harrison, Qiang Chen, Andrew Rutter, and James J. Schauer. 2010. Source apportionment of fine particles at urban background and rural sites in the UK atmosphere. *Atmospheric Environment* 44 (6): 841–851.
75. Zhang, Xiao-jian, Chao Chen, Jian-qing Ding, Aixin Hou, Yong Li, Zhang-bin Niu, Xiao-yan Su, Yan-juan Xu, and Edward A. Laws. 2010. The 2007 water crisis in Wuxi, China: analysis of the origin. *Journal of Hazardous Materials* 182 (1): 130–135.

76. Zhang, X.-F., Xie Quan, J.-W. Chen, Y.-Z. Zhao, Shuo Chen, D.-M. Xue, and F.-L. Yang. 2000. Investigation of polychlorinated organic compounds(PCOCs) in middle and lower reaches of Liao River (in Chinese). *China Environmental Science* 20 (1): 31–35.
77. Zhang, Haijun, Yuwen Ni, Jiping Chen, Su Fan, Lu Xianbo, Liang Zhao, Qing Zhang, and Xueping Zhang. 2008. Polychlorinated dibenzo-p-dioxins and dibenzofurans in soils and sediments from Daliao River Basin China. *Chemosphere* 73 (10): 1640–1648.
78. Zhang, Jing, S.Q. Wang, Yan Xie, X.F. Wang, X.J. Sheng, and J.P. Chen. 2008. Distribution and pollution character of heavy metals in the surface sediments of Liao River. *Environmental Sciences* 29 (9): 2413–2418.
79. Zhang, Zheng, Yue Peng, Lu Yan, and Yuxia Jia. 2011. Analytical procedures to classify organic pollutants in Liao River (in Chinese). *Environmental Science & Management* 36 (1): 19–21.
80. Zhang, Y.L., J. Hu, B.J. Liu, S.L. Li, and J. Guan. 2012. Distribution and sources appointment of polycyclic aromatic hydrocarbons (PAHs) in the Liao River Drainage Basin, Northeast China (in Chinese). *Journal of Earth Environment* 40 (2): 188–194.
81. Zhi, Erquan, Yonghui Song, Liang Duan, Yu. Huibin, and Jianfeng Peng. 2015. Spatial distribution and diversity of microbial community in large-scale constructed wetland of the Liao River Conservation Area. *Environmental Earth Sciences* 73 (9): 5085–5094.
82. Zhihao, Wu, Mengchang He, and Chunye Lin. 2012. Environmental impacts of heavy metals (Co, Cu, Pb, Zn) in surficial sediments of estuary in Daliao river and Yingkou Bay (northeast China): concentration level and chemical fraction. *Environmental Earth Sciences* 66 (8): 2417–2430.
83. Zhou, H.D., and W.Q. Peng. 2005. *Water pollution and water environmental restoration (in Chinese)*. Chemical Industry Press.
84. Zhu, Yingchuan, Wen Liu, Yipin Zhou, and Zexiang Lei. 2008. Reused path of heavy metal pollution in hydro-environment and its research advance (in Chinese). *Guangdong Agricultural Sciences* 8: 143–146.

Chapter 5
Groundwater Risk Sources Identification and Risk Reduction Management in the Song-Liao-River-Basin

Erik Nixdorf, Yuanyuan Sun, Jing Su,
Qiang Wang, Tong Wang, Olaf Kolditz and Beidou Xi

5.1 Groundwater Source Risk Assessment Guideline in SLRB

by SUN Yuanyuan, Erik NIXDORF, SU Jing, WANG Qiang, WANG Tong, Olaf KOLDITZ, XI Beidou

This deliverable reports the establishment of an index system for groundwater source risk assessment in the SUSTAIN H$_2$O project during the project's 1st reporting period. Two demonstration areas in the SLRB were chosen: one in Ashi River Basin and the other in Taizi River Basin. Ashi River is a tributary of Songhua River and Taizi River is a tributary of Liao River. The consortium members have conducted several field surveys in the study areas and acquired first-hand information. Experts from Germany provided technologies and strategies on groundwater risk assessment in EU countries. This deliverable is the outcome of the internal and external collaborations that the consortium members have engaged in the context of the SUSTAIN H$_2$O project.

E. Nixdorf (✉)
Helmholtz Centre for Environmental Research, DE, Leipzig, Germany
e-mail: erik.nixdorf@ufz.de

Y. Sun · J. Su · B. Xi
Chinese Research Academy of Environmental Sciences, Chaoyang , China

Q. Wang
Heilongjiang Provincial Research Institute of Environmental Science, Harbin, China

T. Wang
Liaoning Academy of Environmental Sciences, Shenyang, China

O. Kolditz
Helmholtz Centre for Environmental Research, TU Dresden, DE, Dresden, Germany

© Springer International Publishing AG, part of Springer Nature 2018
Y. Song et al. (eds.), *Chinese Water Systems*, Terrestrial Environmental Sciences,
https://doi.org/10.1007/978-3-319-76469-6_5

5.1.1   Introduction

Groundwater is a valuable natural resource. As the expansion of industrial and commercial, agricultural, and residential land-usage. The demand for groundwater increased accordingly, leading to serious groundwater problems. Once the groundwater is polluted, it is very expensive and time consuming to remedy. Thus it is more significant to protect groundwater by means of preventing it from contacting with pollution. Groundwater risk assessment methods have been put forward to assess potential groundwater vulnerability to pollutants and have been widely used for determining areas with high pollution potential and delineating aquifer protection zones.

In this deliverable, firstly the principles for groundwater risk assessment are outlined, followed by an introduction of groundwater risk assessment methodologies in the EU countries. Based on the data acquisition and field survey, the basic information of the study area in Ashi River Basin and Taizi River Basin are delineated. A risk assessment index system has been established for the SLRB with a detailed description of how the methodology is chosen and how the procedures are preceded. This index system is supposed to be used as a guideline for the groundwater source protection in SLRB.

5.1.2   Principles for Groundwater Risk Assessment

5.1.2.1   Groundwater Resources

Groundwater as a kind of resources plays a significant role in the global water cycle. From the quantitative aspect, it makes up about 20% of the worlds' fresh water supply. From the distributive aspect, it is a pervasive resource and can interact with surface water. Groundwater is widely used for municipal, rural, domestic and industrial water supplies. Although in principle it is a renewable natural resource, groundwater is also very valuable and should be protected from deterioration and contamination.

5.1.2.2   Groundwater Contamination

Groundwater contamination has been observed world-wide and it is obvious that human activity is essential for the contamination of groundwater. In general, there are three main components in the process of groundwater contamination: a potential contamination source, an underlying aquifer and a pathway between these two. Contamination source are often referred to as point sources, diffuse sources, line sources, mobile sources, and so on. Possible pathways are through directly contact with water, air, and soil, either by natural or artificial structures.

5.1.2.3   Groundwater Risk Assessment

Risk assessment is the process of identification of hazards, analysis and evaluation of the risk associated with the hazard, and determination of ways to eliminate or control the hazard. For groundwater risk assessment, it is to identify the contamination sources, analyse the exposure and vulnerability of groundwater resources to these contamination sources and establish risk assessment index system.

5.1.3   Groundwater Risk Assessment Methodologies in the EU Countries

5.1.3.1   Legal Framework

Water Framework Directive

The keystone of modern European water law is the European Water Framework Directive [1] which was adopted in October 2000 and can be considered as the framework of European Water policy. The WFD stipulates its members to protect all water bodies (including inland surface waters, transitional waters and groundwater) from further deterioration and achieve good water status for all waters until 2015.

For Groundwater the following specific objectives are set up by the WFD [2]:

1. To implement measures to prevent or limit the input of pollutants into groundwater and to prevent the deterioration of the status of the groundwater body (groundwater status consists of two parts; quantitative status and chemical status and the overall status of groundwater is taken to be the poorer of the two)
2. To protect, enhance and restore all bodies of groundwater, and ensure a balance between abstraction and recharge of groundwater with the aim of achieving good groundwater status by 2015 in accordance with the provisions laid down in Annex V
3. To reverse any significant and sustained upward trend in the concentrations of any pollutant resulting from the impact of human activity in order to progressively reduce pollution of groundwater.

Groundwater Directive

Although the definition of good quantitative status is set out in WFD Annex V 2.1.2, article 17 of the WFD calls for a daughter directive on groundwater in order to establish criteria for defining a good groundwater status and for the identification of significant and sustained trends in pollution concentration [2].

In consequence, the Groundwater Directive (GWD) was adopted in 2006 to fill this legislative gap [3]. The GWD introduces measures to prevent or limit inputs

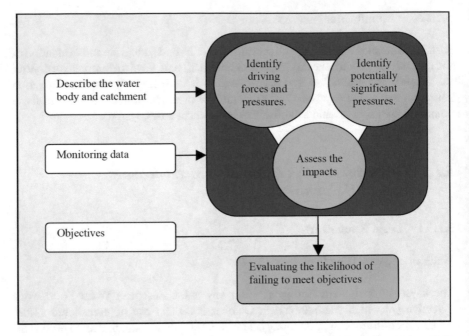

Fig. 5.1 Key components in the analysis of pressures and impacts (European Commission, 2010) [5]

of pollutants into groundwater and requires member states to prevent any inputs of hazardous substances defined by Annex X of the WFD into groundwater and to limit the input of non-hazardous pollutants [4].

In the latter case, The GWD and sets out European-wide minimum groundwater quality standards for Nitrate (50 mg/l) and pesticides ($\sum = 0.5 \,\mu g/l$ and any specific component $< 0.1 \,\mu g/l$). Furthermore, considering local characteristics, the GWD calls the member states to establish thresholds for at least eight additional inorganic ions and two synthetic substances [3].

5.1.3.2   Key Components of Groundwater Risk Assessment in the EU

One of the key points of the WFD implementation is to characterize each groundwater body within all the delineated river basins in order to identify groundwater bodies at risk of not achieving the objectives defined by the WFD.

Apart from the initial delineation description of each groundwater body and its catchment the groundwater risk assessment (Fig. 5.1) is an mandatory part of the water body characterization and is based on a four-step process of pressure and impact analysis [6]:

1. Identifying the driving forces (especially land use, urban development, industry, agriculture and other activities which lead to pressures) without regard to their actual impacts and identifying pressures with possible impacts on the water body and on water uses;
2. Identifying the significant pressures by considering the magnitude of the pressures and the susceptibility of the water body;
3. Assessing the impacts resulting from the pressure;
4. Evaluating the likelihood of failing to meet the objective.

In general, if a water body currently has good status but it is thought that pressures may cause its status to be rendered poor by 2015, then the body is "at risk" and will require further characterization. A water body currently determined to have poor status will automatically be "at risk" [5]. It should be noted that risk assessment differs from status assessment, as the first one's purpose is to make a prediction into the future and therefore should be conducted at the beginning of a management/evaluation period whereas in contrast a status assessment is done at the end of this period and describes the present state of a water body.

Identifying the Driving Forces, Significant Pressures and Potential Impacts

Driving forces are sectors of human activity that have the potential to produce pressure on the water body which may have an impact in the state of the water body itself, connected water bodies or dependent ecosystems [7]. For instance contaminated leachate (pressure) from a landfill (driving force) contaminates the aquifer (impact).

For groundwater bodies the WFD categorizes driving forces of pressures in [8]:

1. Pollution by point sources (e.g. industry, mining, waste water, brownfields)
2. Pollution by diffusive sources (e.g. agriculture, transportation)
3. Abstraction and artificial recharge (e.g. water transfer, private and public supply, land use change).

The first two categories put a risk on the groundwater body to deteriorate its chemical status whereas the remaining category includes all driving forces which may risk the quantitative status.

For a first and coarse evaluation a groundwater body can be considered "at risk" if the sum of similar pollution sources impacts on at least a third of the surface area of the groundwater body [8].

Applying this procedure it is suggested to assume an impact area of $1 \, km^2$ for each identified pollution point source. In case of diffusive sources this most simple approach focusses on the emissions perspective only and can be applied by calculating whether the areas used for agriculture or the areas covered by settlements and roads sum up to be more than 33% of the total site area overlying the groundwater body [8].

The quantitative groundwater body status can be most simply evaluated to be not at "risk" by either showing that assumed abstraction constitute less than 10% of the

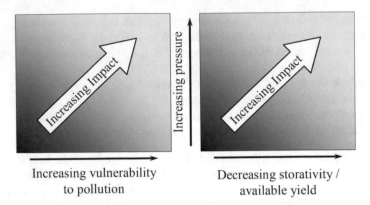

Fig. 5.2 Impact as a consequence of both the magnitude of pollution/abstraction pressure and the vulnerability of the groundwater to that pressure [7]

recharge or by proofing that long-term groundwater level data series don't show a statistically sound reduction in groundwater levels [8].

Among the variety of pressures not all have a significant impact on the groundwater body. To evaluate the significance, specific knowledge of pressures (related to specific pollutants or quantitative status) within the catchment area and a developed conceptual understanding of water flow, chemical transfers and biological functioning of the water body within the catchment system is needed. Firstly correlation assessment between monitored data (chemical and water level) and identified pressures can be used to evaluate significance. If necessary and appropriate, causality assessment e.g. by using numerical modeling can be applied to confirm the impact of significant pressures [7].

For groundwater bodies it has to be taken into consideration that a pressure causing an impact may often be manifested in groundwater quality monitoring data after a considerable time delay. This means that it will not always be possible to accurately measure the impact of actual pressures by monitoring data. Therefore the concept of potential impact is used to describe whether a pressure (typically occurring on the land surface) is likely to bring the groundwater body "at risk" of failing the objectives [7]. Hence, the vulnerability of the groundwater body to pollution and quantitative changes has to be included in the risk assessment (Fig. 5.2) by determining hydrogeological parameters such as conductivity and porosity and by characterizing the overlying strata parameters thickness, layering and cover constitution [7, 8].

Scheme for Identifying Relevant Pollutants

One key question in the context of the analysis of pressures and impacts is choosing adequate specific pollutants for which data on pressures must be collected in order to assess whether there are impacts for the different water bodies in a river basin [7].

As outlined above a number of substances are determined on a European level and must be considered in the pressure and impact analysis. Beside this minimum list specific conditions on the river basin or individual water body level may make it necessary to consider a variety of other substances due to the given driving forces and pressures.

In the optimal case, either a detailed knowledge of the pollution sources or a clear relationship between observed impact and related pollutants is available and determines the number of pollutants relevant for the investigated river basin or water body.

However, given the number of pollutants in general there is most likely a gap of information and data for many pollutants. In this case a generic approach [7] is suggested by the European Union in order to identify relevant specific pollutants (Fig. 5.3).

Starting point is the very comprehensive main pollutants list in the passages 1–9 of Annex VIII WFD.

Secondly available information on pollution driving forces (types of industry, agriculture, etc.), production and usage amounts and discharge of pollutants are collected in order to drive a working list of those pollutants identified as being discharged to the relevant water body [7]. If driving forces are known but no detailed information about produced/used pollutants are available, tables can be used which list the most likely occurring groundwater pollutants to the relevant pollution sources such as industry branches or land use patterns. A comprehensive list of those fingerprints is provided by Länderarbeitsgemeinschaft Wasser [9].

In a third step those pollutants are selected from the list of discharged pollutants which are likely to or already harm the water body or the connected ecosystems. Therefore concentration data in the water body for each contaminant have to be obtained from either measurements or modelling and subsequently have to be compared with thresholds such as the EU environmental quality standards. The subsequent interpretation should include the consideration of natural background concentrations and potential accumulation of substances in sediments or biota.

The fourth step, called safe net is a method to ensure that some pollutants are not incorrectly excluded from the list of specific pollutants by analyzing potential additive effects of smaller pollution source and chemically analog pollutants and by proofing whether trends in concentration indicate a future exceeding of thresholds by several pollutants.

Finally a list of specific pollutants can be presented and used for risk assessment in the relevant water body or river basin.

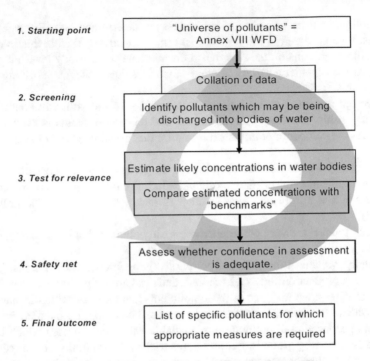

1. Starting point	"Universe of pollutants" = Annex VIII WFD
2. Screening	Collation of data / Identify pollutants which may be being discharged into bodies of water
3. Test for relevance	Estimate likely concentrations in water bodies / Compare estimated concentrations with "benchmarks"
4. Safety net	Assess whether confidence in assessment is adequate.
5. Final outcome	List of specific pollutants for which appropriate measures are required

Fig. 5.3 Steps suggested by the EU to derive a selected list of pollutants [7]

5.1.4 Identification of Typical Groundwater Pollutants in SLRB

Two study areas in the SLRB were chosen: one in Ashi River Basin and the other in Taizi River Basin. Ashi River is a tributary of Songhua River and Taizi River is a tributary of Liao River.

5.1.4.1 Study Area in Ashi River Basin

General Information

The study area locates south of Acheng city, where a drinking water supply work with its surrounded drinking water source protection zone is located (see Fig. 5.4). Ashi River is on the east of the study area, flowing from southeast to northwest.

On the west side of the study area is a plateau with elevations from 162 to 198 m which are higher in the southwest and lower in the northeast of the area. On the fringe of the plateau is a belt-shape, north-south orientated terrace with a width of 200–1500 m and an elevation between 145 and 162 m. The left bank of Ashi River is a 2.5–5 km wide flood plain, with an elevation of 138–145 m. Thus the study area is on the upstream of Acheng city.

Annual precipitation in this area is 515.8 mm (data from 1960–2010), and annual (open-water) evaporation is 1375.2 mm (data from 1967–2003). The annual average temperature is 3.4 °C with the highest in July (22.6 °C) and the lowest in January (−19.5 °C).

In the average and dry period (October to June next year), Ashi River has a width of 30–50 m and an corresponding discharge of 0.64–50 m^3/s. In the wet period (July–Sep), the discharge of Ashi River can increase to up to 50–824 m^3/s. Ashi River is recharged by both precipitation and groundwater.

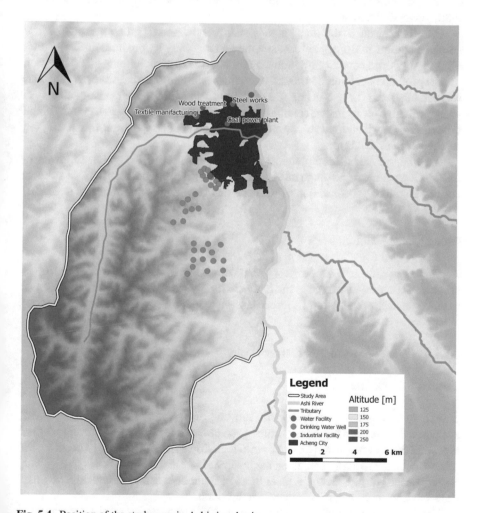

Fig. 5.4 Position of the study area in Ashi river basin

Groundwater Source

The aquifer in this area mainly include 3 types: (1) confined aquifer: in the plateau area, the depth of the confined aquifer is between 19.5 and 46 m, and the distance between the ground surface and the confined aquifer is 9.7–42.3 m. The aquifer is mainly composed of medium and coarse sand, and gravel. (2) Aquitard: in the terrace area, the depth of the aquitard is 28–42 m, and the distance between the ground surface and the aquitard is 7.1–11.4 m. The aquifer is mainly composed of gravel and some layers of clay. (3) Phreatic aquifer: in the flood plain, the depth of the phreatic aquifer is 28–40 m, and the distance between the ground surface and the phreatic surface is 3–5 m. The aquifer is mainly composed of medium and coarse sand, gravel and pebble.

The groundwater in this study area is mainly recharged by groundwater flow from confined aquifer in the upper terrace zone and secondly by precipitation. The drain off is mainly from conduit flow, and also from artificial exploitation and evaporation.

The groundwater source is abundant in this area. The drinking water supply work was established in 1992 and supplied drinking water for the Acheng city ever since using groundwater as its water supply source. The water supply was around 30,000 t/d in 1992, and increased to 45,000 t/d in 2005, as the population in Acheng city reached around 250,000.

The main chemical component in the groundwater is HCO_3–Ca, followed by HCO_3–CaMg. Iron and manganese are showing high concentrations. It is to be noticed that high concentration of iron and manganese are very common in SLRB.

Groundwater Pollution

Through the field survey and investigation, diffusive sources (e.g. agriculture, transportation) could be identified as the main driving force for groundwater pollution risk in this area, followed by point sources (e.g. slaughterhouses, gas stations, factories).

5.1.4.2   Study Area in Taizi River Basin

General Information

The study area locates southwest of Liaoyang city, where a drinking water work, surrounded by a drinking water source protection zone, is located (see Fig. 5.5). It is around 30 km away from Taizi River, which flows from southeast to northwest in this area. Thus the study area is on the upstream of Taizi River.

The terrain in this area is a plain in an altitude between 37.2 and 55.4 m which shows a downward slope in the western direction.

The annual precipitation in this area is 730.6 mm, and annual evaporation is 1627 mm. The annual average temperature is 7.6 °C.

Groundwater Source

The water supply work is currently the main drinking water source for Liaoyang county with a water supply amount of 40,000 t/d.

The groundwater quality is good in this area with the exception of high iron and manganese concentrations. As mentioned above that high concentration of iron and manganese are very common in SLRB.

Groundwater Pollution

The groundwater supply wells are located in the residential areas. Groundwater in the deep confined aquifer is chosen as water supply source. Field investigations found that pollution by point sources (e.g. industry, municipal waste water) is the main driving force of groundwater pollution risk in this area.

5.1.5   Establishment of Risk Assessment Index System for Groundwater

5.1.5.1   Principle for Index System Establishment

The purpose of the risk assessment index system establishment is to assess groundwater resources in SLRB and to classify them by the risk they are exposed to.

The principles for index system establishment include whole process supervision and management, quantitative assessment, and dynamic management.

Whole process supervision and management means that the entire process of groundwater contamination from the contamination source to the pathway in aquifer and eventually to the groundwater should be considered and that the whole process of

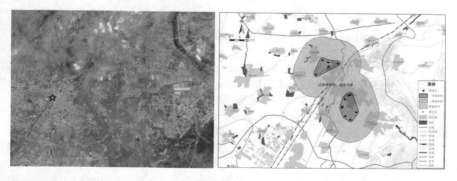

Fig. 5.5   Position of the study area in Taizi river basin

Table 5.1 Distance to the well head

Indicator	Range of the indicator	Assigned rating
W11	Downstream of well head	2
	Upstream >1 km	4
	Upstream 0.5–1 km	6
	Upstream <0.5 km	8
	Capture zone	10

contamination source control, pathway containment, and contamination remediation should be supervised and managed.

Quantitative assessment means that the risk of how the groundwater is exposed to contamination should be quantitatively assessed. Each index in the system should reflect the risk of a certain contamination to the groundwater. Based on this, the risk can be classified to different levels and managed accordingly.

Dynamic management means that the index system is not fixed but varies temporally and spatially. In various hydrogeological and climatic conditions, or after certain time of management, the typical pollutants would be different. Thus the index system should be dynamically managed.

5.1.5.2 Method for Index System Establishment

The index system is intended to serve as a relative indicator of the potential of groundwater contamination. It is utilized for mapping the groundwater pollution potential and to provide a direct evaluation of potential risk from the maps.

There have been many index systems for groundwater vulnerability mapping, including DRASTIC [10], SINTACS [11], GOD [12], AVI [13], PI [14], and GLA [15]. These index assessment systems have become popular due to their little requirement on field data inputs.

5.1.5.3 Procedure of Index System Establishment

First of all, the basic information of the target area (groundwater source) should be acquired by data collection and field investigation. Based on this information, the driving forces and pressures can be identified (Table 5.1).

Accessing the driving forces and pressures and analyzing their impact to groundwater can be done by consulting experts, quantitative screening, hierarchy analysis, etc. Thus the indicators of groundwater source risk assessment can be determined (Table 5.2).

For each indicator, a numerical value is provided. Indicator values are separated into several ranges and each range is assigned a rating. Additionally, each value is

Table 5.2 Existing form

Indicator	Range of the indicator	Assigned rating
W12	Sealed	1
	Partial sealed	5
	Expose	10

weighted by hierarchy analysis. Finally the indicator values are summed to give the index (Table 5.3).

Finally, the index system is compared with the realistic situation to make modifications and adjustments accordingly. Thus, the groundwater resources can be assessed and classified by the risk they are exposed to (Table 5.4).

5.1.6 Index System Establishment in SLRB

5.1.6.1 Determination of Indicators

Based on the aforementioned principles and methodologies, the first step is to identify the driving forces and pressures in groundwater risk assessment. In general, two groups of indicators should be derived, namely pollution source features and groundwater vulnerability (Tables 5.5, 5.6, 5.7 and 5.8).

Pollution source features mainly take into consideration the structures of pollution source and the characteristics of pollutants. The structures of pollution source

Table 5.3 Emission pathway

Indicator	Range of the indicator	Assigned rating
W13	Surface	2
	Surface and underground	5
	Underground	10

Table 5.4 Conductivity

Indicator	Range of the indicator (m/d)	Assigned rating
W37	0.4–4.1	1
	4.1–12.2	2
	12.2–28.5	4
	28.5–40.7	6
	40.7–81.5	8
	>81.5	10

include distance to the wellhead, existing form, emission pathway, probability of emission, duration of emission and amount of pollutant type. The characteristics of pollutants include amount of emission, decay characteristic, migration characteristic and toxicity.

Vulnerability/sensitivity of groundwater source provides an assessment of the need for groundwater protection mechanisms according to the loss of a beneficial use of that groundwater resource. A rating system provides an evaluation of the vulnerability/sensitivity of groundwater source in terms of the physical parameters of the aquifer. The parameters include depth to water table, recharge, aquifer type,

Table 5.5 Probability of emission

Indicator	Range of the indicator	Assigned rating
W14	Once a year	2
	Once a month	4
	Once a week	8
	Once a day	10

Table 5.6 Duration of emission

Indicator	Range of the indicator	Assigned rating
W15	Days	1
	Weeks	4
	Months	8
	Years	10

Table 5.7 Amount of pollutant type

Indicator	Range of the indicator	Assigned rating
W16	<3	4
	3–6	6
	6–10	8
	>10	10

Table 5.8 Amount of emission

Indicator	Range of the indicator	Assigned rating
W21	Low	1
	Moderate low	3
	Medium	5
	Moderate high	8
	High	10

Table 5.9 Decay characteristic

Indicator	Range of the indicator (d)	Assigned rating
W22	<15	1
	15–60	3
	60–180	7
	180–360	8
	>360	10

soil type, topography, impact of vadose zone, and conductivity (Fig. 5.6, Tables 5.9 and 5.10).

5.1.6.2 Rating Assigned to Indicators' Values

Based on the data collection and field investigation as well as experts experiences, each indicator value is separated into different ranges, and each range is assigned a rating (Tables 5.11, 5.12, 5.13, 5.14, 5.15, 5.16 and 5.17).

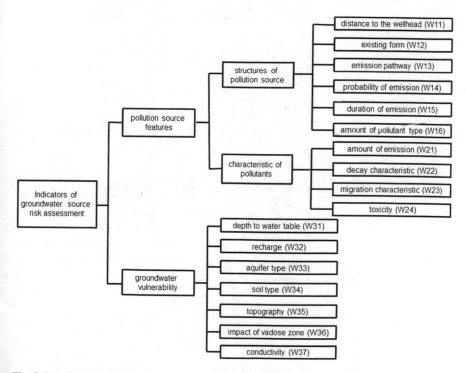

Fig. 5.6 Indicators of groundwater source risk assessment

Table 5.10 Migration characteristic

Indicator	Range of the indicator (Soil adsorption coefficient, K_d)	Assigned rating
W23	>2000	2
	500–2000	4
	150–500	6
	50–150	8
	<50	10

Table 5.11 Toxicity

Indicator	Range of the indicator	Assigned rating
W24	Weak	1
	Moderate weak	3
	Medium	5
	Moderate strong	8
	Strong	10

Table 5.12 Depth to water table

Indicator	Range of the indicator (m)	Assigned rating
W31	0–1.5	10
	1.5–4.6	9
	4.6–9.1	7
	9.1–15.2	5
	15.2–22.9	3
	22.9–30.5	2
	>30.5	1

Table 5.13 Recharge

Indicator	Range of the indicator (mm)	Assigned rating
W32	0–50.8	1
	50.8–101.65	3
	101.65–177.8	6
	177.8–254	8
	>254	9

Table 5.14 Aquifer type

Indicator	Range of the indicator	Assigned rating
W33	Shale	2
	Fine to medium sand	5
	Sandstone	6
	Coarse sand	8

Table 5.15 Soil type

Indicator	Range of the indicator	Assigned rating
W34	Gravel	10
	Sand	9
	Clay	7
	Sandy loam	5
	Loam	5
	Clay loam	2

Table 5.16 Topography

Indicator	Range of the indicator (%)	Assigned rating
W35	0–2	10
	2–6	9
	6–12	5
	12–18	3
	>18	1

Table 5.17 Impact of vadose zone

Indicator	Range of the indicator	Assigned rating
W36	Clay	1
	Shale	3
	Fine and medium sand	6
	Gravel	8

5.1.6.3 Determination of Indicators' Weights

To assess the impacts resulting from the driving forces and pressure, each indicator should be assigned with a weight. The most significant indicators have the highest weight of 9 and the least weight of 1 indicating less significance in the process of groundwater pollution. The weights assigned to each indicator are as follows:

Indicator	W11	W12	W13	W14	W15	W16	W21	W22	W23
Weight	9	6	9	3	3	6	9	2	2
Indicator	W24	W31	W32	W33	W34	W35	W36	W37	
Weight	9	9	7	5	3	1	9	5	

5.1.6.4   Index System

The index is calculated in the following way: each indicator's score is obtained by the multiplication of its assigned ratings with its weights; sum up all the indicator's scores and divide it by the total of the weights.

$$I = \frac{\sum_{i=1}^{n} w_i r_i}{\sum_{i=1}^{n} w_i}$$

where I is the index, n is number of indicators, w is weight, and r is rating.

Thus the index is a number with the scale of 1–10. The higher the number the more possible is the groundwater source at risk.

5.1.7   Summary and Conclusions

The overall objective of the SUSTAIN H_2O project is to develop management tools and practices for pollution reduction in SLRB and to support water quality improvement in the demonstration areas. As part of the project, consortium members of WP4 mainly focused on groundwater protection in SLRB. Groundwater risk assessment methodologies in the EU countries were well studied. Based on the field survey and data acquisition, an index system was established, which could be used as a guideline for groundwater protection in SLRB.

5.2   Sensitive Groundwater Sources Identification and Risk Reduction Management in the Songhua-Liao-River-Basin

by Erik NIXDORF, SUN Yuanyuan, SU Jing, Olaf KOLDITZ, XI Beidou

China has been facing decades of rapid economic development and urban population growth which is in many areas associated with a continuing degradation of natural resources, e.g. aquifer system. In this context, assessing groundwater contamination risk by an index system using verifiable criteria is an import element for efficient water resource management and land use planning.

Most existing risk assessment studies were conducted on small to medium scale catchments and are limited to the evaluation of intrinsic groundwater vulnerability. To our knowledge, this study is the first which established groundwater risk maps for the entire area of Songhua River and Liao River Basin, two of the largest and most contamination endangered water basins in China. Groundwater risk index maps were derived in a spatial resolution of 30 arc seconds with the aid of GIS tools by combining the results of individually conducted groundwater vulnerability and hazard assessment. Groundwater vulnerability was evaluated by a modified DRASTIC method approach; potential pollution pressure was described by two proxy parameters, population density and land cover. Both public datasets at highest available resolution and numerical groundwater modelling results were applied as data sources. Resulting index scores were reclassified into five classes from "very low" to "very high" groundwater contamination risk.

The groundwater risk assessment demonstrated that about 10% of the aquifers in Liao River Basin and 6% in Songhua River Basin are at high or very contamination risk. These areas are mainly located in the vast plain areas of the basins with hotspots of very high groundwater contamination risk in the Liao River Delta, along the Shenyang Economic Zone and in the Harbin Metropolitan area. Moderate groundwater contamination risk areas were predominantly associated with less densely populated agricultural areas in the lowlands of both basins. However, the majority of aquifer area in both catchment was associated with low or very low groundwater contamination risk, particularly in the sparsely populated western mountain ranges.

Although having limitations in resolution and input data consistency, the obtained groundwater contamination risk maps will be beneficial for regional and local decision making process with regard to groundwater protection measure, particularly if other data availability is limited.

5.2.1   Introduction

Since the 1950s, major advances in drilling technology and hydrogeological knowledge had facilitated a massive expansion in groundwater use across the developing world in order to satisfy the needs of irrigation, domestic purposes and industrial demand [16]. In China alone, more than 500 large cities rely on groundwater resources for their drinking water supply [17]. These excessive exploitation and inappropriate activities at the land surface lead to and foster degradation of groundwater resources in many areas [18]. In this context, groundwater contamination risk assessment provides a useful tool to design and implement groundwater protection measures programs [19].

In a risk assessment, where risk is defined as hazard times vulnerability, the combined rating of the potential harmfulness posed by a pollution source (hazard assessment) and the possibility of the spreading into and in the groundwater (vulnerability assessment) could be interpreted as the probability for groundwater contamination, both in quantitative or qualitative terms depending on the used method [20, 21]. At present, index based groundwater risk assessment has be conducted in various regions of the world such as China [22, 23], Italy [24], Brasil [25], Canada [26], Iran [27] and New Zealand [28].

A variety of rating systems have been developed for both integral parts of the groundwater risk assessment, groundwater vulnerability assessment (e.g. DRASTIC [10], GOD [11] and hazard assessment (e.g. [29, 30]) whereas some of the latter group include groundwater vulnerability as integral part of their approach (e.g. [31]). An overview about available methods is given by [19].

For groundwater vulnerability assessment we applied DRASTIC method which, although originally developed by US EPA for aquifers in the USA (e.g. [32–34]), was applied to evaluate groundwater vulnerability for basins with very diverse conditions in many regions of the world such as Europe (e.g. [35, 36]), the Middle East (e.g. [37, 38]) and Northern China (e.g. [23, 39, 40]). Beside the long and diverse application history it has the advantage that it can be easily modified to site-specific conditions (e.g. [41, 42]) and also have a minimum demand of data.

For a detailed hazard assessment, the structure of pollution sources including the distance to the wellhead, emission pathways, probability of emission, duration of emission and amount of pollutant type as well as the characteristic of pollutants including the amount of emission, decay characteristics, migration characteristics and toxicity should be taken under investigation. In fact, detailed information about specific pollution sources and pollutants are often not available, particularly on a larger scale. Nevertheless, proxies can be considered to describe the possible pollution pressure on an aquifer, such as land cover/land use or population density. Land cover/Land use is one principle factor, controlling groundwater contamination as human transformation of the surface changes matter fluxes to the subsurface. Previous studies showed that groundwater from agricultural areas had typically the highest concentrations of Nitrate and pesticides whereas volatile organic compounds and other industrial products are highest in the groundwater underneath of industrial and commercial areas [43]. Consequently several previous studies either directly include land cover/land use in a modified DRASTIC Model [32, 33] or as a parameter describing the human impact on groundwater risk [36].

Population density is a further applicable proxy parameter for assessing hazard. Exemplary, nonagricultural pollution sources of Nitrate in urban areas shows a strong correlation with local population density [44]. In general higher population densities are associated with a higher harmfulness [45].

One of the environmental issues of aquifers in the study area, which is Songhua and Liao River Basin, is an increased pollution of the aquifer system by Nitrate from fertilizer overuse (e.g. [46, 47]). As North-East China is the old industrial heartland of China, discharges and leakages from industrial sources or brownfields put an additional potential pressure on groundwater resources.

In this context the main objectives of this study are: (1) to provide high groundwater contamination risk maps for the entire Songhua River Basin and Liao River Basin based on a combined approach of aquifer vulnerability and hazard potential assessment using public data bases, remote sensing as well as numerical groundwater modeling results for input data generation and (2) to delineate and characterize zones of high contamination risk in order to support efficient groundwater management.

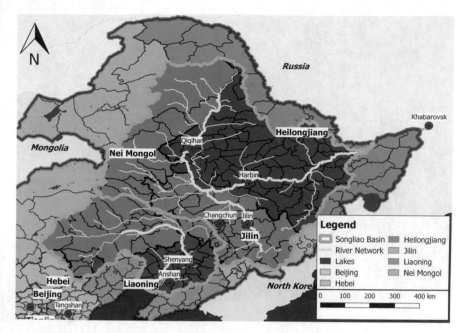

Fig. 5.7 Administrative division of Songliao basin

5.2.2 Materials and Methods

5.2.2.1 Study Area Description

The Songhua River Basin is with an area of more than 550 000 km^2 the third largest Basin in China draining to Songhua River which is the largest tributary of Amur River with a length of more than 2300 km. The main tributaries of Songhua River are Nen River which drains the Northern part of the basin and Second Songhua River coming from Jilin province. The basin covers vast areas of the three Chinese provinces Heilongjiang, Jilin and Inner Mongolia as well as a small part of Liaoning province (Fig. 5.7). It contains mountainous areas in the north-west (Greater Khingan Mountains and Smaller Khingan Mountains) and south-east (Changbai Mountains) with altitudes up to more than 2500 m as well as large plain areas such as the Songnen plain in the north and the Sanjiang plain in the north-east of the catchment (Fig. 5.8). Particularly Songnen plain makes up one of the three largest black soil belts in the world and is an important location for commodity grain and livestock husbandry in China [48]. This study area is located in the northern temperate monsoon climate zone. Continental climate characteristics are very significantly and annual precipitation is between 400 and 900 mm whereas the amount of annual rainfall increases south-eastward and with altitude (Fig. 5.9).

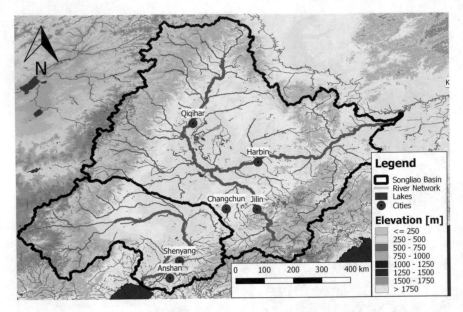

Fig. 5.8 Watershed of Liao and Songhua river with given altitudes and river network

The Liao River is one of the seven largest rivers in China. It is an important river in Northeast China. It is 1345 km long, flowing through Hebei, Jilin, Liaoning Provinces and Inner Mongolia Autonomous Region with a watershed area of about 220 000 km². The upper reaches of the river stretch 882 km, running through loess hills and is composed by two major river systems, an Eastern (Xiliao River) and a Western (Dongliao river) one, which originate from Inner Mongolia and Western Jilin province, respectively. The middle reaches stretches 210 km and the lower reach covers about 300 km of Liao River and forms, together with the lower reaches of the main tributaries Hun and Taizi River the Liao River Delta region before finally draining into Bohai Bay.

The Liao River watershed is located between 40°30′ –45°10′ north latitude and 117°–125°30′ east longitude. The Daxing'an, Qilaotu and Nuluerhu Mountains lie in the west of the watershed with elevations between 500–1500 m; Jilinhada, Longgang and Qianshan Mountains lie in the east with elevations between 500–2000 m. The middle and lower reaches belong to the Liao River plain with elevations below 200 m. The average annual precipitation is between 350–1000 mm, decreasing gradually from southeast to northwest (Fig. 5.10).

5.2.2.2 Groundwater Risk Assessment Index Method

The idea of the of the risk assessment index system is to find a quantitative way of expressing the likelihood that an aquifer be polluted by contaminants which are

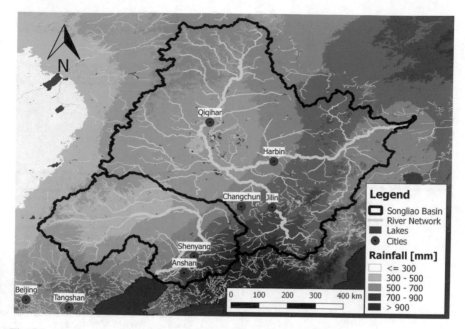

Fig. 5.9 Distribution of average annual rainfall in Songliao basin

introduced into the ground surface. This is determined by two groups of indicators: Firstly by the aquifer intrinsic vulnerability which describes the ease with which a contaminant introduced to the subsurface can reach and diffuse the groundwater [49] and secondly by the hazard which is defined as a potential source of contamination resulting from human activities [50].

In this paper, aquifer vulnerability is assessed using the DRASTIC method. The Acronym DRASTIC stands for the seven parameters which are intended to use for groundwater vulnerability assessment: **D**epth to groundwater, net **R**echarge, **A**quifer media, **S**oil media, **T**opography, **I**mpact of vadose zone and hydraulic **C**onductivity. Each parameter is classified into either ranges of physical quantity or a qualitative description which correspond to a rating varying between 1 (low vulnerability) and 10 (large vulnerability). Finally the DRASTIC index (DI) is computed by sum up the ratings which are each assigned by specific weights in a range from 1 to 5:

$$DI = D_w \cdot D_r + R_w \cdot R_r + A_w \cdot A_r + S_w \cdot S_r + T_w \cdot T_r + I_w \cdot I_r + C_w \cdot C_r \quad (5.1)$$

where r is the variable rating and w is the weight assigned to that parameter.

The calculation of the DRASTIC index can be used to spatially delineate areas of the groundwater which have most prone to contamination. The higher the DRASTIC index, the greater the groundwater pollution potential [10].

Several studies have modified the conventional DRASTIC method by adding or removing factors and changing weights to better reflect specific field site conditions

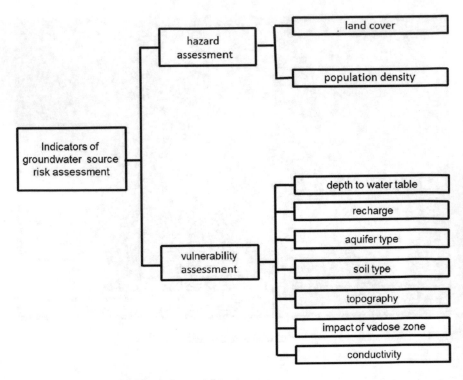

Fig. 5.10 Indicators of the groundwater risk assessment

or research topics [51]. We used a modified DRASTIC rating suggested by the China Geological Survey which provides a linear rating from 1 to 10 for each parameter. Regarding the subsurface parameters aquifer media and vadose zone we adapt the rating scheme and added descriptions from the default DRASTIC rating system for better reflecting non-quarternary geology in the study area (Table 5.18). For the investigation area, the standard weighting scheme was applied as it covered a large variety of different subsurface conditions. Thus, the DRASTIC index could be calculated by:

$$DI = 5 \cdot D_r + 4 \cdot R_r + 3 \cdot A_r + 3 \cdot S_r + 1 \cdot T_r + 5 \cdot I_r + 3 \cdot C_r \qquad (5.2)$$

For hazard assessment, we assigned both used proxy parameters, land cover and population density, with a class based rating system with ratings between 1 and 10 where 1 expresses minimum and 10 maximum hazard potential. For land cover, the corresponding parameter classes quantify the potential harmfulness of different land cover types were adopted from [33]. Hence, for the quantification of potential harmfulness of the parameter population density, we subdivided population density in 10 classes between smaller 100 and more than 900 inhabitants per km^2 and give

Table 5.18 Rating scheme for each parameter of the aquifer vulnerability rating and the hazard rating

| Aquifer vulnerability rating | | | | | | | | Hazard rating | |
Rating	Depth to water (m)	Recharge (mm/year)	Aquifer media	Soil media	Slope (%)	Vadose zone	Conductivity (m/day)	Land cover	Pop. density (Inh/km²)
1	>30.5	<51	–	Nonshrinking clay	>18	–	<4.1	Open water/wetland	<100
2	26.7–20.5	51–71.4	Massive Shale/Silt/Clay	Peat	17–18	Silt/Clay	4.1–12.2	Forests	100–200
3	22.9–26.7	71.4–91.8	Metamorphic/Igneous	Clay Loam	15–17	Shale	12.2–20.3	–	200–300
4	15.2–22.9	91.8–117.2	Weathered/ Bedded rock	Silty Loam	13–15	Metamorphic/Igneous	20.3–28.5	Grassland/Shrub	300–400
5	12.1–15.2	117.2–147.6	Clastic rock/Loam	Loam	11–13	Clastic rock/Loam	28.5–34.6	Settlement	400–500
6	9.1–12.1	147.6–178	Sandstone/Limestone	Sandy Loam	9–11	Sandstone/Limestone	34.6–40.7	–	500–600
7	6.8–9.1	178–216	–	Shrinking Clay	7–9	–	40.7–61.1	–	600–700
8	4.6–6.8	216–235	Sand and Gravel	Peat	4–7	Sand and Gravel	61.1–71.5	Barren land	700–800
9	1.5–4.6	235–254	Basalt	Sand	2–4	Basalt	71.5–81.5	–	800–900
10	1.5	>254	Karst Limestone	Gravel/no soil	<2	Karst Limestone	>81.5	Cropland	>900

each class a rating between 1 for the sparsely populated areas and 10 for the most densely populated areas.

After defining the quantification method for both proxy parameters the harmfulness of a hazard could be expressed with the hazard index by:

$$HI = L_W \cdot L_R + P_W \cdot P_R \tag{5.3}$$

where HI is the hazard index, L_R and P_R are the ratings for land cover and population density and L_W and P_W the specific weights. For this study specific weights were set to 12, meaning that both parameters have an equal impact on HI calculation and that the maximum range of values is the same as the vulnerability index calculation.

Five categories were assigned to both the hazard harmfulness and the intrinsic vulnerability results to describe very low, low, moderate, high and very high vulnerability and harmfulness, respectively. Class boundaries were defined for each catchment separately based on possible maximum span of index values.

Finally, the basic groundwater contamination risk index (RI) was generated by sum up hazard and vulnerability index of each cell:

$$RI = DI + HI \tag{5.4}$$

5.2.2.3 Data Sources

For the calculation of the groundwater contamination risk, primary datasets were obtained from a variety of data sources which are all freely available in the internet (Table 5.19). For the Digital elevation Model (DEM) the dataset of the HydroSHEDS project was used which based on the primary data obtained during a Space Shuttle flight for NASA's Shuttle Radar Topography Mission (SRTM) but were reprocessed for hydrological purposes [52]. Mean monthly precipitation data for the period between the year 1950 and 2000 could be obtained from the WorldClim—Global Climate Database which based on a large set of weather station data from numerous sources [53]. Actual Evaporation (i.e. evapotranspiration) data in a daily resolution were provided by the GLEAM (Global Land Evaporation Amsterdam Model) Version 3.a dataset which includes transpiration, bare-soil evaporation, interception loss, open-water evaporation and sublimation in its calculation algorithm for evaporation [54]. Information on soil cover were taken from the National Soil Database of China which is available from Beijing Normal University (BNU) and provides information about the particle size distribution of Sand, Silt and Clay of Chinese soils [55]. Information on hydrogeology were obtained from hydrogeological maps provided by the Chinese Academy of Geological Science (CAGS). Additionally, water levels from in total 20 wells were supported by Heilongjiang Research Institute for Environment Sciences (HRIES) and Liaoning Academy of Environmental Sciences (LAES).

The two proxy variables describing potential anthropogenic pollution hazard, land cover and population density were available as grid data sets, too. For land

Table 5.19 Data sets for computing DRASTIC index

Dataset	Data source	Type	Domain	Resolution
DEM	HydroSHEDs/SRTM	Grid	Global	3"
Precipitation	WorldClim	Grid	Global	30"
Evaporation	GLEAM	Grid	Global	0.25°
Soil	BNU	Grid	China	30"
Land cover	GlobeLand30	Grid	Global	1"
Population density	GPWv4	Grid	Global	30"
Hydrogeology	CAGS	Figure	Regional	
Water levels	HRIES/LAES	Spreadsheet	Local	

cover data, the GlobeLand30 Land Cover dataset was used which based on images of Landsat and Chinese HJ-1 satellite and classifies land cover into 10 classes based on observations in the year 2010 [56]. Population density data were derived from the Gridded Population of the World, Version 4 (GPWv4) data collection which based on globally-integrated national population data from the 2010 round of the Population and Housing Censuses [57].

5.2.2.4   Data Processing

Prior to index calculations, specific preprocessing steps were needed for most datasets in order to bring them in an adequate structure. Firstly, the data sets which were available in a grid structure were uploaded to a GIS System and clipped to the extent of the particular catchment which was delineated from the DEM data by applying GIS hydrological analysis tools. Furthermore, grids with a cell resolution coarser than 30" were interpolated to 30" cell size by applying the ordinary kriging interpolation scheme.

For the evaluation of topography impact, slopes for each cell were calculated from the DEM using Terrain Analysis tools in GIS. Hereby the rate of maximum change in z-value from each cell was used as slope of the cell and was calculated in percentage.

Recharge was directly computed from the difference between precipitation and evaporation to avoid the errors of other methods of indirect recharge calculation (for a summary see [51]). Prior to calculation, both data sets were summed up to yearly average values. For evaporation, the annual mean of the last 5 years was used for recharge calculation in order to further reduce the impact of climatic outliers.

For depth to water calculation, only a limited number of monitoring wells with a very inhomogeneous spatial distribution were accessible. Furthermore different responsibilities due to the cross-provincial character of the watersheds hinder data availability. Subsequently, numerical groundwater modeling was chosen as an alternative approach to calculate steady-state groundwater levels for both catchments.

Two-dimensional finite element numerical groundwater models with an element resolution of 5 km were set up for Songhua River and Liao River catchment using the modelling software OpenGeoSyS. The water levels of the larger river network in each basin were used as Dirichlet boundary condition and the previously calculated net recharge as Neumann boundary condition of the model. For simplification reasons, subsurface homogeneity was assumed and the hydraulic conductivity for the subsurface of each model was obtained from calibration methods using water level information from the available monitoring wells of the catchment. The computed hydraulic heads were interpolated to a resolution of 30 arc seconds by Kriging ordinary interpolation method and finally subtracted from the elevation data in order to calculate the depth to water of each grid cell.

Soil data were obtained by applying the unified soil classification system (USCS) on the vertically averaged particle size distribution of sand, silt and clay given by the National Soil Database of Chinese soils.

In order to derive information about subsurface parameters for DRASTIC method, vadose zone and aquifer characteristics were estimated from the aquifer's lithological information given in the provincial hydrogeological maps. Prior to estimation these maps were georeferenced, converted to a cell grid structure, merged and clipped to the watershed boundaries. Aquifer and vadose zone categories, which differed among provinces, were unified to a common standard related to the qualitative description by the DRASTIC method.

For aquifer conductivity, a literature study was conducted to derive permeability ranges for each aquifer type. The results were compared with pumping tests submitted by the province environmental agencies HRIES and LAES.

5.2.3 Results and Discussion

5.2.3.1 Parameters Computation

For all resulting data sets, parameter ranges were reclassified to the scaling given in the DRASTIC and the hazard assessment method.

In Songhua River Basin, modelled stationary depth to water ranged between a few meters and more than 50 m and hence covers the entire range of DRASTIC indices (Fig. 5.11a). The simulation showed that a shallow, more vulnerable, groundwater system prevailed mainly in regions with low altitudes such as the area of Songnen and Sanjiang plain as well as along the main rivers draining the catchment. A very different index distribution is given by the recharge conditions (Fig. 5.11b). Low recharge with less than 50 mm per year was calculated for vast parts of the Songnen plain as well as the Greater Khingan mountain range. High recharge rates were calculated for the mountain ranges in the eastern part of the catchment with a peak of about 300 mm/a in the area of the Changbai Mountains in the most southeast and highest part of the catchment. A high topographical gradient with slopes up to 13.3% was found in the mountainous areas of the catchment (Fig. 5.11c). In

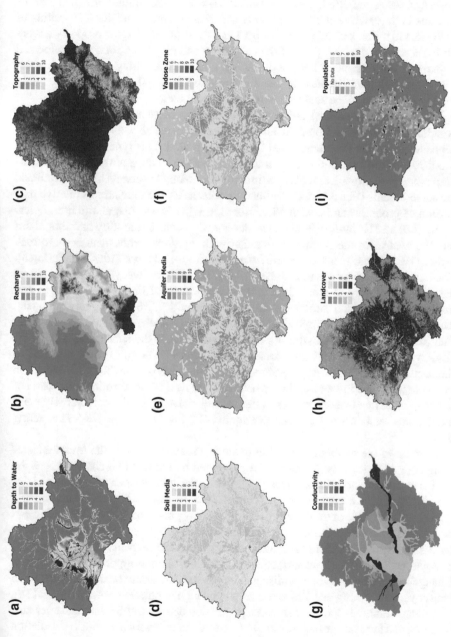

Fig. 5.11 Selected rating maps used on DRASTIC index (**a**, **b**, **c**, **d**, **e**, **f**, **g**) and hazard index (**h**, **i**) computation in Songhua river basin. High and low ratings are visualized by red and green color, respectively

contrast, calculated slopes in the plain areas and the larger river valleys were usually smaller than 2% making potential pollutants more likely to infiltrate the subsurface instead of being run offed. Based on the USCS soil classification more than 72% of the soils in the catchment are loamy soils which have a medium DRASTIC rating of 5 (Fig. 5.11d). The second (about 18.6%) and third (3.6%) largest proportion were silt loam soils in the hilly parts of the catchment and clay loam soils mostly located in parts of the Songnen plain, both having a DRASTIC rating of 4. The highest DRASTIC rating of 9 in the catchment referred to sandy soils, which however, only cover a small area in the southern part of the catchment.

From a hydrogeological point of view, the groundwater in the plain areas of Songhua catchment is stored in an aquifer system which consists of sediments deposited since the Pleistocene and Quaternary period. This type of aquifer covers about 39% of the total aquifer area and has, dependent on the grain size, a DRASTIC rating of either 2 for silt and clay sediment, 5 for loamy sediments or 8 for sand and gravel sediments (Fig. 5.11e, f). Defined by subsurface area the largest relative proportion of groundwater is, with 43%, stored in igneous and magmatite lithological layers (DRASTIC rating = 3 and 4) which mostly occur in the hilly and mountainous headwaters of the catchment together with aquifers within metamorphic rock layers (7% of total area) and sedimentary clastic rock layers (10% of total area). The latter one corresponds to a higher DRASTIC rating of five, which should reflect the potential variety of grain sizes and the occurrence of fissures. Groundwater in massive or thin-bedded carbonate rock aquifers is, with 0.3% in total, very rare in the catchment and limited to small areas in the southern Changbai Mountain headwaters. The pattern of measured hydraulic conductivity reflects the subsurface properties. High hydraulic conductivities, meaning high DRASTIC ratings are associated to the occurrence of Quaternary aquifers in the plain areas of Songhua catchment. On the other hand, the igneous and magmatite lithological layers forming large parts of the catchment headwaters aquifer system usually have hydraulic conductivity magnitudes smaller than 1 m/d, giving these aquifers the lowest possible DRASTIC rating of one.

Coming to the parameters used to quantify the hazard index, the importance of Songhua region for crop production is expressed by the fact that about 42% of the land is cultivated land (hazard rating = 9), particularly in the area of the large plains (Fig. 5.11g). In the mountainous areas forests (hazard rating = 1) were the dominating land cover with 33% of total land cover followed by grassland (DRASTIC rating = 4) with about 19%. Minor parts of the catchments surface of in sum less than 4.5% were covered by open water, bare land, wetlands, etc. Urban settlements (hazard rating = 5) cover about 2.5% of the land surface either as large settlement agglomerations such as Changchun, Harbin or Jilin or as dispersed spots at other parts of the catchment. Urban land cover has a strong interlinkage to areas with high population density (Fig. 5.11h). On average Songhua River Basin has a population density of 98 inhabitants per km^2 meaning that the average hazard rating of the catchment is very low with 1. Population densities went down to less than 10 inhabitants per km^2 in the western and northern regions of the catchment. However, much higher densely populated areas exist in the central region of the catchment in the proximity to the provincial capital of

Heilongjiang province as well as areas along the main rivers of the catchment with a population density in the downtown districts of major cities exceeding 10,000 inhabitants per km².

In Liao River Basin, the results of modelled stationary depth to water showed significantly shallow groundwater levels (DRASTIC rating = 10) both in the plain area south of the Song-Liao watershed and in the lower reach of Liao River (Fig. 5.12a). In contrast, mountainous areas showed, except of the river valley, depths to groundwater which correspond with low DRASTIC rating. Very low annual recharge rates down to 25.6 mm which correspond with a DRASTIC rating of 1 were calculated in the Xiliao River lowlands as well as in the Qilaotu Mountain area where precipitation is low and evaporation is relatively high (Fig. 5.12b). In contrast, high annual recharge rates of up to more than 400 mm were calculated in the south-eastern part of the catchment, particularly in the Changbai Mountains, where major tributaries such as Hun River and Taizi River originate from. Maximum topographical slopes of more than 14% were calculated in parts of the western and south-eastern mountain areas whereas in the vast plain areas and the river valleys slopes were usually smaller than 2% corresponding with a maximum DRASTIC Rating of 10 (Fig. 5.12c). The composition of soil distribution in Liao River Basin slightly differs from Songhua River Basin (Fig. 5.12d). Similar to Songhua River Basin the dominating soil type in Liao River Basin is loam with a proportion of 60% of total soil types. However, soils with a higher ratio of sand material such as sand (2.6%), loamy sand (13.3%) and sandy loam (11.3%) have a larger occurrence in Liao River Basin which is visualized by relatively large areas having DRASTIC rating of 6 and 9 (Fig. 5.12d).

In Liao River Basin, 51% of the subsurface area consists of unconsolidated sand and gravel aquifers that are concentrated in the river valleys as well as in the large Liao River Plain. Although all have intergranular porosity and all contain water primarily under unconfined or water-table conditions, the sorting of aquifer materials and the mount of silt and clay is variable and thus lead to a differences in the DRASTIC rating. Similar to Songhua River Basin, most groundwater in the headwaters of the basin is stored in aquifers in igneous/magmatite (31% of total area), metamorphic (12%) and clastic rocks (5%). Carbonate rock containing aquifer cover less than 1% of the total subsurface area and are concentrated to areas in the South-Eastern headwaters of Taizi and Hun River. Similar to the Songhua region, high hydraulic conductivities occur in the middle and lower part of Liao River Basin, with particularly high hydraulic conductivities either in the Shenyang metropolitan area or close to the watershed to Songhua Catchment. In contrast, small hydraulic conductivities are associated to the dense material forming the aquifer in the headwater of the catchment (Fig. 5.12e, f).

Agricultural areas are the dominating land cover in Liao River Basin, covering more than 43% of the surface (Fig. 5.12g, h). Due to different geographic conditions, forests cover is with about 10% proportionally only one third of the area as in Songhua River Basin. This is compensated by a larger proportion of grasslands (39%), particularly in the western and middle part of the catchment. Additionally urban settlements cover about 4% of available land, particularly in the area of the Central Liaoning City Cluster in the south-eastern lowlands around the megacity of Shenyang. In contrast, larger urban settlements are rare in the entire western part of the basin which is

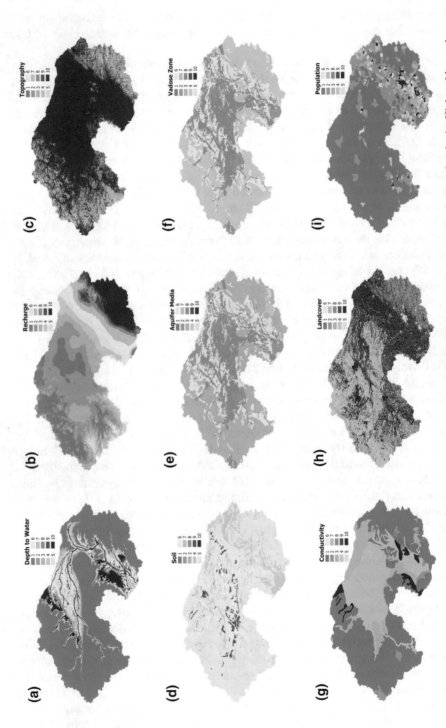

Fig. 5.12 Selected rating maps used on DRASTIC index (**a, b, c, d, e, f, g**) and hazard index (**h, i**) computation in Liao river basin. High and low ratings are visualized by red and green color, respectively

drained by Xiliao River. Similar to the distribution of urban land cover, highest population densities could be found in the south-eastern part of the catchment whereas the north-western areas of the catchment are sparsely populated with typically less than 100 inhabitants per km^2 (Hazard rating = 1). Average population density in Liao River Basin is with 246 Inhabitants per km^2 two and a half times higher than in Songhua River Basin.

5.2.3.2 Groundwater Vulnerability Mapping

After the construction of all DRASTIC parameters maps, a GIS raster calculator was used to combine the seven attributes to compute the DRASTIC vulnerability index for both catchments (Fig. 5.13a, b). Derived DRASTIC Indices (DI) were in a similar range for both catchments with values between 134 and 240 in Songhua River Basin and between 133 and 240 in Liao River Basin.

In Songhua River Basin, more than 81.5% of the aquifers have a low or very low vulnerability, 17% a medium vulnerability and the remaining less than 1.5% a high vulnerability. Zones of high vulnerability were mainly located in the eastern part of Songnen plain, the northwestern part of Sanjiang plain and close to the river banks of the major rivers (Fig. 5.13a). The majority of the aquifers in the basins lowlands is rated by low or medium vulnerability whereas very low groundwater vulnerability was predominant in the mountainous areas. A very high vulnerability could not be obtained for a single cell of the grid. In contrast, the results for Liao River Basin show that the majority of high vulnerability regions were located either in the flat areas in the middle and particularly in the lower reach of the Liao River system as well as in the upstream part of the catchment close to the major water reaches. According to the evaluation, more than 5% of the area was categorized as high vulnerability region, which is triple fold in comparison to Songhua River Basin. Apart from that, medium groundwater vulnerability dominates the eastern and central north region of the catchment with in total 26.5% of the area, whereas mainly the western headwater region and the central south region had a low or very low groundwater vulnerability covering about 68.5% of the catchment (Fig. 5.13c).

5.2.3.3 Hazard Mapping

Potential harmfulness was assessed according to Eq. 5.3. Derived hazard indexes were divided into five classes covering an equal span between the lowest possible index of 24 and the highest which is 240. Areas, for which no population density was available (e.g. larger water surfaces) were excluded from the hazard index calculation. For both catchments, more than half of the area has a very low or low potential harmfulness (Fig. 5.14c). The percentage for areas with very low potential harmfulness is with 52–46% slightly higher for Songhua River Basin than for Liao River Basin which is related to lower population densities and larger areas covered by forest, particularly in the headwaters of the catchment. Similarly, the proportion

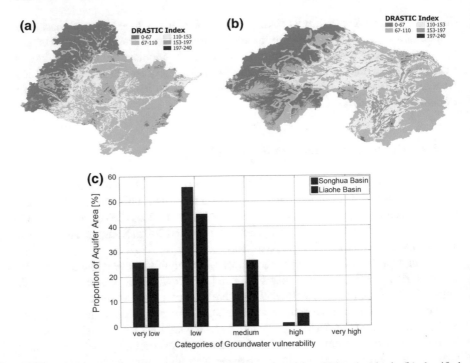

Fig. 5.13 Groundwater vulnerability map of Songhua river (**a**) and Liao river basin (**b**) classified in five classes from very low (dark green color) to very high groundwater vulnerability (red color). The percentage frequency distribution of vulnerability categories for both catchments is given by (**c**)

of areas associated with a low harmfulness is with about 9% almost double as high in Liao River Basin than in Songhua River Basin.

In Songhua River Basin, medium potential harmfulness is to a large part assigned to non-urban areas of Songnen plain and Sanjiang plain as well as to a few clusters in the south-east of the basin (Fig. 5.14a). Although also having large plain areas, medium harmfulness areas in Liao River Basin mainly cover the plains draining to the lower reach of Liao River whereas the plain areas of Xiliao River are to large degree associated with a low harmfulness (Fig. 5.14b). On the other hand, the western mountain ranges of Liao River Basin have a significant amount of areas associated with medium potential harmfulness due to the different land cover (grassland) in comparison to the forested mountain areas of Songhua River Basin.

High and very high potential harmfulness was particularly calculated for urban areas. Considering that Liao River Basin is more densely populated, the percentage of area with high or very high harmfulness is with together about 9% significantly higher than in Songhua River Basin with about 6%. In Liao River Basin, most of the area with high potential harmfulness is clustered around Shenyang metropolitan zone along the lower reach of Hun River as well for areas around Fuxin, Chifeng and along the Dongliao River with zones of very high harmfulness directly at the location

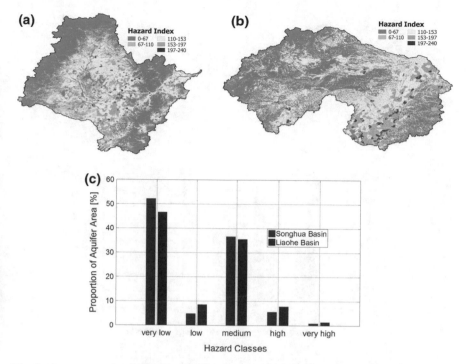

Fig. 5.14 Groundwater hazard map of Songhua (**a**) and Liao river basin (**b**) classified in five classes from low harmfulness (dark green color) to very high harmfulness (red color); **c** shows the percentage frequency distribution of hazard classes for both catchments

of large cities. In Songhua River Basin high harmfulness could be assigned for the more densely populated areas in the Songnen and Sanjiang plain with individual areas of highest hazard class located in the Harbin metropolitan region.

5.2.3.4 Groundwater Risk Mapping

The groundwater risk index was obtained by spatial overlay of the hazard map and the vulnerability map according to Eq. 5.4. The results were reclassified into five equal sized classes between the minimum and maximum index which is 48 and 480, respectively, according to the calculation scheme. These classes can be associated with very low, low, medium, high and very high groundwater contamination risk and are visualized by different colors in Fig. 5.15.

Taking into account the hazard potential and the groundwater vulnerability simultaneously, the aquifer areas with the highest contamination risk class are in both catchments located in the plain areas in the vicinity of large agglomerations. For Songhua River Basin a high intrinsic groundwater vulnerability as well as a very high pollution potential could be obtained for the central Songnen plain region in the surrounding

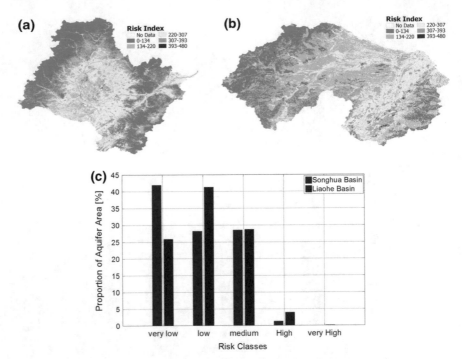

Fig. 5.15 Groundwater risk map of Songhua (**a**) and Liao river basin (**b**) classified in five classes from very low (dark green color) to very high (red color) groundwater contamination risk; **c** shows percentage frequency distribution of risk classes for both catchments

of Harbin due to industrial zones and intensive agricultural use (Fig. 5.15a). Nevertheless, scattered areas associated with high and highest groundwater pollution risk were identified for other plain areas of the catchment, too. In contrast, low groundwater risk is obtained for the mountainous headwaters of Songhua River mainly due to less hazards and favorable vulnerability conditions such as steep slopes and low hydraulic permeability. This is particularly true in the north-western mountain ranges which have also relatively low recharge rates decreasing the load of potential contaminants towards the groundwater. In Liao River Basin, areas with a high and very high groundwater contamination risk are more abundant than in Songhua River Basin (Fig. 5.15c) due to higher recharge rates, lower groundwater tables as well as more area of high potential harmfulness due to a generally higher population density and less forested areas. Most of the plain areas in the Liao River Delta region have at least a high groundwater contamination risk, with a large cluster of areas at highest risk in the metropolitan area around Shenyang city (Fig. 5.15b). In contrast, most of the aquifer in the Xiliao Subbasin as well as in the south-eastern headwaters of Taizi River and Hun River can be considered as being at very low to low pollution risk.

5.2.4 Conclusion

This study developed a method for assessing groundwater contamination risk on a very large scale for the surficial aquifers of Songhua and Liao River Basin. This was achieved by the integration of hazards and intrinsic vulnerability to form a groundwater contamination risk index. We applied the standard DRASTIC method to generate intrinsic vulnerability assessments of the studied aquifers. The potential of anthropogenic hazards was evaluated by using two proxy parameters: land cover and population density. A GIS platform was used to scale and overlay input datasets from different sources and to compute and visualize evaluation results. Two large scale stationary groundwater models were set up using finite element groundwater flow modelling software OpenGeoSys to calculate the groundwater depths within the basins. All indices were estimated for a grid resolution of 30 arc seconds which reflects the resolution limit for most of the currently available datasets.

The final results demonstrated that the flat plain areas of the Songliao River Basin, particularly in the middle reach of Songhua River and the lower reach of Liao River and it south-eastern tributaries, not only are high vulnerable to groundwater contamination due to special surface and hydrogeological conditions but also contain the highest harmfulness to hazards. Focusing on pollution risk from point sources, results can be seen critically for in particular three areas of the Songliao area: Firstly, in the area of Shenyang metropolitan zone which has one of the largest chemical industrial complexes in China producing a large number of chemical compounds hazardous to water. Secondly, the Liao River lowlands where large oil fields with a high potential harmfulness are located and finally Harbin metropolitan area which has a strong but old industrial base posing a large potential source for groundwater contamination. Furthermore intensive agriculture takes place in the chernozem soil belts of the catchments over aquifers with medium to high groundwater contamination risk posing a threat to groundwater contamination by fertilizer residuals and pesticides emission from diffuse sources.

Hence, the derived groundwater vulnerability, potential hazard and groundwater contamination risk maps may support decision-making in the field of future land planning and groundwater management to avoid future contamination of the groundwater by considering the vulnerability of an area before high-risk activities were allowed to take place there. The application of open source software and datasets accessible to public ensures that the methodology of the risk assessment is highly reproducible in other areas of China and global as long as some minimum data requirements are met.

5.3   Groundwater Source Pollution Prevention and Risk Reduction Strategy in SLRB

by SUN Yuanyuan, Erik NIXDORF, SU Jing, WANG Qiang, WANG Tong, Olaf KOLDITZ, XI Beidou

In this deliverable, firstly the European water pollution control strategies were summarized, including strategies of groundwater legislation, integrated groundwater protection, groundwater monitoring, groundwater pollution risk assessment and groundwater source protection. As a comparison, the Chinese Action Plan for Prevention and Control of Water Pollution was introduced with description of the overall consideration, overall framework and overall contents.

Based on the European experience and National Water Action Plan as well as the practices in Songhuajiang-Liao River Basin (SLRB), groundwater source protection strategy in SLRB was composed, including groundwater environment protection and pollution prevention and risk reduction strategies.

This deliverable is a promotion of Deliverable 4.1 "Groundwater source risk assessment guideline in SLRB" and Deliverable 4.2 "Sensitive groundwater source identification technical guideline in SLRB". The strategies proposed in this deliverable will provide some support for the groundwater source protection in SLRB.

5.3.1   Introduction

Groundwater is widely utilized as drinking water in Songhuajiang and Liao River Basin (SLRB) making groundwater sources very crucial for drinking water safety. As SLRB is the old industrial heartland of China, discharges and leakages from industrial sources as well as agriculture pollutions put some potential pressure on groundwater resources. The groundwater pollution risk assessment and the delineation of sensitive groundwater sources are very significant and fundamental for the management of drinking water sources.

European countries have many experiences in the field of water pollution control strategies. One of the most important points is that the European Union takes the water body within river basins as a whole consideration. China issued the Action Plan for Prevention and Control of Water Pollution in 2015, which also learned from the EU strategies' advantages. Groundwater, as a crucial part of water body, has been paid more attention to since the issue of the Water Action Plan.

Groundwater source protection in SLRB could learn from the EU experiences and the Water Action Plan with some consideration of local conditions, which has been described in the form of guidelines for groundwater pollution risk assessment and identification of sensitive groundwater sources.

5.3.2 European Water Pollution Control Strategies

5.3.2.1 Groundwater Legislation

The law-based governance of water is a distinct feature of the EU groundwater resources management. The first wave of EU groundwater protection legislation started with standards for the rivers and lakes as sources of drinking water abstraction in 1975 and ended with the setup of binding targets for drinking water quality in 1980. The second wave began with the 1988 Ministerial Water Forum in Frankfurt, where the existing laws and regulations at that time were reviewed to identify the defects in groundwater legislation. The review facilitated the subsequent groundwater legislation and special legislation, mainly including Directive 91/676/EEC concerning *the Protection of Waters against Pollution Caused by Nitrates from Agricultural Sources* (hereafter referred to as *Nitrates Directive*) and Directive 91/271/EEC concerning *Urban Wastewater Treatment* (hereafter referred to as *Urban Wastewater Treatment Directive*). In this stage, the traditional groundwater legislation was broken through and water policies were combined with policies of other fields. The third wave arrived in 1995 when the European Commission accepted the request of the Environment Committee of European Parliament and the EU Council of Environment Ministers and started to consider reviewing EU's water legislations and policies from a global perspective. Typical legislations include Directive 96/61/EC concerning *the Integrated Pollution Prevention and Control*, Directive 98/83/EC concerning *the Quality of Water Intended for Human Consumption in the EU*, *EU Water Framework Directive* and *EU Groundwater Directive* and their implementation has enriched and improved the EU legal system for groundwater protection.

5.3.2.2 Integrated Groundwater Protection

"Integrated Water Resources Management (IWRM) is a process which promotes the coordinated development and management of water, land, and related resources in order to maximize economic and social welfare in an equitable manner without compromising the sustainability of vital ecosystems," as defined by the Global Water Partnership (GWP) Technical Advisory Committee. Various international organizations are working to put into force the IWRM across the world.

Groundwater management plays an important part in the IWRM of the EU and is featured by integration in typical legislations for the all-round management of water resources, such as *EU Water Framework Directive* and *Groundwater Framework Directive*. *EU Water Framework Directive* attaches importance to the integrated management of "surface water- groundwater- wetland-offshore waters", which means taking groundwater protection as an important part of EU's water resources protection and performing systematic management of it together with rivers, lakes, wetland, and offshore waters. In the meantime, *EU Water Framework Directive* also lays emphasis on the integrated management of "water quantity- water quality- water ecosystems", which refers to all-round and systematic management of the quantity, quality and ecosystems of water through controlling groundwater pollution, protecting groundwater quality and restoring hydrological conditions and aquatic habitats. In addition, it stresses the integrated management of water users and stakeholders from all industries, as EU recognizes the importance of the coordination of policies among industries and sectors within member states and thereby encourages its member states to enact comprehensive groundwater management laws and regulations based on sufficient negotiation with their respective stakeholders. Finally, *EU Water Framework Directive* has also established a scientific assessment system using for reference new achievements of modern ecology and environmental sciences in order to assure the scientific legislation. Goals of the EU Water Framework Directive are to prevent groundwater pollution and reduce or eliminate any damage incurred by the pollution.

5.3.2.3 Groundwater Monitoring

Having built up a nationwide monitoring network, each EU member state is monitoring strictly its groundwater quality and quantity with mature and rigorous techniques. Monitoring stations of each EU member state are classified according to either monitoring objective or management ownership. Seen from monitoring objective, the basic stations of some EU member states such as the Netherlands are used mainly by government organs to provide long-term monitoring data for planning and management. Including state stations and regional stations, they are working to monitor the quantity and quality of water. Special stations are short-term special monitoring stations set up by municipal agencies, NGOs or water users for a specific objective and they are used also to monitor the quantity and quality of water. Temporary stations are established temporarily by scientific research institutions for the purpose of research with monitoring contents determined by actual needs.

5.3.2.4 Groundwater Pollution Risk Assessment

EU Water Framework Directive and *Groundwater Framework Directive* indicate the necessity to assess groundwater risks. It's an assessment on related human activities' influence on groundwater that is performed in the early stage of the watershed management planning cycle. The concept is an extension of groundwater vulnerability. Groundwater risk assessment takes into account pressure and influence and evaluates the chemical conditions of groundwater, but accurate chemical conditions cannot be confirmed until the end of the planning cycle. The estimate can be verified using the recent data of supervision and monitoring and proper trend assessment.

5.3.2.5 Groundwater Source Protection

The EU has been experienced in and achieved progress in the protection of groundwater sources and its member states are working on:

I. Taking legal measures to protect groundwater quality. The zoning-based system of land utilization management is put into force in various natural and social environments.

II. Using computers to simulate and manage groundwater quality. It has been possible to simulate the complicated movement of pollutants, arising from chemical reactions along with the rapid development of groundwater analytics using computers.

III. Surveying the basic conditions of groundwater thoroughly. Because of the great differences in the movement and quality of groundwater, field surveys should be performed when planning groundwater protection zones on: (1) the level, flow rate and flow characteristics of groundwater; (2) the landform, geology, and other hydrogeological characteristics of groundwater-moving areas; (3) the precipitation, evaporation capacity and other meteorological characteristics that determine the balance of groundwater; (4) the strata permeability, clearance rate, and other physical characteristics that control groundwater movement; (5) the stratigraphic distribution that influences the groundwater table; and (6) the chemical characteristics of stratum.

In addition to planning water quality protection zones, it's also very important to further understand the possibility of pollutants removal. Therefore, within the water quality protection zones, surveys should also be performed on (1) the thickness of the unsaturated zone; (2) the property of stratums above main aquifers; and (3) the physical and chemical property of pollutants.

5.3.3 China's Water Protection Strategy: The Action Plan for Prevention and Control of Water Pollution

5.3.3.1 Overall Considerations

Water conversation helps combating pollution. A systemic approach should be adopted and water conversation should be prioritized. Transform extensive water usage to intensive water usage and attach equal attention to emission reduction and water conversation.

Maintain a balance between population, economy and resources as well as strengthening demand side management. Development plans should be dominated by the carrying capacity of water resource and eco-water environment. Water should be a decisive factor while working out urban, land, and population planning.

Both government and market need to play their roles. Managing water is the responsibility of the government. At the same time, market-based measures such as water right/price should be adopted in allocating water resources, so that the government and the market could complement each another.

The ecosystem is a living community. A systemic approach is needed for managing water, mountains, forests, farmlands, lakes, etc. together.

5.3.3.2   Overall Framework

Water Action Plan consists of 4 levels: The target level, the task level, the measures level, and the safeguard level. The target level includes target indicators setup and water eco-environment security guarantee. The task level includes overall control of pollutants, economic structure transformation and upgrade as well as water resource conservation. The measures level includes technical support enhancement, market mechanism exploitation, rigorous environmental law enforcement and supervision and water environment management enhancement. The safeguard level includes specify responsibilities of each party, and strengthen public engagement and social supervision.

5.3.3.3   Concrete Contents

Target Level

There are 3-phased objectives of the Water Action Plan. Phase I is to have improved waterbody quality in stages by 2020. Phase II is to achieve overall water environment quality improvement by 2030. Phase III is to achieve overall eco-environment quality improvement by middle of the 21st century. The main indicators include key watersheds, blackish and smelly water, drinking water, groundwater, coastal water and key regions.

Task Level

The primary task is to reduce the discharge of pollutants from the source. Specify targets for industry, urban households, agriculture and harbors. Secondly, water pollution control forces transformation and upgrading of economic structure. Adapt to the new normality of economic development and environmental protection, coordinate economic/social development with water environment protection in order to realize green development. Thirdly, water conversation helps combating pollution. To address the issues such as conflicts between water supply and demand, low efficiency by the industrial, agricultural, and household sector as well as sever damages to the water eco-environment. Measures will be taken by controlling water consumption volume, improving water utilization efficiency and scientifically protecting and distributing water resources.

Measures Level

Scientific support will be strengthened. In the long run, scientific and technical innovation could safeguard water environment security, promote economic restructuring/upgrading and facilitate sound economic development. Scientific support should be strengthened by means of promoting the demo of applicable technology, R&D on advanced technology and vigorously development of environmental industry.

International cooperation should be strengthened in areas of pollution control of agricultural non-point source, water ecological conservation, early warning for water environment monitoring and technologies and equipment for water treatment process.

Give full play to the role of market mechanism. Cultivate the market with enabling policy, improve the water regime through market forces and let the market play the decisive role in allocating resources.

Promote diversified financing, combine governmental funding and social capital. Encourage social capital investing in water conservation. Establish an incentive mechanism. Develop transboundary water environment compensation mechanism, green loans and "Water-saving champion" system.

Strengthen environmental management. The emission standard and water quality system has not been aligned. Emission permits is crucial to safeguard water quality. Transform the traditional environmental management strategy to the new strategy with water quality as the core.

Safeguard Level

Clearly define the roles and responsibilities of each authority, which lays the foundation for the enforcement of the Action Plan. Local governments have to take more responsibilities. Polluters' responsibilities should be specified. The public is entitled and obligated to supervise and participate in the process of water pollution control. Strengthen the public supervision by information publication and transparency, which helps to enhance the enforcement of the Action Plan.

5.3.4 Groundwater Source Protection Strategy in SLRB

5.3.4.1 Groundwater Environment Protection

As prevention is the primary focus of groundwater pollution control in the principle that "protection should be the priority and prevention should be put first and combined with control", efforts should be made to intensify pollution source control and minimize the occurrence rate of groundwater pollution; to step up groundwater quality monitoring; to establish regional prevention and protection systems; and to develop techniques of emergency response to groundwater risk accidents.

While drinking water source protection is a vital task for drinking water safety assurance in China, this project was piloted in sources of groundwater intended for human consumption. Hence, on the basis of the numerical simulation and risk assessment of groundwater in the early stage, we have analysed the ground pollution sources and the vulnerability of geological conditions of drinking water sources, classified the vulnerability of drinking water sources and directed local protection of sources of groundwater intended for human consumption and related pollution sources management. While the project was implemented, the Chinese government issued the Action Plan for Prevention and Control of Water Pollution. The Action Plan proposes explicitly to survey and assess the drinking groundwater sources at regular intervals and to pilot environmental protection. Therefore, drawing on relevant project results and technical basis, we assisted the Department of Water Environmental Management (the former Department of Pollution Prevention and Control) in completing the *Guidelines for Formulating Implementation Schemes of Groundwater Environment Protection Programs*. It stresses the protection of drinking groundwater sources through problem identification, risk assessment, simulation and evaluation and related projects setup. By reference to *EU Water Framework Directive* and *EU Groundwater Directive*, EU's experience in and methods for groundwater management were applied to Song-Liao River Basin with good achievements obtained. It's a helpful attempt for popularizing these experience and methods across China (Fig. 5.16).

5.3.4.2   Pollution Prevention and Risk Reduction Strategies

Improve the Groundwater Resources Monitoring System

The system, which monitors groundwater quality and quantity, works mainly to analyse data relating to groundwater resources monitoring and to submit them in real time to relevant departments. According to the monitoring objective, monitoring sites can be divided into four groups. The first group of monitoring sites works to serve government management, where government management organs use the information to help formulate water management plans and monitor changes in groundwater level and quality. The second group works to serve water-consuming industries and water supply companies by monitoring the influence incurred by the abstraction of groundwater. The third group works to serve land management, such as the management of natural reserve areas, whose management measures can be formulated according to the monitoring information. The fourth group works to serve scientific research institutions and their primary task is to identify the potential uses of groundwater and they are also intended for research.

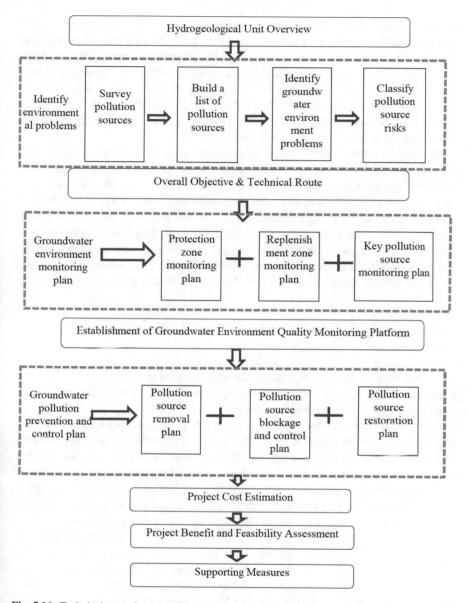

Fig. 5.16 Technical route for groundwater environment protection

Attach Importance to Non-point Source Pollution Control

Groundwater pollution caused by human activities is from either point sources or non-point sources. Point source pollution is resulted from point sources such as storage and transportation facilities of chemicals, pollution treatment sites, industrial sites, accident discharge and refuse landfills, while non-point source pollution is caused when soil and sand grains, nutrient substances such as nitrogen and phosphorus, harmful substances such as pesticides, solid wastes such as straws and agricultural film, fecal sewage from livestock breeding, baits and drugs for aquaculture, rural domestic sewage and wastes and all kinds of atmospheric particulates, which are generated in industrial production and people's life, are going into water, soil or the atmospheric environment through overland runoffs, soil erosion, farmland drainage, eluviation and atmospheric sedimentation.

In comparison with point source pollution, non-point source pollution is much more difficult to prevent and control, because it is derived from dispersed and diversified areas, quite random and with complicated causes and long incubation periods and its geographical boundaries and occurrence positions are hard to identify. Attracting increasing attention from governments, point source pollution has been put under good control and management in many countries, including China. However, non-point source pollution, widely influencing and difficult to control, has been a major source of impact on water environment quality. Therefore, the Chinese government should attach importance to the control of non-point source pollution.

Improve the Monitoring and Management System for Water Source
Protection Zones

There has been already several water source protection zones established in Song-Liao River Basin, but that is far from being enough. We should stick to the principle of "taking protection as the priority and putting prevention first" for groundwater protection, and step up the monitoring and management of established water source protection zones. We should improve the planning of, further perform a general survey on, and designate and adjust in a scientific and reasonable manner drinking water source protection zones. We should survey and assess the current conditions and causes of soil and groundwater pollution, build a standing book of pollution sources, establish environmental quality monitoring systems, prioritize pollution control regions and objects, implement pollution risk assessment, safe zone designation and pollution prevention and control plans and formulate urban and rural water source protection plans. We should establish a pollution incident warning system and an emergency response system for drinking water source protection zones, survey at regular intervals the environment and qualities of drinking water sources and establish a regular water quality information bulletin system for drinking water sources.

5.3.5 Summary and Conclusions

This deliverable introduced the groundwater pollution control strategies in European countries and China, e.g., EU Water Framework Directive, EU Groundwater Directive, Action Plan for Prevention and Control of Water Pollution. From these experiences, it is found that both European countries and China share some ideas in common, e.g. integrated water management, groundwater monitoring network, groundwater pollution risk assessment and groundwater source protection.

By reference to groundwater pollution control strategies in European countries and China, experiences in and methods for groundwater source management were applied to Song-Liao River Basin. The strategy stresses the protection of drinking groundwater sources through problem identification, risk assessment, simulation and evaluation and related projects setup. It's a helpful attempt for popularizing these experience and methods across China.

References

1. Directive 2000/60/EC of the European Parliament and of the Council of 23. Oct 2000. Establishing a framework for community action in the field of water policy, 2000. In *The European parliament and the council of the European union: Technical report*.
2. Technical report on groundwater risk assessment. 2004. Technical report, European Commission.
3. Directive 2006/118/EC of the European Parliament and of the Council of 12. Dec 2006. On the protection of groundwater against pollution and deterioration, 2006. In *The European parliament and the council of the European union: Technical report*.
4. Directive 2013/39/EU of the European Parliament and of the Council of August 12. 2013. Amending directive 2000/60/EC and 2008/105/EC as regards priority substances in the field of water policy, 2013. In *The European parliament and the council of the European union: Technical report*.
5. Guidance document no. 26. 2010. Guidance on risk assessment and the use of conceptual models for groundwater. Technical report, European Commission.
6. Scheidleder, A., J. Grath, and P. Quevauviller. 2007. Groundwater characterisation and risk assessment in the context of the EU Water Framework Directive. In *Groundwater science and policy: An international overview*, ed. P. Quevauviller, 175–192. The Royal Society of Chemistry.
7. Guidance Document No. 3. 2003. Pressures and impacts. Technical report, European Commission.
8. Arbeitshilfe zur Umsetzung der EG-Wasserrahmenrichtlinie (in German). 2003. Technical report, Landesarbeitsgemeinschaft Wasser.
9. Empfehlungen für die Erkundung. 1994. *Bewertung und Behandlung von Grundwasserschäden (in German)*. Länderarbeitsgemeinschaft Wasser: Technical report.
10. Linda Aller, Jay H. Lehr, Rebecca Petty, and Truman Bennett. 1987. Drastic: A standardized system to evaluate groundwater pollution potential using hydrogeologic settings. Worthington, Ohio, United States of America: National Water Well Association.
11. Foster, S.D. 1987. Fundamental concepts in aquifer vulnerability, pollution risk and protection strategy. In *Vulnerability of soil and groundwater to pollutants*, ed. W. van Duijvenbooden, and H.G. van Waegeningh, 69–86. The Hague: TNO Committee on Hydrological Research.
12. Civita, M., and M. De Maio. 1997. Assessing groundwater contamination risk using ARC/INFO via GRID function. In *ESRI Conference*. San Diego, USA.
13. Van Stempvoort, Dale, Lee Ewert, and Leonard Wassenaar. 1993. Aquifer vulnerability index: A GIS-compatible method for groundwater vulnerability mapping. *Canadian Water Resources Journal* 18 (1): 25–37.
14. Nico Goldscheider, Markus Klute, Sebastian Sturm, and Heinz Hötzl. 2000. The PI method—A GIS-based approach to mapping groundwater vulnerability with special consideration of karst aquifers. *Z Angew Geol* 46 (3): 157–166.

15. Hölting, B., T. Haertle, K.H. Hohberger, K.H. Nachtigall, E. Villinger, and W. Weinzierl. 1995. Conception for the evaluation of the protective function of the unsaturated stratum above the groundwater table. *Geol Jahrb* C63: 5–24.
16. Foster, S.S.D., and P.J. Chilton. 2003. Groundwater: The processes and global significance of aquifer degradation. *Philosophical Transactions of the Royal Society of London B: Biological Sciences* 358 (1440): 1957–1972.
17. Jac, van der Gun. 2012. Groundwater and global change: Trends, opportunities and challenges. Technical report.
18. Custodio, E. 2012. Trends in groundwater pollution: Loss of groundwater quality and related services, 74.
19. Zaporozec, A., A. Aureli, J.E. Conrad, et al. 2002. Groundwater contamination inventory: A methodological guide. *Unesco Ihp-Vi, Ser Groundw* 2: 17–21.
20. Johansson, P.-O., C. Scharp, T. Alveteg, and A. Choza. 1999. Framework for ground water protection-the managua ground water system as an example. *Groundwater* 37 (2): 204–213.
21. Varnes, D.J., et al. 1981. The principles and practice of landslide hazard zonation. *Bulletin of the International Association of Engineering Geology-Bulletin de l'Association Internationale de Géologie de l'Ingénieur* 23 (1): 13–14.
22. Li, Yongfang, Da Wang, Yuyan Liu, Quanmei Zheng, and Guifan Sun. 2017. A predictive risk model of groundwater arsenic contamination in China applied to the Huai River Basin, with a focus on the region's cluster of elevated cancer mortalities. *Applied Geochemistry* 77: 178–183.
23. Wang, Junjie, Jiangtao He, and Honghan Chen. 2012. Assessment of groundwater contamination risk using hazard quantification, a modified DRASTIC model and groundwater value, Beijing Plain. *China. Science of the Total Environment* 432: 216–226.
24. Vito, F., Uricchio, Raffaele Giordano, and Nicola Lopez. 2004. A fuzzy knowledge-based decision support system for groundwater pollution risk evaluation. *Journal of Environmental Management* 73 (3): 189–197.
25. Nobre, R.C.M., O.C. Rotunno Filho, W.J. Mansur, M.M.M. Nobre, and C.A.N. Cosenza. 2007. Groundwater vulnerability and risk mapping using GIS, modeling and a fuzzy logic tool. *Journal of Contaminant Hydrology* 94 (3): 277–292.
26. Lavoie, Denis, Christine Rivard, René Lefebvre, and Stephan Séjourné, R. Thériault, Mathieu J. Duchesne, Jason M.E. Ahad, Baolin Wang, Nicolas Benoît, and Charles Lamontagne. 2014. The Utica Shale and gas play in southern Quebec: Geological and hydrogeological syntheses and methodological approaches to groundwater risk evaluation. *International Journal of Coal Geology* 126: 77–91.
27. Neshat, Aminreza, Biswajeet Pradhan, and Saman Javadi. 2015. Risk assessment of groundwater pollution using Monte Carlo approach in an agricultural region: An example from Kerman Plain. *Iran. Computers, Environment and Urban Systems* 50: 66–73.
28. Zhang, Guoqi, Kalyan V. Vasudevan, Brian L. Scott, and Susan K. Hanson. 2013. Understanding the mechanisms of cobalt-catalyzed hydrogenation and dehydrogenation reactions. *Journal of the American Chemical Society* 135 (23): 8668–8681.
29. Foster, S., and R. Hirata. 1988. Groundwater pollution risk assessment: A methodology using available data, Lima.
30. Jarvis, N.J., J.M. Hollis, P.H. Nicholls, T. Mayr, and S.P. Evans. 1997. MACRO-DB: A decision-support tool for assessing pesticide fate and mobility in soils. *Environmental Modelling and Software* 12 (2–3): 251–265.
31. Harry E. LeGrand. 1964. System for evaluation of contamination potential of some waste disposal sites. *Journal (American Water Works Association)* 56 (8): 959–974.
32. Fritch, T.G., C.L. McKnight, J.C. Yelderman Jr., S.I. Dworkin, and J.G. Arnold. 2000. A predictive modeling approach to assessing the groundwater pollution susceptibility of the Paluxy Aquifer, Central Texas, using a geographic information system. *Environmental Geology* 39 (9): 1063–1069.
33. Enrique, Gomezdelcampo, and J. Ryan Dickerson. 2008. A modified DRASTIC model for siting confined animal feeding operations in Williams County, Ohio, USA. *Environmental Geology* 55 (8): 1821–1832.

34. Uddameri, V., and V. Honnungar. 2007. Combining rough sets and GIS techniques to assess aquifer vulnerability characteristics in the semi-arid South Texas. *Environmental Geology* 51 (6): 931–939.
35. Albuquerque, M.T.D., G. Sanz, S.F. Oliveira, R. Martínez-Alegría, and I.M.H.R. Antunes. 2013. Spatio-temporal groundwater vulnerability assessment-a coupled remote sensing and GIS approach for historical land cover reconstruction. *Water Resources Management* 27 (13): 4509–4526.
36. Panagopoulos, G.P., A.K. Antonakos, and N.J. Lambrakis. 2006. Optimization of the DRASTIC method for groundwater vulnerability assessment via the use of simple statistical methods and GIS. *Hydrogeology Journal* 14 (6): 894–911.
37. Muheeb, M., A.A. Awawdeh, and Rasheed A. Jaradat. 2010. Evaluation of aquifers vulnerability to contamination in the Yarmouk River basin, Jordan, based on DRASTIC method. *Arabian Journal of Geosciences* 3 (3): 273–282.
38. Jamrah, Ahmad, Ahmed Al-Futaisi, Natarajan Rajmohan, and Saif Al-Yaroubi. 2008. Assessment of groundwater vulnerability in the coastal region of Oman using DRASTIC index method in GIS environment. *Environmental Monitoring and Assessment* 147 (1): 125–138.
39. Huan, Huan, Jinsheng Wang, and Yanguo Teng. 2012. Assessment and validation of groundwater vulnerability to nitrate based on a modified DRASTIC model: A case study in Jilin city of northeast China. *Science of the Total Environment* 440: 14–23.
40. Xueyan, Y.E., Chuanyu You, D.U. Xinqiang, and Lixue Wang. 2015. Groundwater vulnerability to contamination near the river: A case study of the second Songhua River. *Science and Technology Review* 33: 78–83.
41. Distribution and mass inventory of mercury in sediment from the Yangtze River estuarine-inner shelf of the East China Sea. *Continental Shelf Research*, 132 (Supplement C): 29–37, 2017.
42. Guo, Qinghai, Yanxin Wang, Xubo Gao, and Teng Ma. 2007. A new model (drarch) for assessing groundwater vulnerability to arsenic contamination at basin scale: A case study in Taiyuan basin, northern China. *Environmental Geology* 52 (5): 923–932.
43. Eckhardt, D., W. Flipse, and E. Oaksford. 1989. *Relation between land use and ground-water quality in the upper glacial aquifer in Nassau and Suffolk Counties*. New York: Long Island.
44. Ouedraogo, I., and M. Vanclooster. 2016. A meta-analysis of groundwater contamination by nitrates at the African scale. In *Hydrology and Earth System Sciences Discussions*, 1–43.
45. Masoudi, Masoud, Gholam Reza Zehtabiyan, Reza Noruzi, S. Mohammad Mahdavi, Behruz Kuhenjani, et al. 2009. Hazard assessment of ground water resource degradation using GIS in Mond Miyani basin. *Iran. World Applied Sciences Journal* 6 (6): 802–807.
46. Baojing, Gu, and Ying Ge. 2013. Scott X Chang, Weidong Luo, and Jie Chang. Nitrate in groundwater of China: Sources and driving forces. *Global Environmental Change* 23 (5): 1112–1121.
47. Zhang, W.L., Z.X. Tian, N. Zhang, and X.Q. Li. 1996. Nitrate pollution of groundwater in northern China. *Agriculture, Ecosystems and Environment* 59 (3): 223–231.
48. Wang, Rui, Jian-Min Bian, and Yue Gao. 2014. Research on hydrochemical spatio-temporal characteristics of groundwater quality of different aquifer systems in Songhua River Basin, eastern Songnen Plain. *Northeast China. Arabian Journal of Geosciences* 7 (12): 5081–5092.
49. Vrba, J., and A. Zaporozec. 1994. Guidebook on mapping groundwater vulnerability. In *International association of hydrogeologists. International contributions to hydrogeology*, vol. 16. Heise, Hannover.
50. Andreo, Bartolomé, Nico Goldscheider, Inaki Vadillo, Jesús María Vías, Christoph Neukum, Michael Sinreich, Pablo Jiménez, Julia Brechenmacher, Francisco Carrasco, Heinz Hötzl, et al. 2006. Karst groundwater protection: First application of a Pan-European approach to vulnerability, hazard and risk mapping in the Sierra de Líbar (Southern Spain). *Science of the Total Environment* 357 (1): 54–73.
51. Sharif Moniruzzaman, Shirazi, H.M. Imran, and Shatirah Akib. 2012. GIS-based DRASTIC method for groundwater vulnerability assessment: A review. *Journal of Risk Research* 15 (8): 991–1011.

52. Lehner, Bernhard, Kristine Verdin, and Andy Jarvis. 2008. New global hydrography derived from spaceborne elevation data. *Eos, Transactions American Geophysical Union* 89 (10): 93–94.
53. Robert, J. Hijmans, Susan E. Cameron, Juan L. Parra, Peter G. Jones, and Andy Jarvis. 2005. Very high resolution interpolated climate surfaces for global land areas. *International Journal of Climatology* 25 (15): 1965–1978.
54. Miralles, D.G., T.R.H. Holmes, R.A.M. De Jeu, J.H. Gash, A.G.C.A. Meesters, and A.J. Dolman. 2011. Global land-surface evaporation estimated from satellite-based observations. *Hydrology and Earth System Sciences* 15 (12): 453–469.
55. Shangguan, Wei, Yongjiu Dai, Baoyuan Liu, Axing Zhu, Qingyun Duan, Wu Lizong, Duoying Ji, Aizhong Ye, Hua Yuan, Qian Zhang, et al. 2013. A china data set of soil properties for land surface modeling. *Journal of Advances in Modeling Earth Systems* 5 (2): 212–224.
56. Jun, Chen, Yifang Ban, and Songnian Li. 2014. China: Open access to earth land-cover map. *Nature* 514 (7523): 434–434.
57. Center for International Earth Science Information Network CIESIN Columbia University. 2015. Gridded population of the world, version 4 (GPWv4): Population Density. Palisades, NY: NASA Socioeconomic Data and Applications Center (SEDAC).

Printed in the United States
By Bookmasters